The Shortest Path Problem

Ninth DIMACS Implementation Challenge

DIMACS
Series in Discrete Mathematics
and Theoretical Computer Science

Volume 74

The Shortest Path Problem

Ninth DIMACS Implementation Challenge

Camil Demetrescu
Andrew V. Goldberg
David S. Johnson
Editors

Center for Discrete Mathematics
and Theoretical Computer Science
A consortium of Rutgers University, Princeton University,
AT&T Labs–Research, Alcatel-Lucent Bell Laboratories,
NEC Laboratories America, and Telcordia Technologies
(with partners at Avaya Labs, HP Labs, IBM Research, Microsoft
Research, Georgia Institute of Technology, Rensselaer Polytechnic
Institute, and Stevens Institute of Technology)

American Mathematical Society
Providence, Rhode Island

This DIMACS volume presents the proceedings from the Ninth DIMACS Implementation Challenge held at Rutgers University, Piscataway, NJ, on November 13–14, 2006.

2000 *Mathematics Subject Classification.* Primary 05C90, 05C12, 05C38, 05C85, 68–06, 68W05, 68W40, 90B20, 90B18.

"Engineering Label-Constrained Shortest-Path Algorithms", by Chris Barrett, Keith Bisset, Martin Holzer, Goran Konjevod, Madhav Marathe, and Dorothea Wagner, is reprinted here courtesy of Springer-Verlag

---

Library of Congress Cataloging-in-Publication Data

The shortest path problem : ninth DIMACS implementation challenge / Camil Demetrescu, Andrew V. Goldberg, David S. Johnson, editors.

p. cm. — (DIMACS series in discrete mathematics and theoretical computer science ; v. 74)

Includes bibliographical references.

ISBN 978-0-8218-4383-3 (alk. paper)

1. Combinatorial optimization—Congresses. 2. Graph theory—Mathematics—Congresses. 3. Algorithms—Congresses. I. Demetrescu, Camil. II. Goldberg, Andrew V. III. Johnson, David S., 1945–

QA402.5.S5425 2009
519.6′4—dc22 2009012805

∞ The paper used in this book is acid-free and falls within the guidelines
established to ensure permanence and durability.
Visit the AMS home page at `http://www.ams.org/`

10 9 8 7 6 5 4 3 2 1 14 13 12 11 10 09

Contents

Foreword

The Ninth DIMACS Implementation Challenge is the most recent of a series of highly successful events aimed at narrowing the gap between theory and practice in an area of current research interest. The DIMACS challenges identify an important research area and challenge researchers to find new algorithms appropriate to a challenge problem. The first Challenge event was held in 1991, and this volume represents the fifth volume dedicated to reporting the outcome of such events. The Challenge workshop was held at DIMACS November 13-14, 2006, and was devoted to shortest path problems. Shortest path problems have applications in a wide variety of areas including transportation, robotics, and network optimization, and there has been a huge literature devoted to them. This challenge was intended to revisit the problem in all of its generality and its many variations, with the goal of identifying new algorithmic approaches and finding real-world applications for which good solutions to corresponding shortest path problems are not yet known. A highlight of the activity leading to this volume was a competition involving the point to point shortest path problem, comparing performance and robustness of different approaches.

We would like to express our appreciation to Camil Demetrescu, Andrew V. Goldberg, and David S. Johnson for their efforts to organize and plan this successful event. We especially thank Camil Demetrescu for his efforts in maintaining a website dedicated to the Challenge. We also thank Paolo Dell'Olmo, Irina Dumitrescu, Mikkel Thorup, and Dorothea Wagner for serving as an Advisory Committee for this Challenge.

DIMACS gratefully acknowledges the generous support that makes these programs possible. Special thanks go to the National Science Foundation and Microsoft Research, as well as to DIMACS partners at Rutgers, Princeton, AT&T Labs - Research, Alcatel-Lucent Bell Labs, NEC Laboratories America, and Telcordia Technologies, and affiliate partners Avaya Labs, Georgia Institute of Technology, HP Labs, IBM Research, Microsoft Research, Rensselaer Polytechnic Institute, and Stevens Institute of Technology.

Fred S. Roberts,
Director

Robert Tarjan,
Co-Director for Princeton

Introduction

This volume contains papers arising from the 9th DIMACS Implementation Challenge (`http://www.dis.uniroma1.it/~challenge9`), which was devoted to shortest path algorithms.

DIMACS implementation Challenges are aimed at narrowing the gap between theory and practice for algorithms in specific problem areas. This is achieved by developing a common infrastructure for experimental algorithm evaluation and encouraging researchers to develop and compare state of the art algorithm implementations. The first Challenge was held in 1990-1991 and was devoted to *Network Flows and Matching.* Other addressed problems included: *Maximum Clique, Graph Coloring, and Satisfiability* (1992-1993), *Parallel Algorithms for Combinatorial Problems* (1993-1994), *Fragment Assembly and Genome Rearrangements* (1994-1995), *Priority Queues, Dictionaries, and Multi-Dimensional Point Sets* (1995-1996), *Near Neighbor Searches* (1998-1999), *Semidefinite and Related Optimization Problems* (1999-2000), and *The Traveling Salesman Problem* (2000-2001).

Selection of the problem area is the key to a Challenge's success. A good area is one with active theoretical algorithmic research and a lack of the common infrastructure. In 2004, when the call for the Ninth Challenge was issued, this was the case for the area of shortest path algorithms.

Shortest path problems are among the most fundamental combinatorial optimization problems with many applications, both direct and as subroutines. They arise naturally in a remarkable number of real-world settings. A limited list includes transportation planning, network optimization, packet routing, image segmentation, speech recognition, document formatting, robotics, compilers, traffic information systems, and dataflow analysis. Shortest path algorithms have been studied since the 1950's and still remain an active area of research.

The goal of the Ninth DIMACS Implementation Challenge was to stimulate research on shortest path algorithms by creating a reproducible picture of the state of the art. The main results of the Challenge included:

- Definition of common file formats for several variants of the shortest path problem, both static and dynamic. These comprised extensions of the famous DIMACS graph file format used by several algorithmic software libraries.
- Definition of a common set of core input instances for evaluating shortest path algorithms.
- Definition of benchmark codes for shortest path problems.
- Design of new shortest path algorithms.
- Experimental evaluation of state of the art implementations of shortest path codes on the core input families.

- A discussion of directions for further research in the area of shortest paths, identifying problems critical in real-world applications for which efficient solutions still remain unknown.

The Challenge call for participation was broad, inviting work on most polynomially-solvable variants of shortest paths. The organizers provided input-output file formats and benchmark problem solvers for the most basic problems and indicated that more formats will be developed if needed.

Participants at sites in the U.S. and Europe undertook projects during the period from October 2005 to November 2006, studying several variants of the shortest paths problem. For its relevance in modern applications such as GPS navigation systems, the point-to-point shortest paths problem, which consists of answering multiple online queries about the shortest paths between pairs of vertices, was the most popular problem during the Challenge, attracting about half of the contributions. Other variants under investigation included k-shortest paths and language-constrained shortest paths. Parallel and external memory implementations of shortest paths algorithms were also studied.

In the first stage of the Challenge, the organizers and the participants obtained or developed real-life problems and problem generators, from which benchmark problem families were selected. The participants were encouraged to use these problems whenever practical during the second part of the Challenge, during which algorithm implementations were developed and evaluated. In addition, the participants were free to use any other problems that they found useful in evaluating their algorithm performance.

The Challenge culminated in a workshop held at the DIMACS Center at Rutgers University, Piscataway, New Jersey on November 13–14, 2006. Following the workshop, participants were encouraged to share test instances whenever possible, to rework their implementations and experiments in light of the feedback at the workshop, and to submit a final report for this book. After a careful reviewing process, comparable to that of refereeing for a high-quality journal, twelve full articles were selected for this volume. In view of the practical importance of the problems studied in this Challenge, this volume also contains a survey of applications of shortest path problems.

More than 40 researchers attended the workshop; there were thirteen project presentations, one invited talk, and a panel discussion. The specific application problems presented at the workshop included point-to-point shortest paths, k-shortest paths, parallel shortest paths, external-memory BFS, and language-constrained shortest paths. The invited presentation was given by David A. Bader (Georgia Institute of Technology). The lecture focused on solving massive graph problems on large-scale parallel machines and presented several graph theoretic kernels for connectivity and centrality. The lecture discussed how the underlying parallel architectures affects algorithm development, ease of programming, performance, and scalability.

In addition, many participants took part in a special competition held during the workshop and devoted to the point-to-point shortest path problem. The aim was to compare the performance and the robustness of the different implementations discussed at the workshop. The rules of the competition were announced on the first day of the workshop and the results were due on the second day. The competition consisted of preprocessing a version of the full road network of continental U.S.

CODE	PREPROCESSING		QUERY		
	Time (minutes)	Space (Mbytes)	Node scans	Time (ms)	Benchmark ratio $\times 10^6$
HH-based transit [**5**]	104	3664	n.a.	0.019	4.78
TRANSIT [**1**]	720	n.a.	n.a.	0.052	10.77
HH Star [**2**]	32	2662	1082	1.140	287.32
REAL(16, 1) [**4**]	107	2435	823	1.420	296.30
HH with DistTab [**2**]	29	2101	1671	1.610	405.77
RE [**4**]	88	861	3065	2.780	580.08

TABLE 1. Results of the Challenge competition on the USA graph (23.9 million nodes and 58.3 million arcs) with unit arc lengths. The benchmark ratio is the average query time divided by the time required to answer a query using the benchmark Dijkstra code on the same platform. Query times and node scans are average values per query over 1000 random queries.

with unit edge lengths and answering a sequence of 1,000 random distance queries. Each participating team used their own machine and computational environment. Note that the U.S. graph with lengths corresponding to travel distances and transit times was a part of the benchmark data set, and the unit length function was a natural "new" length function to consider. Some of the participants' codes were unable to deal with the size of the full U.S. graph, or incurred runtime errors. However, six point-to-point implementations successfully preprocessed the graph and answered the queries. These included two codes, *RE* and *REAL*(16, 1), which were not eligible for the official competition because one of the co-authors on the paper was also an organizer. These two codes were run on the problem prior to the competition to ensure that the problem could be solved in 24 hours.

Because results were obtained on different platforms, each participant ran a Dijkstra benchmark code [**3**] on the USA graph to allow machine calibration. The final ranking was made by considering each query time divided by the time required by the benchmark code on the same platform (benchmark ratio). Other performance measures taken into account were space usage and the average number of nodes scanned by query operations.

The results of the competition are reported in Table 1. All six codes that successfully handed the graph had substantially faster query times than the benchmark Dijkstra code, which did no preprocessing. As one can see from the table, the fastest query time was achieved by the *HH-based transit* code of Peter Sanders and Dominik Schultes. (The times reported in the table are for the different platforms used by the participants and are not normalized. They are reasonably comparable, however. Based on the benchmark runs, no pair of platforms differed in speed by more than 22%.)

Work done by the participants during the Challenge improved the state of the art in shortest path algorithms. In addition, the infrastructure developed during the Challenge should facilitate further research in the area, leading to substantial follow-up work as well as to better and more uniform experimental standards.

The expert assistance of the DIMACS staff in hosting and arranging the workshop is gratefully acknowledged. We would like to thank Microsoft Research for providing partial funding for the workshop. We also would like to thank all the Challenge participants, whose hard work and enthusiasm was the key to the success of the Challenge.

Camil Demetrescu
Andrew V. Goldberg
David S. Johnson

March 2009

Bibliography

[1] H. Bast, S. Funke, and D. Matijevic, *Transit: Ultrafast shortest-path queries with linear-time preprocessing*, in 9th DIMACS Implementation Challenge Workshop: Shortest Paths, DIMACS Center, Piscataway, NJ, November 13–14, 2006.

[2] D. Delling, P. Sanders, D. Schultes, and D. Wagner, *Highway hierarchies star*, in 9th DIMACS Implementation Challenge Workshop: Shortest Paths, DIMACS Center, Piscataway, NJ, November 13–14, 2006.

[3] E. Dijkstra, *A note on two problems in connexion with graphs*, Numerische Mathematik, 1 (1959), pp. 269–271.

[4] A. Goldberg, H. Kaplan, and R. Werneck, *Better landmarks within reach*, in 9th DIMACS Implementation Challenge Workshop: Shortest Paths, DIMACS Center, Piscataway, NJ, November 13–14, 2006.

[5] P. Sanders and D. Schultes, *Robust, almost constant time shortest-path queries in road networks*, in 9th DIMACS Implementation Challenge Workshop: Shortest Paths, DIMACS Center, Piscataway, NJ, November 13–14, 2006.

DIMACS Series in Discrete Mathematics
and Theoretical Computer Science
Volume **74**, 2009

Real-World Applications of Shortest Path Algorithms

J.L. Santos

ABSTRACT. The shortest path problem (SPP) is one of the most-studied combinatorial optimization problems in the literature. It has a wide range of applications, both direct and as a subproblem of other problems. Many of these applications require solving very large instances in real time, and recently new algorithmic techniques have been introduced to deal with such instances. This paper presents an overview of the state of the art in shortest path applications and the algorithms used for them.

1. Introduction

The word "network" appears frequently in our vocabulary, associated with many daily situations. In fact, it would be hard to imagine how we would live today without the Internet (World Wide Web), highway systems, train services, air travel, telecommunications, or water and electricity distribution, all of which have underlying networks. In these examples, the network structure is obvious, but there are other applications where a network structure is only implicit, as for example in social relationships, protein-protein interactions, genetics, and linguistics.

In network related problems, we frequently have to determine shortest paths (SPs). In particular, given a graph $G = (V, E)$, we may need to find (1) the shortest path between two specified vertices s and t (the *point-to-point shortest path* (P2PSP) problem), (2) the shortest paths between a given source vertex s and all other vertices (the *single source shortest path* (SSSP) problem), or (3) the shortest paths between all pairs of vertices (the *all pairs shortest path* (APSP) problem). These SP problems have a wide range of (direct and indirect) applications. More than one million results appear if one performs a *Google* search on the term *"shortest path"*. Hundred of thousands matches are returned when using the *Google Scholar* and *Scirus* webpages. Thousands of results can also be found with more specific search engines, namely, *Scopus*, *ScienceDirect* and *Citeseer*. Since *Scopus* clusters the results by date, we can also recognize a significant increase in published papers on SP problems during recent years (about five hundred articles per year in the last four years). Finally, more than 1500 registered patents related to shortest paths bear witness to the practical importance of SP problems.

1991 *Mathematics Subject Classification.* Primary 90B10, 90C27, 90C35.

Key words and phrases. shortest path problem, combinatorial optimization, large real world size network.

Research related to SP problems began in earnest in the 1950s. Dantzig [**Dan51**] developed the network simplex algorithm for the uncapacitated transportation problem and, next, Bellman [**Bel52, Bel53**] established the preliminaries of dynamic programming. In 1956, Ford [**For56**] outlined the first label-correcting algorithm for SP problems. A labeling algorithm assigns to each vertex j an upper bound π_j on the shortest distance from a source vertex s to j, with all values initially infinite except for π_s, which is initialized to 0. The main idea is to update π_j each time one encounters a vertex i such that $\pi_i + c_{i,j} < \pi_j$, where (i, j) is an arc from i to j and $c_{i,j}$ is the associated distance. When updates are no longer possible, one will have determined the lengths of the shortest paths from s to all other vertices, thus solving the SSSP problem. Moore [**Moo57**] and Ford and Fulkerson [**FF62**] studied the general properties of labeling algorithms. In 1958, Bellman applied dynamic programming to SP problems and designed the first-in, first-out (FIFO) implementation of the label-correcting algorithm [**Bel58**].

Some of the above algorithms apply to SP problems where negative edge costs (lengths) are allowed. Assuming we want a path with no repeated vertices, this problem is NP-complete [**GJ79**]. This is essentially because of the possibility of negative-cost cycles. If negative-cost edges are not allowed (the situation in many key shortest path applications), the problem becomes polynomial-time solvable, and it is this special case of the problem on which we concentrate in this article. We note, however, that for the general case there are polynomial-time algorithms that either detect a negative-cost cycle or return a guaranteed shortest path [**AMO93**]. Detecting negative-cost cycles has interesting applications in its own right, for example in determining whether a system of difference constraints with unitary coefficients has a feasible solution [**AMO93**]. However, the algorithmic techniques used in the negative edge cost case can be quite different from those relevant to the case of non-negative costs, and we will not elaborate on them here.

The algorithm for graphs with non-negative arc lengths that is today commonly known as "Dijkstra's algorithm" was proposed independently by Dijkstra [**Dij59**], Dantzig [**Dan60**], and Whiting and Hillier [**WH60**]. It runs in $O(n^2)$ time, where n is the number of vertices, and solves the SSSP problem. This algorithm is a variant of the labeling method that selects to process next a previously unprocessed vertex i with the smallest distance estimate π_i, and then updates π_j for all its unprocessed neighbors. An early observation was that the running time for a graph with n vertices and m edges, given in adjacency list format, could be improved to $O(m \log n)$ using a priority queue implementation.

In 1962, Floyd [**Flo62**] designed an algorithm to solve the APSP problem based on Warshall's work that determined the transitive closure of graphs [**War62**]. Also in the 1960's, Hart *et al.* [**HNR68, HNR72**] introduced the A* Search algorithm (a.k.a. Heuristic or Goal-Directed Search) for the P2PSP problem. This algorithm uses estimates on the distance to the destination to direct – and speed up – the search from the source vertex. Under certain conditions, the method finds the optimal (shortest) path, although in general the optimality is not guaranteed.

In the 1970s and 1980s, several new labeling algorithms for the SSSP problem were proposed. Incremental graph algorithms [**GP88**] solve the problem on a subgraph, then add a vertex to the subgraph, and reoptimize starting from the previous solution. In particular, Pape's algorithm [**Pap74**] processes labeled vertices of the current subgraph in stack (LIFO) order, and Pallottino's algorithm [**Pal84**] processes them in FIFO order. Glover *et al.* [**GGK84**] proposed a hybrid algorithm that that processes vertices with small distance labels in FIFO order, combining the ideas of the Dijkstra and Bellman-Ford methods.

In addition, many researchers worked at improving efficiency of the priority queue implementation of Dijkstra's algorithm. For instance, Dial developed the concept of buckets [**Dia69**], Johnson [**Joh77**] used a d-heap, Denardo and Fox [**DF79**] designed a multi-level bucket data structure, and Fredman and Tarjan proposed the utilization of Fibonacci heaps [**FT87**]. Ahuja *et al.* [**AMOT90**] show that the multi-level buckets in combination with augmented Fibonacci heaps give a better time bound. A later paper shows that the improved bound can be obtained using original Fibonacci heaps [**CGS99**] alone. A variation of the algorithm with linear expected time has been given in [**Gol01**].

In the 1980s, the conventional wisdom was that algorithms with bad worst-case time, such as Pape's and Pallotino's, work best in practice, although there was disagreement about which one was the best. However, in 1996, a comprehensive study [**CGR96**] showed that these algorithms perform poorly on some natural examples, and validated the practical efficiency of the multi-level bucket data structures, with the algorithms that use them currently considered the fastest for the SSSP problem.

As for the P2PSP problem, the practical state of the art has advanced substantially in recent years, as will be evident from some of the other papers in this volume. For the non-negative cost case, the combination of the best current algorithms and data structures with the processing power and storage capacity of modern computers has led to impressive results. With appropriate preprocessing, it is now often possible to solve large-scale instances in real time. This is the case, for example, when we compute driving directions on web mapping sites or on satellite navigation systems using GPS/GIS technology. These systems work with networks having several millions of road segments and provide answers in seconds or less.

For readers interested in reading more about algorithms for SP problems, a good survey of papers published until the middle of the 1980s is presented by Deo and Pang [**DP84**]. This work was later updated by Ahuja *et al.* [**AMO89, AMO91**]. Finally, the most recent improvements on the labeling algorithm for the SSSP problem are reported in [**GH05, GKW07, GW05, SS05, SS07, WW05**]. Developments on the APSP problem can be found in [**AGM97, GM97, Joh77, KKP93, Pet04, Sei95, SZ99, Tak92, Zwi98, Zwi02**]. There has also been considerable study of SP problems restricted to the special case where only integral edge costs are allowed. See [**AMOT90, CGS99, DF79, Dia69, DGKK79, Gol01, GT89, Hag00, Joh82, Mey01, Ram96, Tho04, vEBKZ76, Wag76**].

Many variants of the above SP problems have also been studied. For example, in the minimum non-decreasing path problem, proposed by Minty [**Min58**], we wish to find a path whose consecutive arc weights are non-decreasing and such that the last arc weight is minimized. The problem of ranking the k shortest paths between two given vertices was introduced by Hoffman and Pavley [**HP59**]. Other variants include the stochastic SP problem [**EZ62**], the restricted SPP [**Jok66**], the multi-criteria SPP [**Vin74**], and the robust SP problem [**MG04**]. Of particular note are *dynamic* SP problems, in which our goal is to maintain a representation of the current shortest paths while a sequence of edge length updates is performed [**RR96, BRT08**].

In the remainder of this paper we shall survey some of the many applications that involve SP problems, both directly and indirectly, also covering the types and sizes of the networks involved and the SP algorithms used. In Section 2, we describe the three most widely occurring real-world applications, route-finding services, railroad itinerary generation, and Internet routing. Sections 3 and 4 then briefly cover a selection of additional applications for SP algorithms, as an illustration of the wide variety of such applications that exist. Section 3 covers applications dealing explicitly with shortest paths and shortest path lengths. Section 4 illustrates the general optimization technique known as dynamic programming, which implicitly involves the computation of shortest paths. We conclude in Section 5 with a brief wrap-up of the discussion.

2. Real-world problems where shortest-path codes are in constant use

2.1. Point-to-point route finding services on the Web. In recent years, a variety of free web-based services have sprung up to help people looking for travel directions. These services include the following:

Web page	URL address
GoogleMaps	`http://maps.google.com/`
LiveMaps	`http://www.maps.live.com/`
Map24	`http://www.map24.com/`
Maporama	`http://world.maporama.com/`
MapQuest	`http://www.mapquest.com/`
MultiMap	`http://www.multimap.com/`
ViaMichelin	`http://www.viamichelin.com/`
YahooMaps	`http://maps.yahoo.com/`

Typically one visits the webpage of such a service and enters a starting address and a destination address. The website will then return driving directions, together with maps of the route. It is usually possible to choose between computing the quickest (expected driving time) or the shortest (distance) path. Some services consider additional factors in their routing. For instance, the ViaMichelin webpage allows users to specify preference for most economic or most scenic route among other options. In addition, some services cover more countries than others. For instance, MapQuest accepts locations from Canada, USA and only 12 countries in Europe, while ViaMichelin includes 60 countries.

Tables 1–2 report the results obtained for quickest and shortest routes under two different queries: one in the United States between California City, California, and New York City, New York, and another one in Europe between Lisbon (Portugal) and Wien (Austria). GoogleMaps and YahooMaps do not allow a choice between time and distance metrics, but may be using a linear combination of these two criteria as an objective function. In the tables we put results for these two services by default into the "Quickest path" column.

Note that there is considerable variability between services. This is not surprising. Different services obtain data from different vendors. They estimate travel time in different ways, taking into account several parameters like speed limits, traffic conditions, etc. Furthermore, the applications may use inexact heuristics

	Quickest path		Shortest path	
Web page	Time	Km	Time	Km
GoogleMaps	45:00h	4,784	—	—
LiveMaps	42:26h	4,750	48:15h	4,595
Map24	44:42h	4,410	53:03h	4,351
Maporama	58:58h	5,744	45:55h	5,364
MapQuest	40:35h	4,398	41:59h	4,375
MultiMap	47:05h	4,392	47:39h	4,407
ViaMichelin	42:12h	4,424	51:09h	4,345
YahooMaps	44:42h	4,402	—	—

TABLE 1. Results obtained for the quickest and the shortest path from California City, California, to New York, New York (USA).

	Quickest path		Shortest path	
Web page	Time	Km	Time	Km
GoogleMaps	27:00h	2,965	—	—
LiveMaps	26:25h	2,920	29:08h	2,735
Map24	31:30h	2,915	34:20h	2,755
Maporama	24:59h	2,972	28:53h	2,713
MapQuest	26:00h*	2,979*	31:43h*	2,746*
MultiMap	31:35h	2,929	33:19h	2,838
ViaMichelin	30:19h	2,903	36:52h	2,749
YahooMaps	25:08h	2,982	—	—

TABLE 2. Results obtained for the quickest and the shortest path from Lisbon (Portugal) to Wien (Austria). **MapQuest only considers driving connection among 12 countries in Europe. Consequently, the results presented in this table include the distance/time from Lisbon to Salamanca and then the MapQuest results from Salamanca to Wien (Spain is included in the countries supported by MapQuest but this is not the case of Portugal).*

for computing shortest paths. Maporama's results for the California City to New York City route are particularly anomalous, with the shortest path actually being quicker than the reported quickest path, which is much longer and slower than the results for any of the other services. As might be expected, given the lack of a standard definition of "quickest," there appears to be more variability in the duration of quickest paths than in the lengths of shortest paths. For instance, in the results presented in Tables 1–2, the variation coefficient (that is, the ratio between the standard deviation and the mean) for travel time is around 8%, whereas the variation coefficient for shortest path length approaches 2% (this computation excludes the anomalous results obtained by *Maporama*).

In addition to performing queries using the various services, we also attempted to learn more about the SP algorithms they used, and were able to obtain partial information for three of the services, GoogleMaps, MapQuest, and LiveMaps:

GoogleMaps: Barry Brumitt, Google Software Engineer, has published some notes [**Bru07a, Bru07b, Bru07c**] about features of the GoogleMaps software, such as the use of the mapReduce framework [**DG04**] to process large geographic datasets across thousands of machines simultaneously. The following quote, obtained directly from Brumitt, provides hints as to how the software works: *I'm afraid the technology and algorithms we use are not public, and therefore I can't disclose very much about our process, nor do we have any papers or public documents. We start with a raw graph data structure with on order of 100M vertices. We use a series of mapReduce (MR) passes to transform the data into a (special) graph structure suitable for processing in MR, and then we do a bunch of additional MR work to build a data structure which can support efficient online queries. We didn't use MR when processing the actual live queries. This entire process is exact, in that we do find actual minimum-cost shortest paths.*

MapQuest: Sara Robinson interviewed Marc Smith, MapQuest's chief technical developer, in *SIAM News* in 2004 [**Rob04**]. He did not provide detailed information about how their algorithm worked, but revealed a little about it. Here are some excerpts from the interview: ... *MapQuest uses a "double Dijkstra" algorithm for its driving directions, ... Smith also conceded that the algorithm uses heuristic tricks to minimize the size of the graph that must be searched. For edge weights, Smith said, the algorithm uses estimated driving times, rather than distances, based on classifications of the different types of roads. The MapQuest software also takes into account such factors as left-turns and long-term construction, .. "Over time, we have done such radical things to the algorithm that it totally disresembles the original algorithm" Smith said. MapQuest gets all its road data from about 30 different vendors ...* Smith would not describe how the data was stored, since that was "what makes MapQuest Mapquest."

From this interview, it appears that MapQuest uses a bi-directional Dijkstra algorithm, possibly combined with A* search (the *"heuristic tricks"*). In the interview, Marc Smith also said that the software "runs on about three dozen Sun servers..." but did not reveal whether the software uses a parallel SP algorithm or simply has different machines assigned to different geographical areas.

LiveMaps: At the time of the Challenge, LiveMaps combined A* search and heuristics based on road categories [**Gol08**].

Note that there are very efficient algorithms that exactly solve the sort of point-to-point SP problems considered by these services, where the underlying network may have tens of millions of vertices. These algorithms exploit various ideas, such as landmark-based lower bounds [**GH05, GW05**], reach pruning techniques [**Gut04**], highway hierarchies [**SS05, SS07, BFSS07**], parallel programming [**MBBC07**], geometric containers [**SWW00**], partition-based hierarchical techniques [**HSW06**], etc. Additional references can be found in [**Sch**], as well as in some of the other papers in this volume. These techniques are all based on performing a preprocessing step that then allows individual queries to be processed very quickly, something we presume the web-based services do as well, even if they only compute approximate solutions.

2.2. Constructing public transportation itineraries. For those who would prefer to take public transportation, there are also many online services that will help you plan your route. Now however, the problem is a bit more complicated, since routes must be feasible as well as short – for instance, your next train cannot depart from a station before your current one arrives. There also can be multiple optimization criteria. It may be just as important to have as few train/plane changes as to minimize your total travel time. Price as well may be a consideration. In the airline industry, sites like Orbitz.com or Expedia.com deal with this problem by providing the user with long lists of feasible routes, many of them quite roundabout, from which one can choose. It is not clear that any sophisticated algorithms are used in generating these. For train travel in Europe, however, more sophisticated travel planning sites exist, such a the HAFAS system at `plannerint.b-rail.be`.

A key observation that travel planning systems can exploit is that finding a feasible route with minimum travel time can be reduced to an ordinary shortest path problem. One constructs a "time expanded" version of the underlying railroad network and timetable, in which there is a vertex for each potential event, where an event is the arrival or departure of a particular train from a particular station. For each leg of a scheduled train's itinerary, there is an arc joining the departure event to the next arrival event for that train, whose length is the time that leg is scheduled to take. For each station and train arrival, there are arcs to all departure events for that station that occur afterward (sufficiently long afterward if a train change is required), whose lengths are the times between the arrival and the corresponding departures. If one appropriately augments such a graph with special nodes corresponding to the origin and destination of the planned trip, a shortest path from the former to the latter will correspond to the feasible itinerary with minimum total travel time.

One problem with this approach is that the time-expanded network can be quite large. An alternative approach is the "time-dependent" model where there is only one node per station and the length of an arc depends on the time it is traversed – see [**MHSWZ07**] for the details and additional references. A modified version of Dijkstra's algorithm is still able to compute SSSPs by exploiting the assumption that if two trains travel the same non-stop route between two stations, the first to depart is also the first to arrive. The time-dependent approach was used for instance by Nachtigal [**Nac95**], who computed P2P shortest paths by combining the modified Dijkstra algorithm with A* search techniques and reported computational results for German railways consisting of 26 lines and 37 stations.

Pyrga *et al.* [**PSWZ04**] evaluated both the time-expanded and the time-dependent approaches in studying a bi-criteria version of the problem in which one is interested in minimizing both the travel time and the number of transfers, and so must examine the tradeoffs between the two criteria. They obtained computational results for German timetables from the winter period 2000/01 for both long distance Berlin/Brandenburg and Rhein/Main region traffic. Their biggest instance yielded networks with more than 30,000 vertices and 90,000 arcs for the time-dependent approach and more then 2,250,000 vertices and 4,500,000 arcs for the time-expanded model.

The on-line system used by the Portuguese Railway since 1998 also exploits a time-dependent model and was partially designed by Martins, Pascoal and Santos of the Department of Mathematic of the University of Coimbra, Portugal,

2.3. Routing using the OSPF/IS-IS protocols. Internet routing is another domain in which large numbers of shortest path computations are performed every day. In particular, many Internet Service Providers (ISPs) route packets within their networks along shortest paths, where the edge lengths (link weights) are parameters that can be set by traffic engineers with the goal of minimizing congestion. The two most-common intra-domain routing protocols are Open Shortest Path First (OPSF) [**Moy98**] and Intermediate System to Intermediate System (IS-IS) [**Cal90**]. Under both of these protocols, routers use versions of Dijkstra's algorithm to solve a local SSSP problem every time they learn of a failed link, a new link, or a change in a link weight within the ISP's network [**Cis, Con**]. (OSPF is a bit more complicated because of its use of "Areas," for each of which separate SSSP computations must be performed, and by the special role of Area 0 in routing – interested readers can find more details in [**Moy98**].) The protocols then update their routing tables, which implicitly contain, for each potential destination router within the network, the first link to be followed on a shortest path from the current router to the destination.

Link weights must be integers in the range from 1 to $2^{16} - 1 = 65,535$. They can be set by suggested rules of thumb, such as all equal, proportional to distance, or inversely proportional to link capacity [**Cis97**]. Better results can be obtained in practice, however, if one has a good estimate of the point-to-point traffic in the network. Fortz and Thorup [**FT00**] suggest that weight settings that yield near-optimal congestion can be obtained using metaheuristic search techniques. (It is NP-hard to find the best possible weight settings for minimizing congestion.) The Fortz-Thorup algorithms for determining weights involve repeatedly solving sets of SSSP problems under slight changes in the weight settings, a task which would be hopelessly impractical if one had to perform a full Dijkstra computation at each step. Instead, they make use of dynamic shortest path algorithms, which save extra information so that shortest path information can be updated quickly when a link weight is changed (see for example [**BRT08**]). Note that the updating of OSPF/ISIS routing tables is also a dynamic SSSP problem, but here the updates are sufficiently infrequent that the extra speed of the dynamic approach is not needed, especially given that the networks involved are not large – typically hundreds of vertices or less, compared to the millions of vertices we saw in the trip-routing application of Section 2.1.

3. Other Routing Applications and Variants

3.1. Shortest paths with turn prohibitions. In urban traffic, the movement of vehicles at intersections is often restricted by no-left-turn, no-right-turn, or no U-turn signs – see Figure 1(a) and (b). This give rise to variants on the standard SP problems in which no path is allowed to violate a turn restriction. We can reduce such a problem to a standard SP problem by transforming the network into a directed network that models the turn restrictions using additional vertices and arcs. Such transformations can also allow one to model turn penalties (e.g., the fact that one has to slow down before making a turn).

One way of modeling turn restrictions [**SJK03**] starts by adding, for each arc $A = (u, v)$, two new arc-specific vertices Au and Av and a new arc (Au, Av) with the same length as A. Then, for every triple (u, v, w) of original vertices such that in the original network there are arcs $A = (u, v)$ and $B = (v, w)$, where B can be traversed immediately after A without violating a turn restriction, we add a zero-length arc (Av, Bv). Finally, we delete all the original vertices and arcs. See Figure 1(d). Now the turn-restricted problem can be solved by a standard SP algorithm in the derived network.

A second, more compact way of modeling turn restrictions is to construct a graph whose vertices correspond to the arcs in the original network [**Cal61**]. Suppose (u, v) and (v, w) are arcs in the original network and X and Y are the corresponding vertices of the derived graph. There is an arc (X, Y) in the derived graph if and only if the turn restrictions allow us to traverse (v, w) immediately after (u, v) (see Figure 1(c)). Things are a bit more complicated here, since now it is the vertices that have length. However, Gutirrez and Medaglia [**GM08**] propose an extension of Dijkstra's algorithm that implicitly handles turn restrictions using this approach. In their paper, they present computational results for instances as large as Bogota's road network, which has 69,431 vertices and 213,144 arcs, and for which the network representation has 213,144 vertices and 667,574 arcs.

3.2. Shortest paths in dynamic networks. Huang *et al.* [**HWZ07**] proposed an algorithm for determining the least cost path between a moving object and its destination by continually adapting the dynamic traffic conditions, while making use of the previous search results. This algorithm is based on a variation of the A* algorithm [**KLF04**] with an additional technique that prunes the unnecessary vertices to speed up the dynamic search process. They examined the suitability and performance of the proposed algorithm on the road networks of Calgary (with approximately 8000 vertices and 12500 arcs) and Singapore (with around 7000 vertices and 11800 links).

3.3. K shortest paths and transportation scheduling problems. In many applications, simply computing the shortest path may not be enough. For

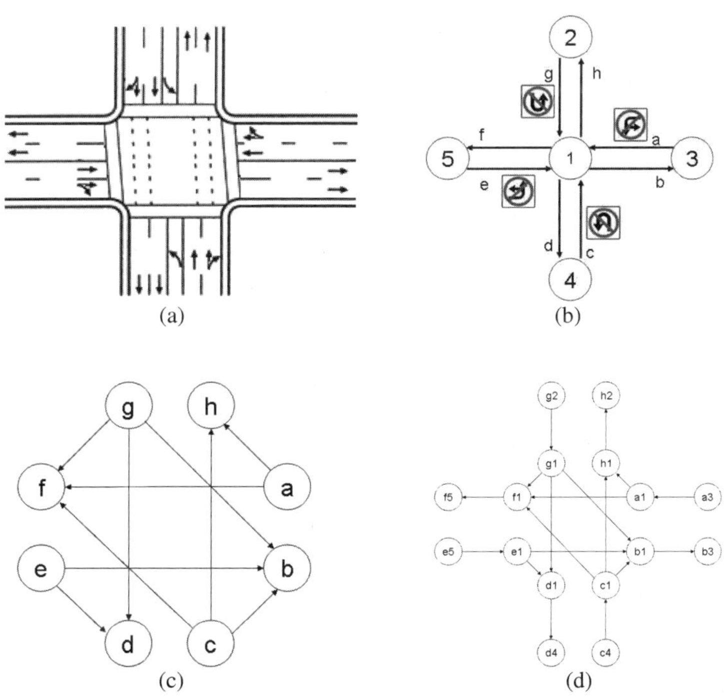

FIGURE 1. Example of turn prohibitions in urban traffic.

example, it can happen that, when we want to use the path, some of its arcs are temporarily unavailable (for example, when driving we might hear on the radio that there is a car accident at some point on our chosen route). In such cases, it might help to have already computed an alternative path. The need for alternative paths can also arise when there are constraints on legal paths that, unlike the turn restrictions mentioned above, are too difficult to include in a simple graph model. In such cases, we may want enumerate the shortest paths in order, and choose the first one that satisfies all the constraints. Similarly, if we confront a problem with multiple objective functions, having a list of the shortest paths with respect to each criterion may enable us to find one that is simultaneously good for all, or more generally to identify all the *Pareto optimal* solutions, where a solution is Pareto optimal if no other solution is better than it under *all* the objective functions [**CM82, San03, PS08**].

The K Shortest Path Problem (K-SPP) was devised to address such situations. Formally, in the K-SPP we are looking for an ordered set of K paths $\{p_1, \ldots, p_K\}$ from an initial vertex s to a terminal vertex t, where the cost (length) of p_i is not greater than the cost of p_{i+1} ($i \in \{1, \ldots, K-1\}$) and no other path from s to t have cost less than the cost of p_K. Currently the theoretically fastest algorithm for the K-SPP is that of Eppstein [**Epp98**], who has also prepared a useful online bibliography on the problem [**Epp01**].

Computing the K shortest paths may also be useful when we want to estimate the relevance of various arcs to the problem of going from s to t, where we can use the number of the K shortest paths that contain a given arc as a measure of its importance. Such information could for instance be used in setting priorities for road maintenance. Another use would be in sensitivity analysis. Knowledge of the K shortest paths lets us determine whether the shortest path is substantially better than its successors, or differs substantially from them in the arcs it contains.

As a final, and more specific, type of application for the K-SPP, consider the problem of column generation for crew scheduling problems. Column generation is a technique that is used to solve implicitly formulated linear or integer programs that potentially contain an exponential number of variables (columns). For such LPs, one typically starts by solving a version restricted to a relatively small collection of variables, thus obtaining an upper bound on the optimal solution (assuming we are minimizing). We then look for promising variables to add in hopes of finding a better solution. For crew scheduling applications, where the variables correspond to the possible routes that a pilot, driver, or other crew member may follow throughout a day or week, the column generation process often reduces to finding shortest paths in a reduced-cost graph, and it can be much more efficient to add multiple variables at a time, as can be done using K-SPP algorithms. Here are two real-world examples of this approach from the literature.

Airline crew scheduling: Medard and Sawhney [**MS07**] considered the problem of scheduling crews for airlines, and provided algorithms to the Swedish company Carmen Systems (subsequently acquired by the Jeppesen navigation services subsidiary of Boeing). Their approach was based on an integer programming model where the integer variables corresponded to crew itineraries selected to cover all the flights' needs. In their column generation scheme, they needed to find attractive paths that respected additional restrictions, and they used a K-SPP routine based on the work [**MS00**] as a tool. There test instances, derived from a Swedish airline database, required applying the K-SPP algorithm to a network with around 4,000 vertices and 60,000 arcs. In the first stage, the procedure needed to rank almost one thousand paths, but then, as the column generation proceeded, this number fell to a few hundred, and finally ended with less than 10.

Bus driver scheduling: Pais and Paixão [**Pa06**] use a K-SPP routine for column generation in a bus driver scheduling problem. The problem is formulated as a set covering model, where columns correspond to driver itineraries and rows to continuous driving periods. They adopted a branch-and-price approach based on a proposed branching rule and a combination of state space relaxation with column generation. The columns of the set covering problem correspond to paths in an associated network and are generated as needed by solving shortest path problems with resource constraints using a k-shortest paths routine, [**Mar84**]. Pais and Paixão present computational results for real instances concerning several mass transit companies operating in Portugal. The biggest set covering model has 96 rows and 2015 columns.

3.4. Distances in social networks. In a *social network*, the vertices typically represent people, telephone numbers, companies, computers, or similar objects, with edges joining vertices that communicate or are associated with each other, and the length of an edge reflecting the intensity of the association (shorter edges representing stronger associations). In this section we discuss two examples of social networks where shortest path length is of key concern.

Six degrees of separation: (This discussion is adapted from the Wikepedia article on the topic [**Wik**].) The classic example of a social network is the one in which the vertices are people and there is an edge between two people if they know each other (all edges of length 1). This network has become a part of popular culture as a result of the famous "Six Degrees of Separation" conjecture, apparently originally stated by Frigyes Karinthy. This Hungarian author published a volume of short stories in 1929 including one titled "Chain-Links" where he essentially asserted that the maximum distance between any two people in the world-wide social network was 6.

In recent years, this conjecture and its variants have caught the attention of the public. John Guare, an American playwright, wrote the play Six Degrees of Separation in 1990, which was turned into a movie starring Will Smith in 1993. Soon thereafter, the game "*Six Degrees of Kevin Bacon*" was devised for the social network whose vertices are movie actors, with an edge between two actors if they have appeared in a movie together. The challenge here is to find a path of length 6 or less from any other given actor to Kevin Bacon. The related concept of the *Bacon Number* of an actor is simply the length of the shortest path in this graph from the actor to Kevin Bacon. This is an adaptation of the earlier concept, devised by mathematicians, of the *Erdös Number* of a researcher. Here the underlying social network consists of authors of research papers, with an edge between two researchers if they are co-authors of some paper. A researcher's Erdös Number is simply the length of the shortest path in this graph from the researcher to Erdös, with low numbers highly prized.

Given complete lists of the vertices and edges of a social network (which unfortunately have not yet been compiled for the above examples), it would be a simple algorithmic task to verify the 6 Degrees hypothesis, and if 6 is not the correct number, to determine what that correct number should be. Simply run an APSP algorithm on the graph and report the longest shortest path length found, i.e., the diameter of the graph. Such a computation has been made for a large related social network. In 2007, Jure Leskovec and Eric Horvitz [**LH07**], examined a data set of 30 billion messages and constructed a communication graph with 180 million vertices (Microsoft Messenger users) and 1.3 billion undirected edges. They found the average path length among Microsoft Messenger users to be 6.6, close to the original conjecture. Note, however, that this is the average, not the maximum. The maximum was 29.

Fighting organized crime: Xu and Chen [**XC04**] considered a social network built by extracting noun phrases from crime reports. The vertices of the network are records of entities (persons, organizations, vehicles, properties, and locations) stored in crime databases. There is an edge between a pair of entities if they appear together in the same criminal incident report. The more frequently they appear together, the stronger the association weight (shorter the edge length). The network is undirected and usually has high vertex-degrees. Given a subset N of vertices, Xu and Chen wished to identify criminal associations by determining the shortest paths between all pairs of vertices in N. This could be accomplished for instance by a straightforward "one-directional" approach in which, for each vertex i in N, one runs Dijkstra's algorithm until it had determined the shortest path from i to each other vertex in N.

Computational results were obtained for two networks based on real data taken from kidnapping and narcotics reports. In the first case, the network had 280 vertices and 25,862 arcs (average vertex-degree: 92.4) and in the second one there was 4,257 vertices and 733,572 arcs (average vertex-degree: 172.3). The abovementioned one-directional approach worked best for the narcotics network, but for the kidnapping network was outperformed by a bi-directional approach.

4. Dynamic Programming

Dynamic programming is a method for efficiently solving optimization problems that can be broken down into a partially ordered set of sub-problems, where the optimal solution for one subproblem is determined by the optimal solutions for its predecessors. The name "dynamic programming" was introduced by Richard Bellman [**Bel52, Bel53**] who first systematized the approach. Most dynamic programming problems can be reformulated as shortest path problems on directed acyclic graphs. As an illustration, consider Knuth's paragraph formatting problem.

In this problem we wish to determine the break points for separating a string of words into lines to obtain a formatted paragraph. When formatted text is required to be aligned with both the left and right margins, the choice of break points greatly affects the quality of the formatted document. The document processing program TeX uses a dynamic programming algorithm [**KP81**] to do its line breaking. Such a program might work as follows:

Suppose we have a paragraph formed by n words $w_1, \ldots, w_n$, where the word w_k has length $|w_k|$ and L is the maximum length allowed for lines in the paragraph, with $|w_k| \leq L$ for all k. For $1 \leq i \leq j \leq n$, let $W(i,j)$ denote the line that starts with word w_i and ends with word w_j. If we assume equal-length characters, including spaces, then the length of line $W(i,j)$ is $L(i,j) = (j-i) + \sum_{k=i}^{j} |w_k|$. Suppose that we have a demerit function that assigns a penalty value $c(i,j)$ for each such line. As an example, one could use

$$c(i,j) = \begin{cases} \infty & \text{if } L(i,j) > L \\ (L - L(i,j))^2 & \text{otherwise} \end{cases}$$

To find the collection of linebreaks that minimizes the total penalty for the paragraph (assuming there is no penalty for the last line if it has length L or less, since it need not be right justified), all we need do is find a shortest path in the following graph.

There are $n+1$ vertices, $v_0, v_1, \ldots v_n$, where vertex v_i represents the text through word w_i. We then have an arc for each possible line. Specifically, for each pair i, j with $1 \le i \le j \le n$, there is an arc (v_{i-1}, v_j) representing $W(i,j)$. The length of this arc is $c(i,j)$ unless $j = n$ and $L(i,j) \le L$, in which case it is 0. It is not difficult to confirm that if $v_0 = v_{i_0}, v_{i_1}, v_{i_2}, \ldots, v_{i_k} = v_n$ is a shortest path from v_0 to v_n in this graph, then there is a minimum-cost version of the paragraph consisting of the lines $W(i_h + 1, i_{h+1})$, $0 \le h < k$.

There are thousands of applications of dynamic programming that can be solved by shortest path algorithms, as here, from multiple sequence alignment in DNA reconstruction [**LR00**] to parsing continuous speech [**Ney91**] to database query optimization [**Cha98**]. The interested reader can find further examples for in [**AMO93**], for instance. Note that because the graphs that arise are typically directed and acyclic, one can dispense with the priority queue in implementing Dijkstra's algorithm and simply treat the vertices in breadth first order, yielding a running time that is linear in the number of edges.

5. Conclusion

This paper has surveyed algorithms and applications for the standard shortest path problems (P2PSP, SSSP, APSP, K-SPP) and some of their variants. Of necessity, we have been highly selective, particularly in the area of dynamic programming applications. However, there can be no doubt of the importance of shortest path computations, and the interested reader is encouraged to read the other papers in this volume and follow the pointers in our bibliography to learn more.

Acknowledgments

The author would like to express his sincere thanks to the referees for their valuable suggestions which helped him in modifying and improving the contents of the manuscript, and to David Johnson who helped in organizing and clarifying the presentation.

References

[AGM97] N. Alon, Z. Galil, and O. Margalit, *On the exponent of the all pairs shortest path problem*, Journal of Computer and System Sciences **54** (1997), no. 2, 255–262.

[AMO89] R. K. Ahuja, T. L. Magnanti, and J. Orlin, *Handbooks in operations research and management science. vol i: optimization*, ch. Network Flows, pp. 211–369, North-Holland, Amsterdam, 1989.

[AMO91] R. K. Ahuja, T. L. Magnanti, and J. B. Orlin, *Some recent advances in network flows*, SIAM Rev. **33** (1991), no. 2, 175–219.

[AMO93] ———, *Network flows – theory, algorithms, and applications*, Prentice-Hall, Inc., New Jersey, 1993.

[AMOT90] R. K. Ahuja, K. Mehlhorn, J. Orlin, and R. E. Tarjan, *Faster algorithms for the shortest path problem*, J. ACM **37** (1990), no. 2, 213–223.

[Bel52] R. E. Bellman, *On the theory of dynamic programming*, Proceedings of the National Academy of Sciences, vol. 38, 1952, pp. 716–719.

[Bel53] ———, *Dynamic programming and a new formalism in the calculus of variations*, Proceedings of the National Academy of Sciences, vol. 39, 1953, pp. 1077–1082.

[Bel58] ———, *On a routing problem*, Quarterly Applied Mathematics **16** (1958), 87–90.

[BFSS07] H. Bast, S. Funke, P. Sanders, and D. Schultes, *Fast routing in road networks with transit nodes*, Science **316** (2007), no. 5824, 566.

[BRT08] L. S. Buriol, M. G. C. Resende, and M. Thorup, *Speeding up dynamic shortest-path algorithms*, INFORMS J. Comput. **20** (2008), no. 2, 191–204.

[Bru07a] B. Brumitt, http://igniteseattle.com/index.php?s=brumitt, February 2007.

[Bru07b] ———, http://googleblog.blogspot.com/2007/11/road-to-better-path-finding.html, September 2007.

[Bru07c] ———, *The road to better path finding*, http://googleblog.blogspot.com/2007/11/road-to-better-path-finding.html, June 2007.

[Cal61] T. Caldwell, *On finding minimal routes in a network with turn penalties*, Com. ACM **4** (1961), no. 2, 107–108.

[Cal90] R. W. Callon, *Use of osi is-is for routing in tcp/ip and dual environments*, 1990.

[CGR96] B. V. Cherkassky, A. V. Goldberg, and T. Radzik, *Shortest paths algorithms: Theory and experimental evaluation*, Mathematical Programming **73** (1996), no. 2, 129–174.

[CGS99] B. V. Cherkassky, A. V. Goldberg, and C. Silverstein, *Buckets, heaps, lists, and monotone priority queues*, SIAM Journal on Computing **28** (1999), no. 4, 1326–1346.

[Cha98] S. Chaudhuri, *An overview of query optimization in relational databases*, Proc. ACM Symp. on Principles of Database Systems, 1998, pp. 34–43.

[Cis] Cisco, *OSPF design guide*, http://www.cisco.com/warp/public/104/2.html.

[Cis97] ———, *Configuring OSPF, 1997*, 1997.

[CM82] J. N. Clímaco and E. Q. Martins, *A bicriterion shortest path algorithm*, European Journal of Operational Research **11** (1982), 399–404.

[Con] Data Connection, *Is-is protocol: Intermediate system - intermediate system*, http://www.dataconnection.com/iprouting/isisprotocol.htm.

[Dan51] G. B. Dantzig, *Activity analysis of production and allocation*, ch. Application of the simplex method to a transportation problem, pp. 359–373, NY, John Wiley & Sons, 1951.

[Dan60] ———, *On the shortest route through a network*, Management Science **6** (1960), 187–190.

[DF79] E. V. Denardo and B. L. Fox, *Shortest route methods: reaching, pruning and buckets*, Operations Research **27** (1979), 161–186.

[DG04] J. Dean and S. Ghemawat, *Mapreduce: Simplified data processing on large clusters*, Proceedings of the Sixth Symposium on Operating System Design and Implementation (OSDI) 2004, 2004, San Francisco, CA, 2004.

[DGKK79] R. Dial, F. Glover, D. Karney, and D. Klingman, *A computational analysis of alternative algorithms and labeling techniques for finding shortest path trees*, Networks **9** (1979), no. 3, 215–248.

[Dia69] R. Dial, *Algorithm 360. shortest path forest with topological ordering*, Communications of ACM **12** (1969), 632–633.

[Dij59] E. W. Dijkstra, *A note on two problems in connexion with graphs*, Numerische Mathematik **1** (1959), 269–271.

[DP84] N. Deo and C. Pang, *Shortest path algorithms: taxonomy and annotation*, Networks **14 (2)** (1984), 275–323.

[Epp98] D. Eppstein, *Finding the k shortest paths*, SIAM J. Computing **28** (1998), no. 2, 652–673.

[Epp01] ———, *K shortest paths and other "k best" problems*, 2001, `http://www.ics.uci.edu/~eppstein/bibs/`.

[EZ62] J. H. Eaton and L. A. Zadeh, *Optimal pursuit strategies in discrete-state probabilistic systems*, Trans. ASME Ser. D, J. Basic Eng **84** (1962), 23–29.

[FF62] L. R. Ford and D. R. Fulkerson, *Flows in networks*, Princeton University Press, Princeton, NJ, 1962.

[Flo62] R. W. Floyd, *Algorithm 97: Shortest path*, Commun. ACM **5** (1962), no. 6, 345.

[For56] L. R. Ford, *Network flow theory*, Tech. Report Paper P-923, The Rand Corporation, Santa Monica, 1956.

[FT87] M. L. Fredman and R. E. Tarjan, *Fibonacci heaps and their uses in improved network optimization algorithms*, J. ACM **34** (1987), no. 3, 596–615.

[FT00] B. Fortz and M. Thorup, *Internet traffic engineering by optimizing ospf weights*, in Proc. IEEE INFOCOM, 2000, pp. 519–528.

[GGK84] F. Glover, R. Glover, and D. Klingman, *Computational study of an improved shortest path algorithm*, Networks **14** (1984), no. 1, 25–36.

[GH05] A. V. Goldberg and C. Harrelson, *Computing the shortest path: A* search meets graph theory*, SODA '05: Proceedings of the sixteenth annual ACM-SIAM symposium on Discrete algorithms (Philadelphia, PA, USA), Society for Industrial and Applied Mathematics, 2005, pp. 156–165.

[GJ79] M. R. Garey and D. S. Johnson, *Computers and intractability: A guide to the theory of NP-completeness*, W. H. Freeman, San Francisco, 1979.

[GKW07] A. V. Goldberg, H. K., and R. F. Werneck, *Better landmarks within reach.*, WEA (Camil Demetrescu, ed.), Lecture Notes in Computer Science, vol. 4525, Springer, 2007, pp. 38–51.

[GM97] Z. Galil and O. Margalit, *All pairs shortest paths for graphs with small integer length edges*, Journal of Computer and System Sciences **54** (1997), no. 2, 243–254.

[GM08] E. Gutirrez and A. L. Medaglia, *Labeling algorithm for the shortest path problem with turn prohibitions with application to large-scale road networks*, Ann Oper Res **157** (2008), 169–182.

[Gol01] A. V. Goldberg, *A simple shortest path algorithm with linear average time*, Proceedings of the 4th. Annual European Symposium Algorithms, vol. 230–241, 2001.

[Gol08] ———, 2008, Personal communication.

[GP88] G. Gallo and S. Pallottino, *Shortest Paths Algorithms*, Annals of Oper. Res. **13** (1988), 3–79.

[GT89] H. N. Gabow and R. E. Tarjan, *Faster scaling algorithms for network problems*, SIAM Journal on Computing **18** (1989), no. 5, 1013–1036.

[Gut04] R. J. Gutman, *Reach-based routing: A new approach to shortest path algorithms optimized for road networks*, ALENEX/ANALC, 2004, pp. 100–111.

[GW05] A. V. Goldberg and R. F. Werneck, *Computing point-to-point shortest paths from external memory.*, ALENEX/ANALCO (Camil Demetrescu, Robert Sedgewick, and Roberto Tamassia, eds.), SIAM, 2005, pp. 26–40.

[Hag00] T. Hagerup, *Improved Shortest Paths in the Word RAM*, 27th Int. Colloq. on Automata, Languages and Programming, Geneva, Switzerland, 2000, pp. 61–72.

[HNR68] P. E. Hart, N. J. Nilsson, and B. Raphael, *A formal basis for the heuristic determination of minimum cost paths*, IEEE Transactions on Systems Science and Cybernetics **4** (1968), no. 2, 100–107.

[HNR72] ———, *Correction to "a formal basis for the heuristic determination of minimum cost paths"*, SIGART Bull. (1972), no. 37, 28–29.

[HP59] W. Hoffman and R. Pavley, *A method for the solution of the th best path problem*, J. ACM **6** (1959), no. 4, 506–514.

[HSW06] M. Holzer, F. Schulz, and D. Wagner, *Engineering multi-level overlay graphs for shortest-path queries*, Proceedings of the Eighth Workshop on Algorithm Engineering and Experiments (ALENEX), SIAM, 2006, pp. 156–170.

[HWZ07] B. Huang, Q. Wu, and F. B. Zhan, *A shortest path algorithm with novel heuristics for dynamic transportation networks*, International Journal of Geographical Information Science **21** (2007), no. 6, 625–644.

[Joh77] D. B. Johnson, *Efficient algorithms for shortest paths in sparse networks*, Journal of the ACM **24** (1977), no. 1, 1–13.

[Joh82] ———, *A priority queue in which initialization and queue operations take o(log log d) time*, Mathematical Systems Theory **15** (1982), no. 4, 295–309.

[Jok66] H. Joksch, *The shortest route problem with constraints*, Journal of Mathematical Analysis and Applications **14** (1966), 191–197.

[KKP93] D. R. Karger, D. Koller, and S. J. Phillips, *Finding the hidden path: Time bounds for all-pairs shortest paths*, SIAM Journal on Computing **22** (1993), no. 6, 1199–1217.

[KLF04] S. Koenig, M. Likhachev, and D. Furcy, *Lifelong planning a**, Artif. Intell. **155** (2004), no. 1–2, 93–146.

[KP81] D. E. Knuth and M. F. Plass, *Breaking paragraphs into lines*, Software: Practice and Experience **11** (1981), no. 11, 1119–1184.

[LH07] J. Leskovec and E. Horvitz, *Planetary-scale views on an instant-messaging network*, Technical Report MSR-TR-2006-186, Microsoft Research, June 2007, Also available at `http://arxiv.org/abs/0803.0939v1`. A shorter version appeared at *WWW 2008*.

[LR00] M. Lermen and K. Reinert, *The practical use of the a* algorithm for exact multiple sequence alignment*, Journal of Computational Biology (2000), 655–671.

[Mar84] E. Q. Martins, *An algorithm for ranking paths that may contain cycles*, European Journal of Operational Research **18** (1984), 123–130.

[MBBC07] K. Madduri, D. A. Bader, J. W. Berry, and J. R. Crobak, *An experimental study of a parallel shortest path algorithm for solving large-scale graph instances*, 2007 Proceedings of the Ninth Workshop on Algorithm Engineering and Experiments (ALENEX), New Orleans, Louisiana, January 2007.

[Mey01] U. Meyer, *Single-Source Shortest Paths on Arbitrary Directed Graphs in Linear Average Time*, 12th Symp. on Discr. Alg., 2001, pp. 797–806.

[MG04] R. Montemanni and L. M. Gambardella, *An exact algorithm for the robust shortest path problem with interval data*, Computers and Oper. Res. **31** (2004), no. 10, 1667–1680.

[MHSWZ07] M. Müller-Hannemann, F. Schulz, D. Wagner, and C. Zaroliagis, *Algorithmic methods for railway optimization*, Lecture Notes in Computer Science, vol. 4359, ch. Timetable Information: Models and Algorithms, pp. 67–90, Springer Berlin / Heidelberg, September 2007.

[Min58] G. J. Minty, *A variant on the shortest-route problem*, Operations Research **6** (1958), no. 6, 882–883.

[Moo57] E. F. Moore, *The shortest path through a maze*, Proceeding of the International Symposium on the Theory of Switching (Part II), vol. 30, Harvard University Press, 1957, pp. 285–292.

[Moy98] J. T. Moy, *OSPF: Anatomy of an internet routing protocol*, Addison-Wesley Longman Publishing Co., Inc., Boston, MA, USA, 1998.

[MS00] E. Q. Martins and J. L. Santos, *A new shortest paths ranking algorithm*, Investigação Operacional **20 (1)** (2000), 47–62.

[MS07] C. P. Medarda and N. Sawhneyb, *Airline crew scheduling from planning to operations*, European Journal of Operational Research **183** (2007), no. 3, 1013–1027.

[Nac95] K. Nachtigal, *Time depending shortest-path problems with applications to railway networks*, European Journal of Operations Research **83** (1995), 154–166.

[Ney91] H. Ney, *Dynamic programming parsing for context-free grammars in continuous speech recognition*, IEEE Trans. Signal Processing **39** (1991), no. 2, 336–340.

[Pa06] A. Pais and J. M. Paix ao, *A branch-and-price approach for the bus driver scheduling problem*, Working Paper 3, Centro de Investigação Operacional da Faculdade de Ciências da Universidade de Lisboa, 2006.

[Pal84] S. Pallottino, *Shortest-path methods: Complexity, interrelations and new propositions*, Networks **14** (1984), no. 2, 257–267.

[Pap74] U. Pape, *Implementation and efficiency of moore-algorithms for the shortest route problem*, Mathematical Programming **7** (1974), 212–222.

[Pet04] S. Pettie, *A new approach to all-pairs shortest paths on real-weighted graphs*, Theoretical Computer Science **312** (2004), no. 1, 47–74.

[PS08] J. M. Paixão and J. L. Santos, *A new ranking path algorithm for the multi-objective shortest path problem*, Tech. Report 08-27, Department of Mathematics of the University of Coimbra, 2008, Submitted for publication (revision phase).

[PSWZ04] E. Pyrga, F. Schulz, D. Wagner, and C. Zaroliagis, *Experimental comparison of shortest path approaches for timetable information*, Proceedings of the Sixth Workshop on Algorithm Engineering and Experiments (ALENEX), SIAM, 2004, pp. 88–99.

[Ram96] R. Raman, *Priority Queues: Small, Monotone and Trans-Dichotomous*, 4th Europ. Symp. on Algo., Springer-Verlag, Lect. Notes in CS 1136, 1996, pp. 121–137.

[Rob04] S. Robinson, *Mapping magic*, SIAM news (2004).

[RR96] G. Ramalingam and T. Reps, *An incremental algorithm for a generalization of the shortest path problem*, J. Algorithms **21** (1996), no. 2, 267–305.

[San03] J. L. Santos, *Optimização vectorial em redes*, Ph.D. thesis, Departamento de Matemática, Universidade de Coimbra, 2003.

[Sch] F. Schulz, *Shortest path algorithms: Some references*, http://i11www.iti.uni-karlsruhe.de/ fschulz/shortest-paths/.

[Sei95] R. Seidel, *On the all-pairs-shortest-path problem in unweighted undirected graphs*, Journal of Computer and System Sciences **51** (1995), no. 3, 400–403.

[SJK03] L. Speicys, C. S. Jensen, and A. Klygis, *Computing data modeling for network-constrained moving objects*, Proc. 11th ACM Int. Symp. on Adv. in Geo. Info. Sys., ACM Press, 2003, pp. 118–125.

[SS05] P. Sanders and D. Schultes, *Highway hierarchies hasten exact shortest path queries.*, ESA (Gerth Stlting Brodal and Stefano Leonardi, eds.), Lecture Notes in Computer Science, vol. 3669, Springer, 2005, pp. 568–579.

[SS07] ———, *Engineering fast route planning algorithms.*, WEA (Camil Demetrescu, ed.), Lecture Notes in Computer Science, vol. 4525, Springer, 2007, pp. 23–36.

[SWW00] F. Schulz, D. Wagner, and K. Weihe, *Dijkstra's algorithm on-line: an empirical case study from public railroad transport*, J. Exp. Algorithmics **5** (2000), 12.

[SZ99] A. Shoshan and U. Zwick, *All pairs shortest paths in undirected graphs with integer weights*, FOCS '99: Proceedings of the 40th Annual Symposium on Foundations of Computer Science (Washington, DC, USA), IEEE Computer Society, 1999, pp. 605–614.

[Tak92] T. Takaoka, *A new upper bound on the complexity of the all pairs shortest path problem*, Information Processing Letters **43** (1992), no. 4, 195–199.

[Tho04] M. Thorup, *Integer priority queues with decrease key in constant time and the single source shortest paths problem*, J. Comput. Syst. Sci. **69** (2004), no. 3, 330–353.

[vEBKZ76] P. van Emde Boas, R. Kaas, and E. Zijlstra, *Design and implementation of an efficient priority queue*, Theory of Computing Systems **10** (1976), no. 1, 99–127.

[Vin74] P. Vincke, *Problèmes multicritères*, Cahiers du Centre d'Études de Recherche Opérationelle **16** (1974), 425–439.

[Wag76] R. A. Wagner, *A shortest path algorithm for edge-sparse graphs*, J. ACM **23** (1976), no. 1, 50–57.

[War62] S. Warshall, *A theorem on boolean matrices*, J. ACM **9** (1962), no. 1, 11–12.

[WH60] P. D. Whiting and J. A. Hillier, *A method for finding the shortest route through a road network*, Operations Research Quarterly **11** (1960), 37–40.

[Wik] Wikipedia, *Six degrees of separation*, http://en.wikipedia.org/wiki/Six_degrees_of_separation.

[WW05] D. Wagner and T. Willhalm, *Drawing graphs to speed up shortest-path computations.*, ALENEX/ANALCO (Camil Demetrescu, Robert Sedgewick, and Roberto Tamassia, eds.), SIAM, 2005, pp. 17–25.

[XC04] J. J. Xu and H. Chen, *Fighting organized crimes: using shortest-path algorithms to identify associations in criminal networks*, Decis. Support Syst. **38** (2004), no. 3, 473–487.

[Zwi98] U. Zwick, *All pairs shortest paths in weighted directed graphs - exact and almost exact algorithms*, FOCS '98: Proceedings of the 39th Annual Symposium on Foundations of Computer Science (Washington, DC, USA), IEEE Computer Society, 1998, Los Alamitos, CA, November 8-11, 1998, pp. 310–319.

[Zwi02] ———, *All pairs shortest paths using bridging sets and rectangular matrix multiplication*, Journal of the ACM **49** (2002), no. 3, 289–317.

Department of Mathematics, University of Coimbra, Portugal

Current address: Department of Mathematics, Largo D. Dinis, Apartado 3008, 3001 454 Coimbra, Portugal. Fax number: 00 351 239793069

E-mail address: zeluis@mat.uc.pt

DIMACS Series in Discrete Mathematics
and Theoretical Computer Science
Volume **74**, 2009

An Experimental Evaluation of Point-To-Point Shortest Path Calculation on Road Networks with Precalculated Edge-Flags

Ulrich Lauther

ABSTRACT. An efficient algorithm for fast and exact calculation of shortest paths in graphs with geometrical information in nodes (coordinates), e.g., road networks, is presented. The method is based on preprocessing and therefore best suited for static graphs, i.e., graphs with fixed topology and edge weights.

In the preprocessing phase, the network is divided into regions and edge flags are calculated that indicate whether an edge belongs to a shortest path into a given region.

In the target application (say, a navigation system) where we want to calculate shortest paths, only those edges that carry the appropriate flag need to be investigated.

We compared this method to a classical Dijkstra implementation using USA road networks with three different edge weight metrics (travel times, distance, and unit edge weights) and report on speedup, preprocessing time, and memory needed to store edge flags.

1. Introduction

A huge amount of research work has been done concerning fast shortest path algorithms, but - if we disregard preprocessing methods for a moment - still efficient exact algorithms are based on Dijkstra's algorithm [**Dij59**], combined with efficient data structures for implementing the priority queue [**Dial79, Tar83, AhMaOr93**], which is a central part of an efficient implementation.

Given a directed or undirected graph $G = (V, E)$ with n nodes V and m edges E, edges $e = (v, w)$, positive edge weights $w(e)$, and two special nodes s (source) and t (sink), calculating a shortest path (i.e., a path with minimum total edge weight) between s and t can be solved efficiently using Dijkstra's algorithm. Depending on data structures used and additional assumptions on the density of the graph and the distribution of edge weights, the worst case complexity lies between $O(m + n\ log\ n)$ (for general graphs) and $O(m)$ (for a sparse graph with limited, integral edge weights [**Dial79**]).

1991 *Mathematics Subject Classification.* Primary 54C40, 14E20; Secondary 46E25, 20C20.
Key words and phrases. shortest path, preprocessing, navigation system.

In a typical geographical application, e.g., finding a shortest path in a road map, the *expected* running time of a careful implementation will be $O(d^2 \log d)$ if d is the number of edges in the shortest path; as we observe a circular expansion of the algorithm around the source node, $O(d^2)$ nodes will be touched and each one will be inserted into and retrieved from a priority queue with size $O(d^2)$ once. (Updates of the priority queue are of the same order, as the node degree in road networks is limited and small). "Careful implementation" implies for instance that - other than usually seen in textbooks - an initialization phase processing *all* nodes in the network is avoided.

Often these running times are prohibitive, e.g., in autonomous automotive navigation systems with limited resources (memory and CPU power). Note that the CPU speeds of current car navigation systems are - due to cost limitations - 10 to 20 times lower then that of today's typical personal computers that have been used for the experiments reported in this paper [**Sch08**]. Also, the memory access times of these systems (CD-ROM or flash memories) impose a serious bottleneck.

In practice, some speed-up is often achieved by compromising accuracy: heuristics are employed that limit the search space and hopefully find reasonable solutions. However, the latter is not guaranteed and not always achieved in practice. Typical examples of such heuristics are the layer concept found in autonomous car navigation systems [**Fli04**] and heuristic variants of the A^*-algorithm [**Do67, HaNiRa86, Po71, AhMaOr93**].

Much less effort has been spent on solutions which use preprocessing, i.e., trade off-line preprocessing time for running time in the target application. Of course, preprocessing is not trivial: we cannot afford to calculate (and store!) all shortest paths that might be requested in the application.

This paper describes a very successful preprocessing strategy and its efficient implementation. Here, *successful* means a big speed-up in the target application and acceptable memory needs, *efficient* implies acceptable preprocessing running time. Moreover, the resulting algorithms are exact, i.e., the shortest (cheapest) path is guaranteed to be found.

At least in our implementation, the suggested method needs not just an abstract graph, but also some geometrical information: coordinates of nodes. In addition, we assume positive length (or weight) of edges. No further assumptions (e.g., planarity) are made. (As will be seen, we use coordinates for partitioning a road network into regions; other ways of partitioning are possible, but not discussed in this paper.)

2. Previous Work

The work reported here was done in the late nineties already, but published at that time only in the form of patent applications [**LrEn03, LrEn98**]. A scientific publication was made available in 2004 [**Lr04**], but contained quite limited experimental results. More experimental results have been given by Köhler, Möhring, and Schilling [**KoMoSc05**].

Other preprocessing approaches include Ertl's work [**Er98**] using edge radii, the work by Goldberg and Harrelson [**Go05**], who use so called *land marks* that give upper bounds for the travel time to the target node and can be used in A^* search, Wagner et. al.'s [**Wa05**] work on using geometric containers enclosing subtrees of shortest paths trees to reduce the search space, and finally hierarchical methods [**Fli04, SaSc05**] that are similar to the heuristics currently used in production navigation systems, but - unlike these heuristics - guarantee optimum path calculations.

To the best of our knowledge, most of these approaches are - in comparison to the method described here - inferior in terms of speed-up achieved and/or memory needed to store results of preprocessing. One competitive method seems to be the hierarchical approach presented by Sanders and Schultes [**SaSc05**], but this method needs major modifications of the shortest path implementation in the target application that has to deal with the hierarchy imposed on the network.

3. Basic Idea: Edge Flags

When we drive through a road network in real life, we usually do not calculate shortest paths at all; we follow signposts. These do not need to be very specific: driving from Munich to Hamburg we do not (initially) care whether we are heading for the Binnenalster or the Reeperbahn, we just follow the signs to Hamburg.

This principle can be transferred to a software solution.

Let the road map be represented by a graph $G = (V, E)$ with n nodes and m edges. Edges carry two weights (e.g., length or estimated travel time), one for the forward direction, one for the backward direction. (A one-way street would have an infinite weight in one direction).

We first partition the whole network into r regions i. These regions should be geometrically connected, but the only hard requirement is that each node v of the network belongs to exactly one region, and that (in the application) we have fast access to a node's region (for instance by way of a region-id per node). We then define two edge flags for each edge $e = (v, w)$ and region i, $vflag_{e,i}$ and $wflag_{e,i}$. Let us (for now) assume that node v does not belong to region i. Then edge flag $vflag_{e,i}$ is set iff there is a shortest path from node v over edge e into region i. $wflag_{e,i}$ is defined in a similar way. Thus we have $m * r * 2$ different flags, and we need 1 bit to store each of these flags. These flags are our road signs; how they can be calculated will be discussed later

4. How to Use Edge Flags

Utilization of edge flags can be added to any existing shortest path algorithm; in Fig. 1 we show the Dijkstra algorithm with edge flag enhancements printed in bold face. We use pseudo-code, which is very near to the actual implementation based on our C++ class library TURBO [**Lr00**]. To utilize edge flags we need just two additional lines, one to retrieve the region index of the target node, the other one for skipping edges which cannot be on the shortest path to the target. This simple trick gives an enormous reduction of nodes that need to be scanned, and thus a corresponding speed-up for the shortest path calculation.

(Note that in a real application the initialization of all nodes would be done outside the Dijkstra-procedure only once; inside the procedure, we would keep track of touched nodes and re-initialize only these before the procedure returns. Or we may use an even faster method described in Section 10.1.)

5. The Problem of Cones

Coming back to our previous example, as we approach Hamburg we need to make up our mind, whether to head for Binnenalster or Reeperbahn. The signs

```
int dijkstra(Graph& g, Node* source, Node* target) {

  int target_region = target->region();

  // initialize all nodes:
  Node* v;
  forall_nodes(v,g) v->dist = ∞;

  // initialize priority queue:
  priority_queue q;
  source->dist = 0;
  q.insert(source);

  while ((v = q.del_min())) { // while q not empty
    if (v == target) return v->dist;
    Node* w;
    Edge* e;
    forall_adj_edges(w,v,e,g) { // scan node v
      if (! flagged(v,e,target_region)) continue;
      if (w->dist > v->dist + e->length) {
        w->dist = v->dist + e->length;
        if (q.member(w)) q.update(w);
        else             q.insert(w);
      }
    }
  }
  return ∞;
} // dijkstra
```

FIGURE 1. Edge-flags-enhanced Dijkstra code.

pointing to Hamburg are not very helpful anymore. When we use edge flags as described above and animate the implementation, we observe that in the initial part of the route very few nodes are touched that do not belong to the final shortest path. This is so because usually only *one* edge out of a node v has its flag on for a particular region which is far away from v's region. However, when we are near the target region, the situation changes. As edge flags describe shortest routes to *all* nodes within the region, now many edges out of a node will have its flag set and we observe a broad cone of touched nodes in front of the target region. Fig. 2 shows an example of such a cone.

This cone can easily be avoided: if we had started the shortest path calculation at the target (now looking for the shortest path to the source node), we would not have had a cone there, but one in front of the source region. We can combine the two approaches (and their benefits) by using a bidirectional shortest path algorithm, expanding from source and target simultaneously. The two expansions will meet somewhere in the middle between source and target and hopefully before the cones start to develop. This is at least true when source and target are far from each

FIGURE 2. Cone in Front of the Target Region

other; if they are near to each other, we do not have a problem with computation time in the first place.

There is a small problem with this approach: when we expand from the target, we have to account for the fact that we are driving (one-way) roads in the wrong direction. To compensate for this, we have to use the "wrong" weight value for each edge traversed. However, the edge flags as defined above have not been calculated for this situation. As a consequence, we need *four* flags per bidirectional edge and region, two for the nominal driving direction and two for the wrong one.

We also have to make sure that the two paths developing from source and target meet in a common node. One way to guarantee this, is to make sure that *all* edges (v, w) on a shortest path from v into region i are flagged, not just one out of v (which would be sufficient in a unidirectional implementation). Another way would be to define canonical shortest paths in a way that the shortest path from node v to node w uses the same edges as the shortest path from node w to node v (calculated using the "wrong" edge weights), but we did not find an efficient way to achieve this.

A further complication arises, when we are inside the target region. For a thorough discussion of this special case see Section 9.

6. How to Calculate Edge Flags

The key question is, of course, how to calculate the edge flags in an efficient way. This is where the problem gets interesting (or where the fun begins). In this paper, we will only give a rough idea, without going into all details.

6.1. A Simple but dead slow method. The simplest method is as follows: for each node v in a region i we calculate a tree of shortest paths, with node v as its root. We can use Dijkstra's algorithm for this purpose. When a tree has been calculated, each node carries a distance label, which gives the node's distance to the root. Then we look at all edges $e = (v, w)$: if the difference between the two distances $distance(v)$ and $distance(w)$ corresponds to the length (or weight) of e, this edge is in some shortest path out of region i and we can set the corresponding edge flag. Flags from different trees within the same region are or-ed together. (Actually we are interested in shortest paths *into* region i, not out of it; we fix this by using the "wrong" edge weights).

Note that - as discussed in Section 5 not only edges which form the shortest-path tree are flagged, but all edges without "slack"; thus we have to inspect *all* edges after each tree calculation.

If we have done all these tree calculations (two for each node, as we need four flags, see above) and edge inspections (about 20 weeks later for a road map of Germany containing 1.2 million edges and using an older CPU, see below) we can write the collected flags to disk and are done.

6.2. A somewhat faster method. We can classify the edges into two types: *internal* edges and *interface* edges. Internal edges are those that connect two nodes which belong to the same region. Interface edges connect nodes that belong to different (usually geometrically adjacent) regions. Based on the classification of edges we can classify nodes into *interior* nodes and *exported* nodes. Exported nodes are nodes which are incident to at least one interface edge. All others are interior nodes. Obviously, when we try to reach a target node in some region i (the target region) from the outside, we have to traverse some interface edge and thus cross an exported node of that region. Therefore, it is sufficient to apply the tree calculation process just discussed to exported nodes. After nodes have been assigned to regions (see Section 8 for details), it is easy to identify exported nodes and to store them in a list for each region. We just look at all nodes of a region and check outgoing edges; if the edge connects into another region, both nodes can be marked as exported nodes and put into the respective list. If we now apply our flag calculation method to exported nodes only, we can cut down the running time for the mentioned data set from 20 weeks (estimated) to about 35 hours. (Running times in this paragraph have been measured on a somewhat outdated 1 GHz Pentium III).

6.3. A fast but wrong method. One might be tempted to suggest the following clever way to achieve an even faster solution: Instead of expanding from one exported node of a region at a time, expand from all exported nodes of a region simultaneously. This is easy to implement: we initialize the priority queue of Dijkstra's algorithm with all the exported nodes, after setting their distance to zero. Then we keep expanding nodes until the queue is empty. This is equivalent to expanding from some virtual inner node of the region.

This method is fast, but unfortunately wrong. Resulting flags do not reflect shortest paths to *any* interior node of a region (or equivalently to *any* exported node of the region), but to the *nearest* exported node. Thus, in the target application, we would generate paths that pass trough the exported node of the target region that is nearest to the source node and from there run along a shortest path to the target node. This is not necessarily a shortest path between source and target.

6.4. The Fast Way. An other obvious attempt to achieve faster tree calculations uses the observation that trees rooted at neighboring exported nodes will usually show much similarity. A method that exploits this fact has been developed and gives indeed another big speed-up. (For the example and hardware mentioned, we achieve a preprocessing time of about 3.75 h).

Preprocessing times for the DIMACS-provided USA road networks using modern hardware are reported in the result section.

Unfortunately, the algorithms actually used (exploiting similarity of trees) are proprietary and thus cannot be discussed in this paper. However, a similar approach to this problem has been discussed in [**HiKoMo06**].

7. Saving Space

As described, we would need $4r$ bits per edge when we use r regions. Some of these flags are redundant (in case of one way streets). For some edges all flags are zero and need not be stored individually. And different edges may share the same set of flags. Using these observations we can greatly reduce the space needed for edge flags: In a first step we calculate four bit strings per edge, each containing r bits. Then we eliminate duplicates by inserting for each edge its four bit strings into a dictionary (only if it is not yet in the dictionary) and let the edge point to its bit strings. (One could also use sorting to eliminate duplicates.) For the road maps considered in the next section, this technique reduces the number of bit strings to be stored to between 0.2 % to 5.9 % (average 1.4 %) of the initially calculated strings. This implies that preprocessing needs considerably (here up to 500 times) more memory than is finally needed to store as result and to be used by the target application.

8. How to Define Regions

So far we have not considered how regions are defined. This may be application specific. Many applications use - due to main memory shortage - anyhow a parcelized data structure; these parcels or groups of parcels could be used to define regions.

In our program, two different methods have been implemented for region definition, a very simple rectangular grid method and a slightly more sophisticated square covering method. Both methods take an approximate number of regions, r, as input parameter.

8.1. Grid based region definition. We first calculate the bounding box of the whole network; this takes one pass over all nodes. From the area of the bounding box and the number r of regions to be generated we calculate the area and the width w of one square region. Relating w to width and heights of the bounding box gives us two numbers nx and ny, which are then rounded up. Now the bounding box is cut into nx columns and ny rows, resulting in nearly square regions, whose addresses are stored in a two-dimensional array. In a second pass nodes are assigned to regions; horizontal and vertical region indices can be calculated from a node's coordinates and the final grid width and height derived from nx and ny. Finally, empty regions are weeded out. Usually, the number of regions created will be smaller than the initially requested number, due to some empty regions.

8.2. Square covering. Here again we start by finding the bounding box and the width of a square region. Based on this width, we imagine the whole area to be partitioned into horizontal stripes and create one empty list of nodes for each stripe. Now each node is inserted into the appropriate list based on its y-coordinate. List are then sorted by x-coordinates. Scanning the sorted lists we can assign nodes to regions; whenever the next node does not fit into the current region, a new region is started. Other than in the grid-based method, gaps develop between horizontally consecutive regions and regions are in general not vertically aligned. The number of regions (of limited size) needed to cover the whole set of nodes is smaller than in the grid based method.

We can reduce this number further by extending a region vertically as soon as its lower and upper x-coordinate is known by "stealing" nodes from the next stripe that fit into this x-interval. This will lead to geometrically overlapping rectangular regions, but of course each node belongs to exactly one region.

A similar algorithm [**Go91**] is well known for covering a set of points with circular disks of a specified maximum size.

This method is slightly slower than the simple grid based method. No detailed investigation on its impact on preprocessing results has been carried out.

8.3. Further methods for region definition. Regions could also be formed by clustering algorithm, e.g., by using a combination of a minimum spanning tree algorithm and tree cutting [**KuMi77**] as described in [**Lr03**]. In this case, we could form regions without needing node coordinates. However, we did not investigate this method.

9. How to Use Edge Flags, A Closer Look

When discussing the flag-enhanced shortest path algorithms it was assumed that the node v being scanned is outside the target region. Here we discuss the general case. In the following, the term "target region" is used relative to the mode of expansion in the bidirectional Dijkstra algorithm, i.e., if we are expanding from the target node, the source node's region is the relative target region. There are two main cases to consider:

- **1:** Source and target nodes belong to different regions.
  - **1.1:** node v is outside the target region i. We follow edges out of v only if the v-flag for region i has been set.
  - **1.2:** node v is inside the target region i. When investigating edge $e = (v, w)$, there must be a shortest path from some exported node of region i trough v over edge e to node w; otherwise, this edge needs not to be considered. We can check this by inspecting the edge flag of e for node w. As we are now driving in the opposite direction relative to the exported node passed earlier, we need to use flags for another direction than outside the target region.
- **2:** Source and target nodes belong to the same region.
  - **2.1:** node v is outside the target region i. This is a totally legal situation; a shortest path between two nodes of one region may run trough one or more other regions. However, we know that we need to get eventually back to the target region and we must do this along a shortest path. So case 1.1 applies. Thus, even when source and

target are within the same region we achieve some speed-up, as we limit the investigation of nodes which are outside this region.

2.2: node v is inside the target region i. Here we cannot use any flags.

Let us reconsider case 1.1: a shortest path from node v over edge e to the target node must exist; the respective flag is checked. But, there must also be a shortest path from node w over edge e back to the source region (driving in the wrong direction); otherwise w would be reached along a shortest path over another edge. We can use this observation to construct an unidirectional algorithm that avoids the cones problem discussed in Section 5; however, the bidirectional version is slightly faster.

For the bidirectional method to work, it is essential to flag all tight edges, not just those in the shortest path tree, as was pointed out in Section 5; this is the case in our implementation.

10. Experimental Results

Experiments have been made with road networks provided by the 9th DIMACS challenge. We report on running times for preprocessing, speed-up, and efficiency (see definition below) for point-to-point queries, and memory needs for storing edge flags. For comparison we use our own Dijkstra implementation, described in the next subsection.

As preprocessing needs considerable intermediate memory (much more than needed to store the results), the CTR-instance with about 14 million nodes was the largest that we could run in 32-bit mode. (Under Linux, we can address 4 GBytes in 32-bit mode).

10.1. How to Make the Dijkstra-Algorithm Run Fast. There are two issues with implementing Dijkstra's algorithm efficiently: Firstly, we need an efficient priority queue implementation. Secondly, we should avoid re-initialization of all nodes prior to each path calculation.

For the first issue, we experimented with leftist trees, splay trees, the radix-queue [**AhMaOr93**], and the bounded priority queue from Dial's [**Dial79**] implementation. For the USA road-maps, we found the latter to be best suited; it is also the simplest one.

Avoiding re-initialization of the whole network is crucial when source and target nodes are close to each other; re-initialization of the whole network (i.e., resetting the distance labels of nodes) when only a small part of it was touched would be a waste of time and dominate the running time. One obvious way to avoid this is to keep track of touched nodes in a list and to use this list for re-initialization. But there is a better way: for each node we have a field "mark", initially set to zero, and we have a global variable "visited", initially set to 1. Both, "mark" and "visited" are integers. A node is considered unvisited, when its *mark* is unequal to *visited*. When a node is reached during the path calculation and it is unvisited, its distance label is considered invalid and will be set according to the current path length; its mark is set to *visited*. If it is visited, its distance label is considered valid and will be updated if appropriate. When the search for a path terminates, the value of *visited* is incremented by one. This will make all nodes unvisited in $O(1)$ time. Using this technique, we need one additional simple operation (the marking step) for each visited node; the check for being visited comes for free, as we need

TABLE 1. Comparison of conventional Point-to-Point implementations, using networks with travel times

	nodes	edges	running time		own/D	touched		own/D
			DIMACS	own		DIMACS	own	
NY	264346	733846	37.7	44.7	1.18	132871.7	132872.7	1.0
BAY	321270	800172	43.9	55.3	1.26	160053.1	160054.1	1.0
COL	435666	1057066	63.4	85.7	1.35	210653.1	210654.1	1.0
FLA	1070376	2712798	178.3	194.2	1.09	511662.6	511663.6	1.0
NW	1207945	2840208	188.6	255.1	1.35	600689.7	600690.7	1.0
NE	1524453	3897636	270.6	319.6	1.18	745720.3	745721.2	1.0
CAL	1890815	4657742	336.4	396.1	1.18	924878.7	924879.7	1.0
LKS	2758119	6885658	533.9	654.5	1.23	1399986.8	1399987.8	1.0
E	3598623	8778114	749.9	933.1	1.24	1807251.0	1807252.0	1.0
W	6262104	15248146	1444.7	1688.4	1.17	3168381.6	3168382.6	1.0
CTR	14081816	34292496	5157.8	5274.3	1.02	7189658.1	7189659.1	1.0

to distinguish first time visited nodes and revisited nodes anyhow, in order to do the appropriate priority queue operation (insert or update). With the list based method, on the other hand, we need one insert- and one remove-operation per visited node which leads to the same time complexity but is considerably slower, as was also verified by experiments. (One referee pointed out that the technique discussed above is well known under the name of $time - stamping$.)

In Table 1 we compare the running times and number of touched nodes of our own and the DIMACS-provided implementation for road networks with travel times and random point-to-point pairs as generated by the provided software. Running times are given in milliseconds/path and were measured - as all other results reported in this section - on an AMD Opteron Processor 252 with 2.6 GHz running in 32-bit mode under Linux. We used the GNU-compiler gcc version 4.0.2 with optimization flag -O4.

We see that our Dijkstra-implementation is between 2 % and 35 % slower than the DIMACS-provided software - possibly due to our list based data structure used for storing and traversing the graphs. As the implementation using edge flags uses the same graph data structure, we will use our own Dijkstra-implementation for comparisons in what follows.

10.2. Preprocessing. Table 2 shows preprocessing time and memory needs for different edge weights, namely travel times, distance, and unit edge weights.

Running times are given in CPU-seconds, memory is the space needed per edge in Bytes which includes a 4-Byte pointer to a bit string and the average size of the bit string itself. This is the additional memory needed in the application, resulting from preprocessing; memory needs *during* preprocessing are much higher, cf. Section 7.

The number of regions was set to 200 for all runs and the simple grid based scheme described in Section 8.1 used to define regions. (The actual number of non-empty regions may be smaller, e.g., if a region covers the sea or a large lake.)

We see that running times and memory needs are lower for the travel time metric than for the distance metric, probably because there are more similar shortest

TABLE 2. Running times and memory needs for preprocessing and for different edge weight metrics

		travel time		distance		unit weights	
Instance	nodes	run time	memory	run time	memory	run time	memory
NY	264346	123.0	7.5	124.9	9.5	97.4	9.1
BAY	321270	127.0	6.1	160.4	7.2	103.6	7.3
COL	435666	179.4	6.3	195.8	7.5	125.3	7.7
FLA	1070376	289.7	4.9	462.9	5.7	222.5	5.5
NW	1207945	666.4	5.9	800.9	6.9	452.5	6.6
NE	1524453	937.7	5.7	954.9	6.9	611.4	6.5
CAL	1890815	1091.6	5.4	1291.5	6.4	712.0	6.1
LKS	2758119	1475.7	5.5	2372.3	6.5	1077.8	6.1
E	3598623	2095.8	5.2	2108.5	6.1	1309.3	5.8
W	6262104	6126.6	5.3	7200.9	6.3	4860.6	6.0
CTR	14081816	25808.1	5.4	25975.6	6.6	16488.5	6.2

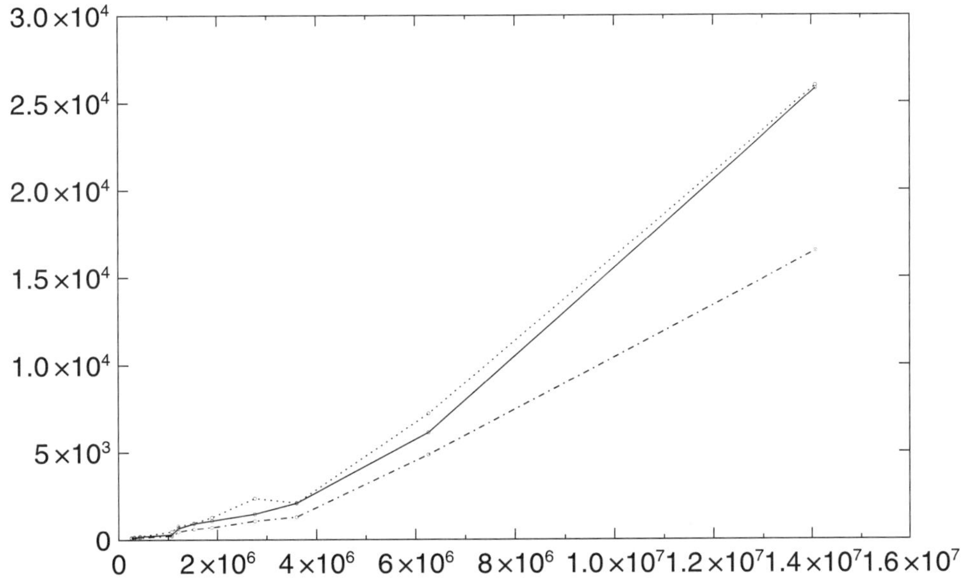

FIGURE 3. Running times [seconds] for preprocessing as a function of the number of nodes (solid: travel time metric, dotted: distance metric, dot-dashed: unit edge weights)

paths that can be handled together and produce similar edge flags. The memory need for unit edge weights is similar to that of the distance metric (for the same reason), but running times are lower, due to faster priority queue operations.

The graph in Fig. 3 shows how the running time increases with the number of nodes.

TABLE 3. Preprocessing time [sec] and average path calculation times [msec] with (fast) / without (slow) preprocessing for roadmaps with travel times.

Instance	nodes	pre	slow	fast	slow/fast	pre/slow	eff
NY	264346	123.0	44.7	0.3	141.5	2750.4	66.9
BAY	321270	127.0	55.3	0.4	146.2	2298.7	70.1
COL	435666	179.4	85.7	0.8	109.1	2092.8	54.3
FLA	1070376	289.7	194.2	1.2	161.9	1491.6	41.2
NW	1207945	666.4	255.1	1.4	182.4	2612.0	60.9
NE	1524453	937.7	319.6	1.3	249.1	2934.3	51.9
CAL	1890815	1091.6	396.1	2.1	190.6	2755.9	41.1
LKS	2758119	1475.7	654.5	2.4	272.6	2254.6	57.1
E	3598623	2095.8	933.1	3.5	269.8	2246.2	39.9
W	6262104	6126.6	1688.4	4.4	385.5	3628.5	57.3
CTR	14081816	25808.1	5274.3	7.5	701.0	4893.2	43.1

10.3. Random Point Pairs. In this paragraph we show results (for different edge weight metrics) for the 1000 random point pairs per network that were provided with the DIMACS benchmark data.

For the evaluation of preprocessing methods, not only the speed-up achieved is interesting, but also the total cost (in terms of running time) for preprocessing plus path calculations. Preprocessing pays off when the sum of preprocessing time t_{pre} and production time $n\ t_{\text{fast}}$ for calculating n shortest paths is lower than that of just using the slow algorithm $n\ t_{\text{slow}}$. The break even point is achieved for

$$n = \frac{t_{pre}}{t_{\text{slow}} - t_{\text{fast}}}$$

or

$$n \approx \frac{t_{\text{pre}}}{t_{\text{slow}}} \qquad \text{for} \qquad t_{\text{fast}} \ll t_{\text{slow}}$$

(This assumes that preprocessing and shortest path calculations are done on the same host. If the target host is slower, break even is reached earlier).

Another interesting measure is the efficiency as defined in [**Go05**], the number of nodes on the shortest path divided by the number of nodes scanned, given in percent. This number would ideally be 100, and actually we achieve this value for long paths.

The following tables show - for the three different edge weight metrics - running times measured for our own plain vanilla Dijkstra-implementation and those achieved using edge flags, the speed-up resulting from preprocessing, the break-even path number, and the efficiency in percent.

Table 3 contains the results using the travel time metric. Preprocessing time is given in seconds, path calculation times in milliseconds per path.

Table 4 shows the results when a distance metric is used instead of travel times. And finally, Table 5 shows results for unit edge weights.

For the travel time metric, we see a speed up factor between about 110 and 700, increasing with the size of the network. The break even is around 2500 paths for the smaller networks and goes up to about 4900 for the largest one, due to the high preprocessing time.

With the distance metric, results are not so good; speed up is lower and break even is higher. This result was to be expected: when faster roads are not preferred, shortest routes use more different roads instead of concentrating on the fast ones. Consequently, edges will carry more on-flags and more edges need to be investigated during the average path calculation.

With unit edge weights, results are difficult to interpret. Due to the faster priority queue operations, both slow and fast path calculation are faster than for the distance metric, but speed up is still higher than for the distance metric.

In the following we will concentrate our considerations mostly on the travel time metric, which is most relevant in real life (at least for road networks).

The plot of Fig. 4 shows how path calculation time and path length are related. Path lengths are measured as Dijkstra rank of the target node, which is equal to the number of scanned nodes when the target node is about to be scanned in a plain vanilla Dijkstra implementation.

We see that most times are below 10 msec (the average is 3.7 msec) and that higher times are needed for path lengths in the middle range. Short paths are trivially fast and long paths gain most from preprocessing.

10.4. Local Point Pairs. Next we show results in Table 6 for more local point-to-point path calculations. Here we give also the number of touched nodes (per path) for the two versions.

The *locality* shown in the table is the logarithm to the basis 2 of the Dijkstra rank.

As to be expected, the speed-up factor grows with the length of paths calculated, with the exception of the most local paths where the edge flags help to keep the path calculations local to a region (see Section 9 for explanation). Accordingly, the efficiency is low for local paths and grows up to 80.8% as paths get longer.

We see a similar dependency between path length and number of touched nodes.

For long paths, the reduction in the number of processed nodes is higher than the reduction in running time. This is so because we have to spend more time on each visited node in the fast version: looking up and processing of edge flags.

Again, results for distance metric and unit edge weights are inferior; nevertheless we achieve considerable speedup.

For the travel time version, we show a plot of running time versus path length in Fig. 5. The same data are shown as a box-and-whiskers plot in Fig. 6. We see again how running times increase towards medium path lengths and decrease as long paths are calculated. Then finally there is again a slight increase for very long paths, as their running time must be proportional to the number of nodes along the path.

11. Application Scenarios

There are two main applications for our algorithm in the context of route planning:

Firstly, in autonomous car navigation systems, the preprocessing can be done once and results are stored on the system's CD-ROM or flash memory. This way we can immensely speed up route calculations (including slow seeks on the CD-ROM or slow access to flash memory). However, when we want to be able to react to changing traffic situations, say traffic jams or blocked roads, we have to fall back to slow, exact algorithms doing the calculations from scratch or to fast heuristics. (At

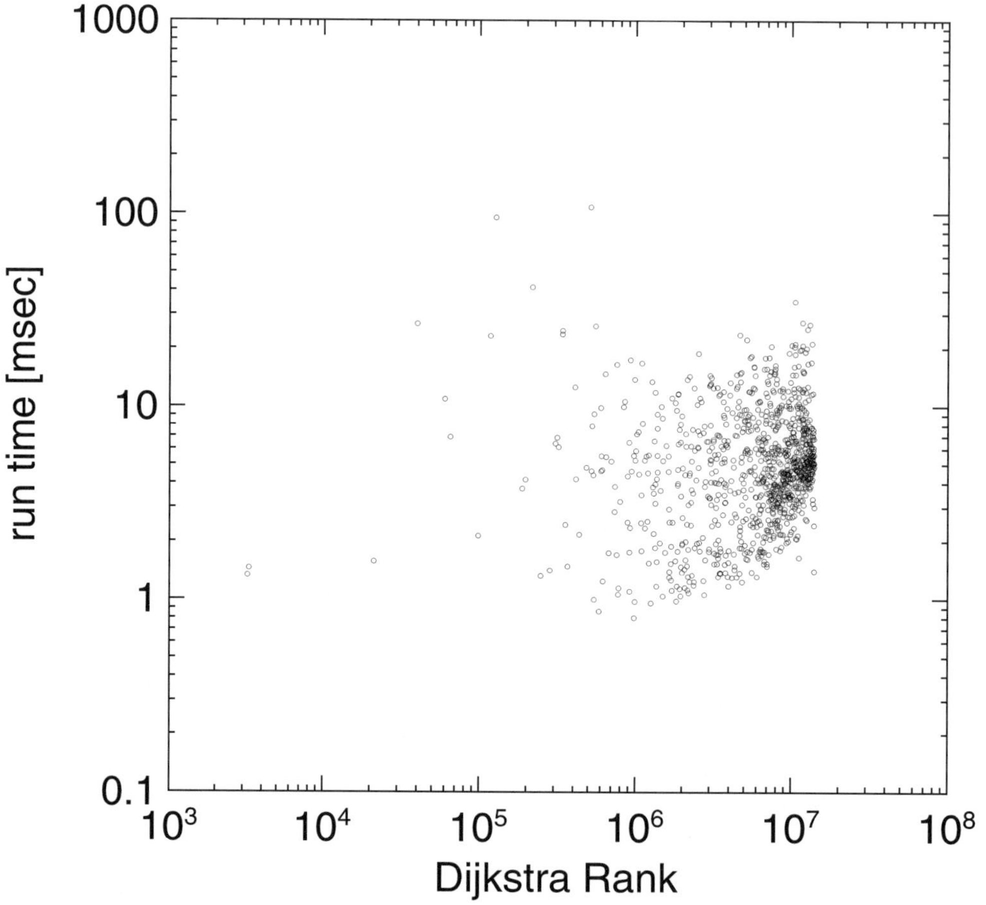

FIGURE 4. Path calculation times vs. Dijkstra rank for instance CTR with travel-time metric; results for 1000 random point-to-point pairs.

least at the time when the methods discussed in this paper were developed, CPU's used in cars were much slower than those used for the experiments presented in this paper. Also memory was a scarce resource forcing developers to store the network on a CD-ROM and load only small parcels of data into main memory when the data were needed during calculation of a path. Today, there is still a big gap in efficiency between cheap navigation systems and current personal computers.)

Secondly, in a navigation system with client server architecture (a central route planning server with the cars as clients), we can do the preprocessing at the server in a cyclic fashion (say, every 15 minutes) based on the current traffic situation. The server would then have to answer a large number of routing requests within a short time span.

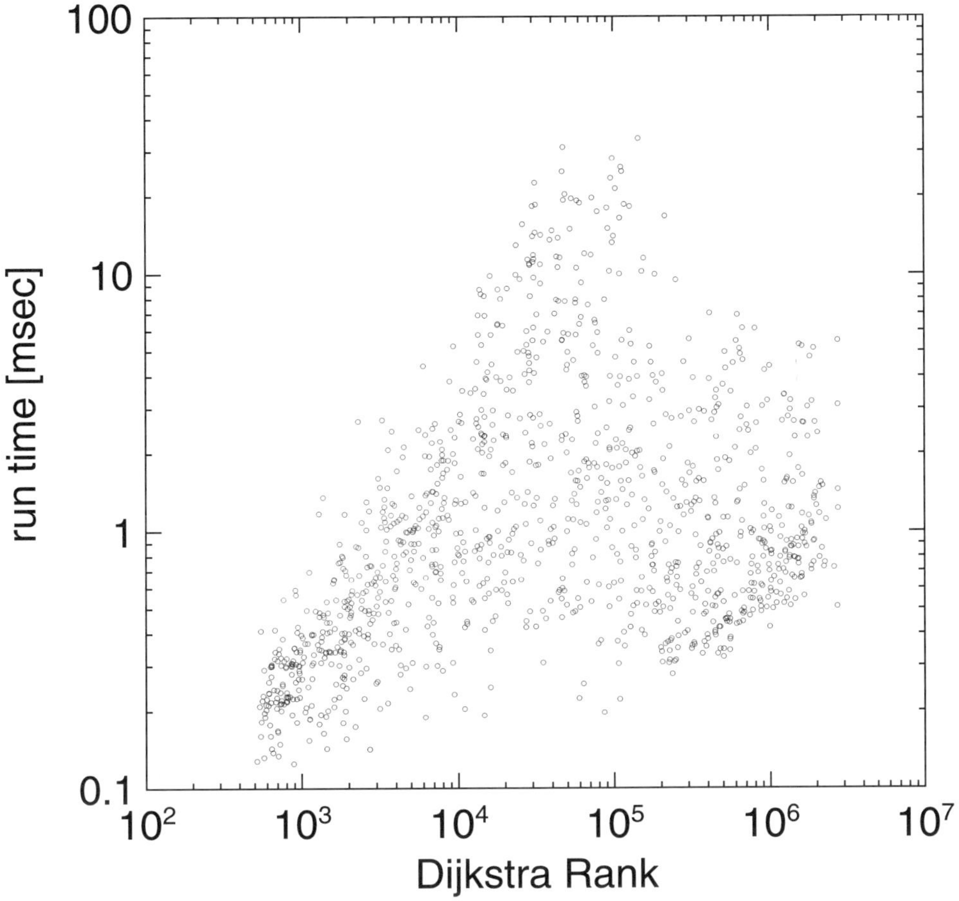

FIGURE 5. Path calculation times vs. Dijkstra rank for 100 random node pairs per locality range in the LKS road-map with travel time metric

12. Availability

The concept of edge flags and associated algorithms are patent protected. Thus commercial use of this concept is not possible without first acquiring a license. Source code licenses are available.

13. Conclusions

Our preprocessing algorithm using edge flags has been evaluated using road networks from the 9th DIMACS challenge. We achieve high speed-ups with a memory overhead of about 6 Bytes per edge and a break-even point for amortization of preprocessing around 2000 path calculations.

In comparison to other methods, the modification needed in the target application is very small and the additional operation is just one bit-lookup per edge to be traversed, whereas container based algorithms need to answer a containment query

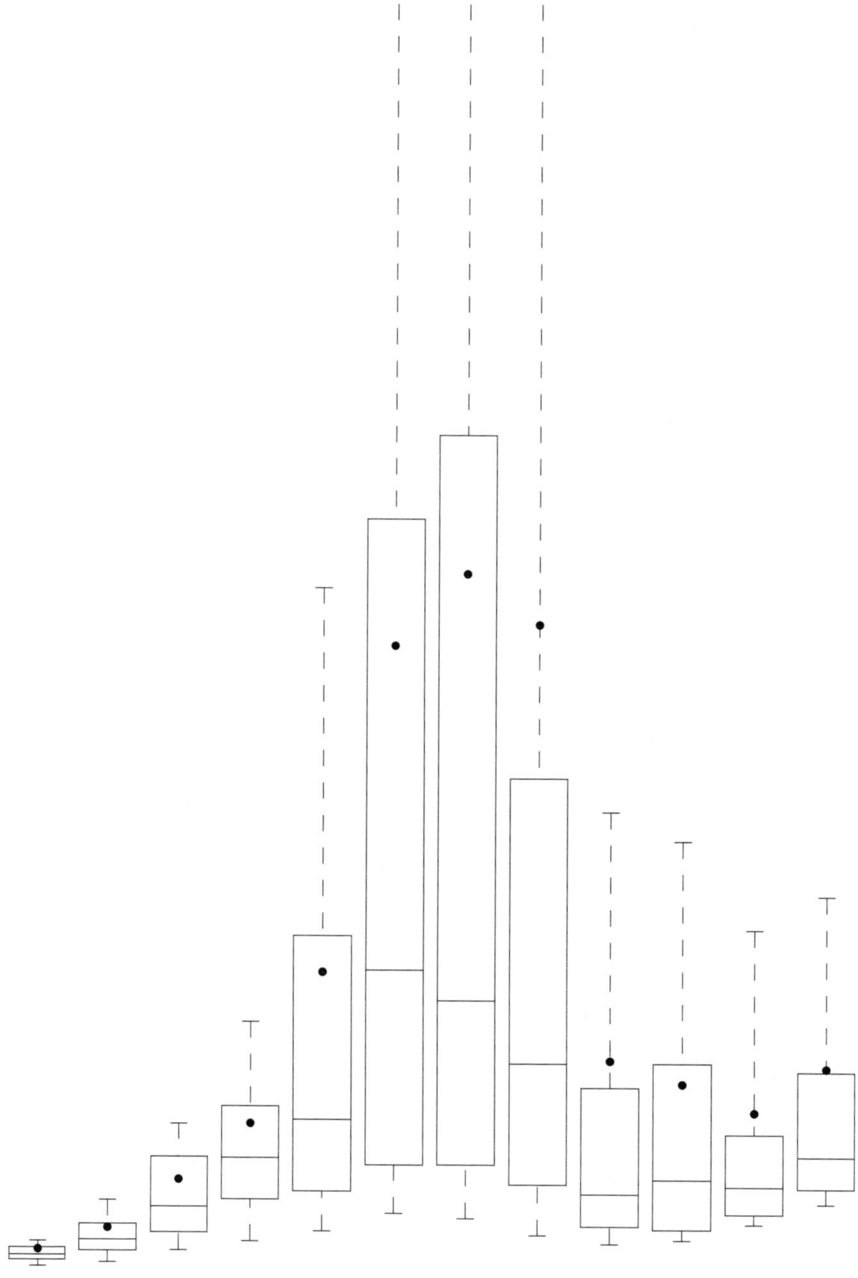

FIGURE 6. Box and whiskers plot of path calculation times vs. Dijkstra rank for 100 random node pairs per locality range in the LKS road-map with travel times, showing 10%, 25%, 75%, and 90% quantiles, median, and average running time (dot).

between a geometric container and a node's coordinates and hierarchical methods need to use the hierarchy in the target application.

Though the basic methods discussed here date ten years back, they are still competitive. Subsequent work that builds on this approach has been done since then, improving for instance the partitioning step in way that reduces preprocessing time [**HiKoMo06**].

There are faster methods (e.g. [**Sa06**], when just the length of the shortest path is to be found. But if we need to report the whole path, $\Omega(l)$ with l the number of edges in the path is a lower bound that cannot be broken and is - for long paths - achieved by our method, i.e. just the edges that belong to the shortest path are investigated during the shortest path calculation.

References

[AhMaOr93] R. K. Ahuja, T. L. Magnanti, and J. B. Orlin, *Network Flows: Theory, Algorithms, and Applications*, Prentice Hall, 1993.

[Dial79] R. Dial, F. Glover, D. Karney and D. Klingman, *A computational analysis of alternative algorithms for finding shortest path trees*, 1979, Networks, vol. 9, pp. 215–248.

[Dij59] E. W. Dijkstra, *A note on two problems in connexion with graphs*, 1959, Numerische Mathematik, vol. 1, pp. 269–271.

[Do67] J. Doran, *An Approach to Automatic Problem-Solving*, Machine Intelligence, vol. 1, pp. 105–127, 1967.

[Er98] G. Ertl, *Shortest Path Calculation in Large Road Networks*, OR Spectrum, vol. 20, 1998, pp. 15–20.

[Fli04] I. C. M. Flinsenberg, *Route planning algorithms for car navigation*, PhD Thesis, Technische Universiteit Eindhoven, 2004.

[Go05] A. V. Goldberg and C. Harrelson, *Computing the shortest path: A* search meets graph theory*, SODA, 2005, pp. 156–165.

[Go91] T. F. Gonzales, *Covering a Set of Points in Multidimensional Space*, Inf. Proc. Letters 40 (1991), pp. 181–188

[HaNiRa86] P. E. Hart, N. J. Nilsson, and B. Raphael, *A Formal Basis for the Heuristic Determination of Minimum Cost Paths*, IEEE Transactions on System Science and Cybernetics,SSC-4(2), 1986.

[HiKoMo06] M. Hilger, E. Köhler, R. H. Möhring, and H. Schilling, *Fast Point-to-Point Shortest Path Computations with Arc-Flags*, 9th DIMACS Implementation Challenge - Shortest Paths, Rutgers University, Nov. 2006.

[KoMoSc05] E. Khler, R. H. Mhring, H. Schilling, *Acceleration of Shortest Path and Constrained Shortest Path Computation*, WEA 2005: 126-138

[KuMi77] S. Kundu and J. Misra, *A linear tree partitioning algorithm*, SIAM Journal on Computing, 6(1977), pp. 151–154

[Lr00] U. Lauther, *The C++ Class Library TURBO - A Toolbox for Discrete Optimization.*, Software@Siemens, 2000, pp. 34–36.

[Lr04] U. Lauther, *An Extremely Fast, Exact Algorithm for Finding Shortest Paths in Static Networks with Geographical Background*, Geoinformation und Mobilität - von der Forschung zur praktischen Anwendung, Editors M. Raubal, A. Sliwinski, and W. Kuhn, vol 22, 2004, ISBN 3-936616-22-1, pp. 219–230.

[LrEn03] U. Lauther, and R. Enders, *United States Patent No. US 6,636,800 B1: Method and Device For Computer Assisted Graph Preprocessing*, 2003.

[LrEn98] U. Lauther, and R. Enders, *Europäische Patentschrift EP 1 027 578 B1: Verfahren und Anordnung zur Rechnergestützten Bearbeitung eines Graphen* 1998.

[Lr03] U. Lauther, T. Winter, and M. Ziegelmann, *Proximity Graph based Clustering Algorithms for Optimized Planning of UMTS Access Network Topologies*, 10th International Conference on Telecommunications ICT 2003, Vol. 2, pp. 1329–1334.

[Po71] I. Pohl, *Bi-directional search*, In Machine Intelligence, vol. 6, pp. 124–140, Edinburgh Univ. Press, Edinburgh, 1971.

[Sa06] P. Sanders and D. Schultes, *Robust, Almost Constant Shortest-Path Queries in Road Networks*, 9th DIMACS Implementation Challenge - Shortest Paths, Rutgers University, Nov. 2006.

[SaSc05] P. Sanders and D. Schultes, *Highway Hierarchies Hasten Exact Shortest Path Queries*, ESA, 2005, 568–579.

[Sch08] H. Schilling, personal communication.

[Tar83] R. E. Tarjan, *Data Structures and Network Algorithms*, 1983, Society for Industrial and Applied Mathematics.

[Wa05] D. Wagner, T. Willhalm, and C. D. Zaroliagis, *Geometric containers for efficient shortest-path computation*, ACM Journal of Experimental Algorithms, vol. 10, 2005.

SIEMENS AG, CT PP 7, D-81730 MÜNCHEN, GERMANY
E-mail address: ulrich.lauther@t-online.de

TABLE 4. Preprocessing time [sec] and average path calculation times [msec] with (fast) / without (slow) preprocessing for road-maps with distance metric.

Instance	nodes	pre	slow	fast	slow/fast	pre/slow	eff
NY	264346	124.9	39.0	0.8	48.5	3199.6	33.3
BAY	321270	160.4	48.7	0.8	58.9	3292.5	35.0
COL	435666	195.8	79.7	1.6	51.1	2455.8	27.1
FLA	1070376	462.9	165.6	2.6	64.2	2796.1	24.9
NW	1207945	800.9	216.4	2.1	104.8	3701.4	33.0
NE	1524453	954.9	275.3	3.6	75.5	3468.7	21.1
CAL	1890815	1291.5	325.7	4.8	67.4	3965.0	16.1
LKS	2758119	2372.3	545.8	6.7	82.0	4346.5	26.4
E	3598623	2108.5	737.1	12.8	57.7	2860.7	11.8
W	6262104	7200.9	1337.3	11.3	118.3	5384.7	17.9
CTR	14081816	25975.6	4071.5	28.0	145.5	6379.9	10.7

TABLE 5. Preprocessing time [sec] and average path calculation times [msec] with (fast) / without (slow) preprocessing for road-maps with unit edge weights.

Instance	nodes	pre	slow	fast	slow/fast	pre/slow	eff
NY	264346	97.4	36.2	0.4	101.8	2689.2	34.8
BAY	321270	103.6	40.0	0.3	116.1	2587.3	38.2
COL	435666	125.3	69.8	0.7	104.8	1795.5	29.0
FLA	1070376	222.5	199.2	0.6	341.7	1117.2	38.8
NW	1207945	452.5	195.5	1.1	177.9	2314.3	29.1
NE	1524453	611.4	275.7	1.1	242.4	2217.8	30.3
CAL	1890815	712.0	310.1	1.4	220.7	2295.9	23.8
LKS	2758119	1077.8	452.4	1.8	248.4	2382.5	35.0
E	3598623	1309.3	620.2	3.1	198.5	2111.0	17.9
W	6262104	4860.6	1214.0	3.3	364.8	4003.8	23.5
CTR	14081816	16488.5	3570.0	12.3	290.6	4618.6	12.0

TABLE 6. Average path calculation times [msec] and number of touched nodes with (fast) / without (slow) preprocessing for the LKS road-map with travel time metric tabulated by locality.

	running time			touched nodes			
locality	slow	fast	slow/fast	slow	fast	slow/fast	eff
9	5.879	1.270	4.6	765.7	494.3	1.5	7.9
10	6.179	1.320	4.7	1551.1	867.9	1.8	6.1
11	6.909	1.750	3.9	3014.2	1698.9	1.8	4.4
12	8.069	2.180	3.7	6211.2	2507.5	2.5	4.0
13	11.258	3.300	3.4	12564.1	4535.6	2.8	3.0
14	15.078	5.259	2.9	24821.6	8142.8	3.0	2.5
15	23.836	5.999	4.0	49038.4	9205.4	5.3	3.0
16	41.384	6.079	6.8	97731.6	8991.4	10.9	4.2
17	78.218	3.379	23.1	196369.9	4078.1	48.2	13.8
18	158.706	2.370	67.0	399665.6	2351.4	170.0	34.1
19	311.983	2.670	116.8	773964.6	2685.4	288.2	46.6
20	706.273	2.630	268.5	1541386.1	2613.4	589.8	80.8

TABLE 7. Average path calculation times [msec] and number of touched nodes with (fast) / without (slow) preprocessing for the LKS road-map with distance metric tabulated by locality.

	running time			touched nodes			
locality	slow	fast	slow/fast	slow	fast	slow/fast	eff
9	2.920	1.280	2.3	765.7	463.5	1.7	7.8
10	3.319	1.310	2.5	1551.1	888.7	1.7	5.5
11	3.809	1.760	2.2	3014.2	1741.4	1.7	3.9
12	5.409	2.300	2.4	6211.1	2791.9	2.2	3.3
13	6.939	3.270	2.1	12564.2	4783.6	2.6	2.6
14	10.988	4.879	2.3	24821.6	7518.8	3.3	2.4
15	18.847	7.279	2.6	49038.4	11830.9	4.1	2.2
16	34.465	6.759	5.1	97731.5	10315.3	9.5	3.6
17	66.710	8.099	8.2	196369.9	12051.6	16.3	4.4
18	134.530	7.929	17.0	399665.6	11214.1	35.6	7.7
19	262.090	7.629	34.4	773964.5	10553.2	73.3	13.1
20	557.165	7.499	74.3	1541386.0	9950.6	154.9	24.6

TABLE 8. Average path calculation times [msec] and number of touched nodes with (fast) / without (slow) preprocessing for the LKS road-map with unit edge weights tabulated by locality.

	running time			touched nodes			
locality	slow	fast	slow/fast	slow	fast	slow/fast	eff
9	1.510	1.190	1.3	1013.0	430.6	2.4	7.6
10	1.770	1.090	1.6	2051.8	758.5	2.7	5.6
11	2.210	1.280	1.7	3896.3	1486.3	2.6	3.7
12	3.020	1.510	2.0	7323.6	2265.8	3.2	3.2
13	4.579	2.020	2.3	14015.4	4035.7	3.5	2.3
14	8.529	2.670	3.2	29805.7	6029.2	4.9	2.2
15	15.168	3.569	4.2	55455.5	8830.9	6.3	1.9
16	28.676	3.160	9.1	106137.3	6740.7	15.7	3.6
17	58.821	3.100	19.0	208332.9	5968.7	34.9	5.8
18	127.181	2.630	48.4	421416.4	4246.3	99.2	11.9
19	249.752	2.260	110.5	771541.3	3109.0	248.2	25.1
20	489.996	2.610	187.7	1520322.5	3673.6	413.8	35.7

DIMACS Series in Discrete Mathematics
and Theoretical Computer Science
Volume **74**, 2009

Fast Point-to-Point Shortest Path Computations with Arc-Flags

Moritz Hilger, Ekkehard Köhler, Rolf H. Möhring, and Heiko Schilling

ABSTRACT. In this paper, we conduct a detailed study of the *arc-flag approach* introduced in [**Lau97, Lau04**]. Arc-flags are a modification of Dijkstra's algorithm to accelerate point-to-point (p2p) shortest path computations. The usage of arc-flags avoids exploring unnecessary paths during shortest path query computations. We present two improvements of the original arc-flag method that reduce the pre-calculation times significantly, tweak the efficiency of queries, and cut down the space requirements. First, we improve the preprocessing of the arc-flags by introducing a *centralized shortest path algorithm* and thereby we overcome the main drawback of the arc-flag method: for the first time it is now possible to apply the pure arc-flag method to large networks such as continental road networks (US network: 24M nodes, 58M edges). Second, we improve the partitioning used in our revised arc-flag method by using multi-way arc separators. Thereby we almost doubled the efficiency of queries on some instances compared to [**Lau06**] and further reduced the space requirements. This achievement stresses the vital importance of the right choice of partitioning for the performance of the arc-flag method. On the US network, our revised arc-flag method requires only between 4.1 and 15.5 bytes of additional preprocessed data per arc and p2p queries can be answered within 3 to 9 milliseconds (depending on the metric and the underlying partitioning). These results make our revised arc-flags method the currently best-performing purely goal-directed approach. Without having to use combinations with other advanced acceleration techniques or any network compression method we have kept the modification of Dijkstra's original algorithm minimal—literally only one additional line of code in the route search algorithm needs to be implemented. Its simplicity suggests its usage in existing code bases of route finding applications—whether that be website routers, mobile phones, or personal navigation devices.

1. Introduction

In the present work we investigate the *point-to-point (p2p) shortest path problem* where one has to find a shortest path between two specified nodes in an input graph. Dijkstra's algorithm is the standard for this problem [**Dij59**]; it runs in $\mathcal{O}(m + n \log n)$ time [**FT87**]. For a long time the main focus in developing shortest path algorithms has been on finding algorithms with good theoretical time-bounds; overviews are given, for instance, in [**Sch08, Sch06, Wil05, GH05**]. Although fast in theory, the corresponding algorithms are often not fast enough for applications in large networks, e.g. continental road networks.

In our study we assume that the shortest path problem has to be solved repeatedly for different node pairs of a fixed underlying network (static case) and therefore, preprocessing

Heiko Schilling was supported by the Deutsche Forschungsgemeinschaft (DFG) as part of the Research Cluster "Algorithms on Large and Complex Networks" (1126).

of the network data is possible. We work on large but sparse directed graphs with given arc weights and a given 2D layout arising from road networks. Nevertheless, the presented acceleration of shortest path computations works equally fast on higher dimensional layouts (time-expanded traffic networks) and even on graphs with no layout at all [**BDW07a**].

More precisely, we consider a generalization of a partition-based arc labelling approach that we refer to as the *arc-flag approach*. The basic idea of the arc-flag approach using a simple rectangular geographic partition was suggested in [**Lau97, Lau04**] and patented by [**EL99**]. The arc-flag approach divides the graph $G = (V, A)$ into regions and gathers information for each arc $a \in A$ and for each region $r' \in R$ ($V = \bigcup_{i=1}^{|R|} r_i \in R$) on whether the arc a is on at least one shortest path leading into region r'. For each arc $a \in A$ this information is stored in a flag (bit) vector f_a. The vector f_a contains a flag (`true` or `false`) for each region $r' \in R$ indicating whether the arc a can contribute to answering shortest path queries for nodes in region r' or not. Thus, the size of each flag vector is determined by the number $|R|$ of regions and the number of flag vectors is bounded by the number $|A|$ of arcs. Since the actual number of unique flag vectors can be much smaller than the number of arcs, storing the flag vectors at one point and adding an index (or pointer) to each arc can reduce the extra amount of memory below the obvious $|A||R|$ bits. The number of regions depends on the input graph size, but can be kept to a moderate size: 200 regions already lead to considerable speedups on instances with 24M nodes and 58M edges.

We use arc-flags in a slightly modified Dijkstra computation to avoid exploring unnecessary paths. This means that we check the flag entry of the corresponding target region (the region where the target node t belongs to) each time before the Dijkstra algorithm tries to traverse an arc. Thus, implementing the arc-flags is one of the simplest acceleration modifications of the standard Dijkstra algorithm and therefore suggests itself for straightforward usage in existing code bases.

We evaluate the quality of our method by measuring *speedup* factors. A speedup factor is the ratio between the cost of Dijkstra's algorithm and the cost of the accelerated algorithm. The cost is either measured by overall running time or by the size of the *search space*. The search space is the set of nodes that are scanned during a run of Dijkstra's algorithm.

The choice of the underlying partition is crucial for the speedup of the arc-flag acceleration of Dijkstra's algorithm. In [**KMS05**] we suggested a multi-way arc separator as an appropriate partition for the arc-flags. This improvement achieved much better speedups compared to the original arc-flag version in [**Lau04**]. For instance, we were able to reach acceleration factors 10 times higher than with Lauther's version of the arc-flags (on networks with up to 0.3M nodes, 0.5M arcs and 278 bits of additional information per arc). Together with Birk Schütz, Dorothea Wagner and Thomas Willhalm a computational study to find out which partitions achieve the best speedups for the arc-flag method [**MSS$^+$06**] was conducted. Both partitions from computational geometry and a multi-way arc separator partition were studied. The multi-way arc separator partition suggested in [**KMS05**] proved to be the best choices for the arc-flags. This is the partition method we concentrate on in the present paper.

When combining the arc-flags with a multi-way arc separator partition and a bidirected search, the overall performance of the method is competitive to those of other acceleration techniques like the highway hierarchy method [**SS06**]. An extensive study of the success of the different acceleration methods on a variety of graphs can be found in [**BDW07a**]. In comparison to the arc-flag method, preprocessing the highway hierarchy can be done faster. Yet, in [**BDW07a**] it is shown that the arc-flag method outperforms other methods on most graph classes in terms of speed-up. Thus, the arc-flag method is a very robust acceleration technique for shortest path computations. Note that the preprocessing times reported in [**BDW07a**] do not take advantage of the new approach presented in this paper.

The additional memory necessary for storing precomputed information is comparably small for the arc-flag method. However, its space requirements can even be reduced by using two or more level partitions: a coarse one for far-away nodes and a finer one for close-by nodes; the idea has been suggested in [**KMS05**] and further studied in [**MSS$^+$06**]. The multi-level approach is the basis for SHARC ([**BD08**]) which combines arc-flags with the iterative shrinking and shortcut ideas of the highway hierarchies. What makes the multi-level arc-flags remarkable is that this idea leads to the currently fastest uni-directional acceleration approach [**BD08**]. [**BD08**] report on query times which are comparable to the currently fastest bi-directional approaches [**BDS$^+$08**]. Furthermore, a fast uni-directional shortest path algorithm can, for instance, be applied to networks with historical speed profiles per arc (i.e. time-dependent arc weights) or to time-expanded networks as they are used for timetable information system. Introducing a time parameter to static networks makes bi-directional searches prohibitive since the exact start time for the backward search is not known.

For shortest path applications on devices with limited processing power or limited battery charge such as PNDs or mobile phones, it is of utmost importance that the shortest path algorithm does not rely on complicated and expensive data structures or subroutines. To keep the implementation of the arc-flag query as simple as possible we do not use any kind of reduction or contraction of the original input graph (e.g. attached trees or shortcuts as used in [**SS06, GKW06**]). Thus, the results we present reflect running times and speedups relative to the input graph sizes rather than to the reduced sizes. Even without using any kind of reduction, the arc-flag method yields speedups comparable to the speedups of [**SS06**] and [**GKW06**].

Our contribution. We conduct a detailed study of the pure arc-flag approach introduced by Lauther [**Lau97, Lau04**]. We were able to reduce the pre-calculation times significantly, to improve the efficiency of queries and to cut down the space requirements compared to previous versions of the arc-flags [**Lau97, Lau04, KMS05, MSS$^+$06**]. Our revised arc-flag approach is currently the best performing purely goal-directed acceleration technique [**BDS$^+$08**]. Our improvements were achieved without having to combine the arc-flags with more advanced acceleration techniques such as hierarchical approaches [**SS06, BFM$^+$07, SS07**] and without performing queries on reduced networks, e.g. by cutting off attached trees or introducing shortcuts [**SS06**]. By keeping our revised arc-flag approach plain and simple it is one of the easiest modifications of Dijkstra's algorithm: in fact, only one line of code needs to be added in the search algorithm.

For the preprocessing of the arc-flags we present two new approaches. On continental road networks like the US network (24M nodes, 54M edges) our method only needs a few hours to complete the pre-calculation. Our approach is the first to apply the arc-flag accelerated shortest path method to networks of this size. This is achieved by a new *centralized shortest path* algorithm which computes distances simultaneously from a set of starting vertices instead of one starting vertex. Another improvement on the preprocessing time is achieved by removing small attached structures for which the arc-flags can be calculated in a much easier way. Note that this reduction is only performed during preprocessing, queries are calculated on the unmodified graph using the pre-calculated information.

We use improved multi-way arc-separator partitionings of our test networks which reduce both the query times and the pre-calculation times. Our partitionings make excessive use of the graph-theoretic structure of the network and furthermore do not depend on any embedding of the network in the plane. Choosing the right partition is vital for the performance of the arc-flag method. The efficiency (see Section 5.3) of our queries are much higher than the ones for the grid partitionings, i.e. we achieve an efficiency two times better on the E instance compared to results presented in [**Lau06**].

As for combinations with other advanced acceleration techniques we suggest a way to combine arc-flags naturally with hierarchical approaches, see also [**KMS05**]. [**BD08**] excessively tested this idea and further improved it by using shortcuts [**SS06**].

Outline. After a brief review of recent related results in the field (Section 1), Section 2 starts with basic definitions and a precise description of the p2p problem. Furthermore, Section 2 explains the pruning of the search space of Dijkstra's algorithm with arc-flags. The preprocessing is described in Section 3. In Section 4 we present the selection of partition algorithms that we used for our analysis. We discuss the two-level variant of the arc-flags in Section 4. Section 5 describes our experiments and computational results and we discuss the results and ongoing work in Section 6.

Previous Work

Transit nodes. Bast et al. [**BFM06**] show that a very natural idea leads to extremely fast computations for vertices that are not too close to each other. The basic idea of their algorithm TRANSIT is as follows. They use a geometric subdivision of the given map into basic quadratic grid cells of length a. Then they precompute for every vertex v in cell C a set of *transit nodes* $T(C)$, which are nodes at distance approximately $d := 2a$ from C such that every shortest path from $v \in C$ to a vertex w at distance more than $2d$ from C passes through one of C's transit nodes (and, by symmetry, also through one of the transit nodes of the cell containing w). In addition, they precompute for every vertex $v \in C$ the distances $dist(v, t_v)$ from v to every transit nodes $t_C \in T(C)$ and all distances $dist(t_C, t_D)$ between any two transit nodes of different cells C, D at a distance of at least $2d$ apart. A shortest path length computation between vertices $v \in C$ and $w \in D$ (whose cells are at least $2d$ apart) then reduces to finding the minimum over all sums $dist(v, t_C) + dist(t_C, t_D) + dist(w, t_D)$ with $t_C \in T(C)$ and $t_D \in T(D)$, which can be done by simple table look-ups. This approach works because transit nodes can be computed very efficiently by sweep line techniques on the geometric subdivision, and the number of transit nodes per vertex turns out to be small (between 17 and 8 on average for a 64×64 and 1024×1024 subdivision of the US road network, respectively). A recursive application of this idea is used to keep the total number of transit nodes small. TRANSIT achieves an average query time of 12 microseconds for all long (i.e. $> 4d$) queries which is 99% of all queries. The shortest path itself can also be obtained through such queries by successively adding the best edge to the already computed initial path segment. Bast et al. [**BFM$^+$07**] combine transit nodes with highway hierarchies [**SS05, SS06**] and deal also with short queries. They obtain query times between 5 microseconds for long queries and 20 microseconds for short queries.

Highway hierarchies. Sanders and Schultes [**SS05, SS06**] introduced the notion of a *highway hierarchy*. The *highway hierarchy method* is based on the idea that only a *highway network* needs to be searched outside a fixed size neighborhood around source and target. This approach can be iterated to generate a hierarchy of *highway networks*. A hierarchy preserving all shortest routes can be constructed very efficiently: preprocessing the European road network (24M nodes, 58M arcs) takes 15 minutes. Because of the fast preprocessing step, highway hierarchies can be computed on very large networks and on such large networks the method achieves large speedup factors. Furthermore, the highway hierarchies can be adapted to speed up the computation of shortest path distances between all pairs of nodes from given sets of sources and targets, see [**KSS$^+$06**]. In comparison, preprocessing highway hierarchies can be done faster than preprocessing arc-flags, but the arc-flags deliver higher query speedups with the same amount of additional information; see [**BDW07a**] for corresponding computational results. Also [**BDW07a**] demonstrated for different graph classes that arc-flags are more robust than highway hierarchies in terms of achieved query times. Recently the highway hierarchies approach was extended to *highway node routing* [**SS07**] an idea similar to tranist nodes. Highway nodes further reduce memory needs and are suitable for a dynamic setup.

Landmarks. Goldberg and Harrelson [**GH05**] (see also [**GW05a**]) have shown that the performance of A^* search can be significantly improved if landmark-based lower bounds are used instead of Euclidean bounds. This leads to the ALT (A^* search, landmarks, and triangle inequality) algorithm for the problem. In [**GH05**], it was noted that the ALT method could be combined with reach pruning (see next paragraph) in a natural way. Not only would the improved lower bounds direct the search better, but they would also make reach pruning more effective.

Reach. Gutman [**Gut04**] defines the notion of *vertex reach*. Informally, the reach of a vertex is a number that is big if the vertex is in the middle of a long shortest path and small otherwise. Gutman shows how to prune an $s - t$ search based on vertex reaches (upper bounds) and vertex distances (lower bounds) from s and to t. He uses Euclidean distances for lower bounds, and observes that the idea of reach can be combined with Euclidean-based A^* search to improve efficiency. Goldberg et al. [**GKW06**] improved the reach-based approach of [**Gut04**] in two ways: they introduced a bi-directional version of the algorithm that uses implicit lower bounds and they added shortcut arcs to reduce vertex reaches. These modifications improve both preprocessing and query times. The resulting algorithm is as fast as the highway hierarchies introduced in [**SS05**]. However, the combination of landmarks with reach is a simpler modification of Dijkstra's algorithm than the highway hierarchies.

Geometric containers. Schulz et al. [**SWW00**] used the concept of enriching the graph with arc labels that mark for each arc $a \in A$ geometric regions of the given layout. The geometric regions contain all possible target nodes of a shortest path that start with the arc a. This labelling approach was done for the special case of a timetable information system. In their work, arc labels are angular sectors in the given layout of a train network. Wagner and Willhalm [**WW03**] studied this approach for general weighted graphs. Instead of the angular sectors, different types of convex geometric objects are implemented and compared with them. The arc-flags also use such a labelling approach, but there are three crucial differences between arc-flags and the geometric containers. Using arc-flags results in a much smaller search space than using geometric containers. The reason for this is that the partition used by arc-flags approximates in a geometrical sense much better for each arc a the set of nodes for which a is useful in a shortest path computation (see Figure 3). Even more important is that the arc-flags allow a considerably faster preprocessing, that can be done without the computation of all-pairs shortest paths. With the geometric containers one geometric object per arc needs to be computed by a shortest path tree computation. Since the geometric objects are different for different arcs, the shortest path tree computation has to be done for each arc. This is not the case with the arc-flags, since the same partition can be used for all flag vectors. See Sections 3 for details of the arc-flag preprocessing. Finally, the geometric containers rely on a given layout of the graph. Such a layout is not necessary for the arc-flags. Therefore, arc-flags provide a more general approach than the geometric containers.

2. Problem Description and Dijkstra's Algorithm with Arc-Flags

Graphs. A directed simple *graph* G is a pair (V, A), where V is a finite set of *nodes* and $A \subseteq V \times V$ are the *arcs* of the graph G. Throughout this paper, the number $|V|$ of nodes is denoted by n and the number $|A|$ of arcs is denoted by m. A *path* in G is a sequence of nodes $u_1, \ldots, u_k$ such that $(u_i, u_{i+1}) \in A$ for all $1 \leq i < k$. A path with $u_1 = u_k$ is called a *cycle*. A graph (without multiple arcs) can have up to n^2 arcs. We call a graph *sparse*, if $m \in \mathcal{O}(n)$. If we are given a *layout* $L : V \to \mathbb{R}^2$ of the graph in the Euclidean plane, then we will identify a node $v \in V$ with its location $L(v) \in \mathbb{R}^2$ in the plane. Furthermore, we introduce *arc weights* by a function $\ell : A \to \mathbb{R}$. We interpret the weights as *arc lengths* in the sense that the *length of a path* is the sum of the weights of its arcs. The *reverse graph* G_{rev} of a directed graph $G = (V, A, \ell)$ with arc weights ℓ is defined as $G_{\text{rev}} = (V, A_{\text{rev}}, \ell_{\text{rev}})$ with $A_{\text{rev}} = \{(u, v) \mid (v, u) \in A\}$ and $\ell_{\text{rev}}(u, v) = \ell(v, u)$. Hence, the reverse graph is the graph G with all arcs reversed. It is easy to see that $s, \ldots, t$ is a shortest path from s to t in G, iff $t, \ldots, s$ is a shortest path in G_{rev} with the same arcs reversed.

The P2P Shortest Path Problem. Let $G = (V, A, \ell)$ be a directed graph whose arcs are *weighted* by $\ell : A \to \mathbb{R}$. The goal of the *point-to-point (p2p)* shortest path problem is to find a path of minimum length from a given source $s \in V$ to a given target $t \in V$. The problem is only well defined for all $s - t$ pairs, iff G does not contain negative cycles. If there are negative arc weights but no negative cycles, it is possible, by using Johnson's algorithm [**Joh77**], to convert in $\mathcal{O}(nm + n^2 \log n)$ time the original arc weights ℓ to non-negative arc weights $\ell' : A \to \mathbb{R}^+$ that result in the same shortest paths. Hence, throughout the paper, we assume that arc weights are non-negative.

Dijkstra's Algorithm with Arc-Flags. The classical algorithm for computing shortest paths in a directed graph with non-negative arc weights is the one developed by Dijkstra [**Dij59**] with $\mathcal{O}(m + n \log n)$ worst-case running time [**FT87**]. However, in practice, speedup techniques can reduce the running time and often result in a sub-linear running time. They crucially depend on the fact that Dijkstra's algorithm is label-setting and that it can be terminated when the destination node is settled. Therefore, the algorithm does not necessarily search the whole graph.

If we allow for a preprocessing step, the running time can be further reduced with the following insight: consider, for each arc a, the set S_a of nodes that can be reached by a shortest path starting with a. It is easy to verify that Dijkstra's algorithm can be restricted to the subgraph with those arcs a for which the tail node t is in S_a. However, storing all sets S_a requires $\mathcal{O}(n^2)$ space which is prohibitive for large graphs. We will therefore use a partition of the set of nodes V into p $(:= |R|)$ regions for an approximation of the set S_a. Formally, we will use a function $r : V \to \{1, \ldots, p\}$ that assigns to each node the number of its region. We now use a flag vector $f_a : \{1, \ldots, p\} \to \{\texttt{true}, \texttt{false}\}$ with entries, each of which corresponds to a region. For each arc a, we set the entry $f_a(i)$ to `true`, iff

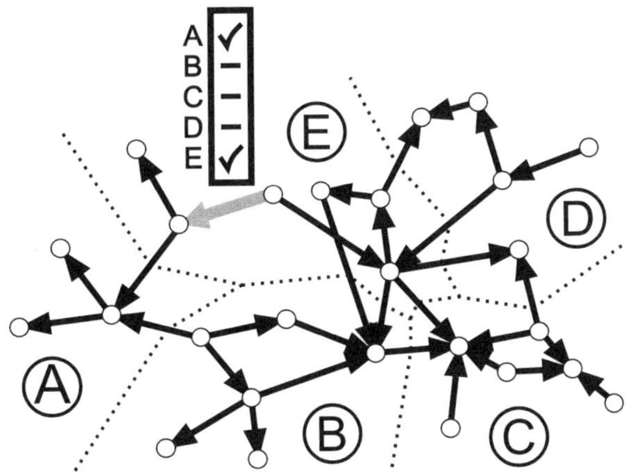

FIGURE 1. The arc-flag method together with a separator partition. The labelled (gray) arc only leads to nodes in region A and E. A search with targets in regions B, C or D can ignore this arc.

a is the beginning of any shortest path to at least one node in region $i \in \{1, \ldots, p\}$ (see Figure 1). Additionally, for each arc (v, w) with $v, w \in V$ we set the flag entry $f_{(v,w)}(r_w)$ to `true`.

For a specific shortest path query from s to t, Dijkstra's algorithm can be restricted to the subgraph induced by those arcs where the flag entry corresponding to the target region (the region where t belongs to) is `true` (see Lemma 2.2); we will call this subgraph G_t in the following.

DEFINITION 2.1 (Consistent target set). Let $G = (V, A)$ be a weighted graph with length function $\ell : E \to \mathbb{R}^+$. We call a set of nodes $V'_{(u,v)} \subseteq V$ a *consistent target set* for an arc (u, v), if for all shortest paths from u to some t that start with the arc (u, v), the target node t is in $V'_{(u,v)}$. A minimum consistent target set is the one with minimum size among all such sets.

In the following, we always mean minimum consistent target sets and therefore omit the word 'minimum'.

LEMMA 2.2 (Dijkstra's algorithms with arc-flags). *Let $G = (V, A)$, $\ell : A \to \mathbb{R}^+$ be a weighted graph and for each arc $a \in A$ let V'_a be a consistent target set. Then Dijkstra's algorithm with arc-flags finds a shortest path from s to t, $s, t \in V$, if one exists.*

PROOF. Consider the shortest path P from s to t that is found by Dijkstra's algorithm. If for all arcs $a \in P$ the target node t is in V'_a, then the path P will also be found by Dijkstra's algorithm with arc-flags. This is because the arc-flags for each $a \in A$ describe by construction a superset of the consistent target set V'_a. Therefore, no necessary arc will be missed when Dijkstra's algorithm traverses the graph, ('necessary' means useful for the shortest path P). Furthermore, the arc-flags do not change the order in which the arcs are processed. Since a subpath of a shortest path is again a shortest path, it follows by the definition of the consistent target sets that $t \in V'_{(u,v)}$ for all arcs $(u, v) \in P$. □

FIGURE 2. The search space of an arc-flag-accelerated Dijkstra search: the search started in s and the region containing the target node t is highlighted.

Bi-directed Search. In bi-directed search, two Dijkstra runs start simultaneously from s and from t and compute shortest path distances $dist_s(u)$ from s in the common (forward) graph and $dist_t(u)$ from t in the *reverse graph*, the graph with every arc reversed. The bi-directed search algorithm alternates between running the forward and reverse search version of Dijkstra's algorithm and stops with an appropriate stopping criterion when the two searches meet. Note that any alternation strategy will correctly determine a shortest path. More precisely, the bi-directed search stops if one direction is about to scan a node v from the priority queue that is already labelled by the other direction: then the shortest path between s and t is already found. The node v is not necessarily on that shortest path. In order to avoid searching for the connector-node v of the two searches, we determine the shortest path "on the fly": every time we consider a node which is labelled in both directions, we update the minimal sum of the shortest paths to source and target. The bi-directed search leads to speedup factors of up to two in the unaccelerated case. In principle, this speedup method can be combined with almost any other one. It improves the arc-flag based search by tackling one major problem of uni-directional arc-flag search, labelled the "problem of cones" in [**Lau04**]. It stems from the fact that the arc-flags only point into the direction of the target region. As soon as this region is reached by the search, no further information is provided and therefore the search space spreads (see Figure 2). With bi-directional search the two searches are likely to meet *before* their search spaces spread. However, application of an arc-flag accelerated bi-directional search requires some care to be taken during preprocessing, as will be explained in the following section. In [**KMS05, MSS$^+$06**] we suggested two and more level partitionings as another way to cope with the spreading and in order to improve the arc-flag method even further. This is discussed in more detail in Section 4.

The subgraph G_t (see page 7 for the definition) can be computed "on the fly" during a run of Dijkstra's algorithm. In a shortest path search from s to t, while scanning a node u, the modified algorithm takes all those outgoing arcs of u into account for which the flag entry corresponding to the target region is `true`. All other outgoing arcs will be ignored.

The space requirement of the preprocessed data is $\mathcal{O}(pm)$ for p regions because we have to store one flag for each region and arc. However, one can observe that many arcs share a common flag vector. Using this fact allows for a significant reduction of space requirement, depending on the partition used: Instead of storing for each arc its flag vector separately, one simply stores each occurring flag vector once and just points each arc to its corresponding flag vector (see Tables 3, 4 and 5 for examples of resulting space requirements).

There is a clear trade-off between achieved speedup factors and space requirements. Depending on the chosen partition, one can regard the arc-flag acceleration of shortest path computation as an interpolation between no precomputed information at all ($p = 1$) and full pre-computation by determining all possible shortest paths of the graph ($p = n$). Thus, in theory, we can get as close as possible to the ideal shortest path search by increasing the number of regions in the partition ('ideal' means that the shortest path algorithm visits only arcs that belong to the shortest path itself). Obviously, an increase in the number of regions also entails an increase in preprocessing time and space consumption. However, in practice, even for $p \ll n$ we achieve an average search space that is three orders of magnitude smaller than the one of plain Dijkstra. In fact, the number of regions can be kept of moderate size while one still achieves good speedups: already about 200 regions on continental road maps deliver enormous speedup factors with little extra space.

It is common practice in many applications to cache the shortest paths to the most important nodes in the given graph. Note that it is possible within the framework of the arc-flag speedup technique to do this by simply using a single region for each of the most important nodes. Storing all shortest paths to important nodes can therefore be realized without any additional implementation effort.

3. The Preprocessing

During the preprocessing phase calculations have to be performed to supply all arcs with the necessary arc-flags, i.e. for any pair of vertices s and t the arcs on a shortest path connecting the two have to be supplied with the correct flags for the region containing t. Note that for uni-directional search only *one* shortest path needs to be flagged. For bi-directional search however this does not suffice. Bi-directional search requires two sets of arc-flags, one for forward and one for backward search. In fact, forward and backward search can be accelerated independently, even the underlying partition can differ for the two directions. So every arc has to carry two flag-vectors, one indicating the flags needed for forward search, one for backward. In order to guarantee that the two searches meet, both need to explore *all* shortest paths connecting any pair of vertices. Therefore, the preprocessing has to supply the arcs on *all* shortest paths with the corresponding flags. Then the preprocessing for the bi-directional search in directed graphs simply consists of two independent runs of the preprocessing for the uni-directional search. Therefore we will only refer to the uni-directional case in the following. For undirected graphs preprocessing has to be performed only once, since the corresponding directed graph is equivalent to its reverse. Therefore the backward search can utilize the same set of arc-flags as the forward search.

All preprocessing methods require vast amounts of calculations of shortest path trees. The running time of these calculations highly depends on the number of vertices that need to be processed. Therefore a reduction of the input graph can improve running times substantially. For road networks it has been observed that the majority of instances consist of one large bi-connected component, called the "core", and small attached structures. It suffices to calculate the set of arc-flags on the core due to the following observation: All attached structures G_i are connected to the core G_c by exactly one root vertex v_{r_i}. Therefore all shortest paths to vertices in G_i as well as all shortest paths from vertices in G_i have to pass through v_{r_i}. So if all vertices in G_i are assigned to the same region as v_{r_i}

calculating the flags for arcs in G_i is straightforward. All arcs on a shortest path tree in G_i rooted at v_{r_i} need the flag for the region of v_{r_i}, since only vertices of that region can be reached. All arcs on a reverse shortest path tree, i.e. considering in-edges instead of out-edges, in G_i edges rooted at v_{r_i} need the flag for all regions, since those form the first part of shortest paths to any vertex in the graph.

3.1. The Preprocessing with All-Pairs Shortest Path. We have to calculate the arc-flag vectors for all arcs. This can be done by computing two shortest path trees for every arc $a \in A$: a one-to-all shortest path computation both from the head and from the tail node of arc a. The computation is done by a standard Dijkstra algorithm which stops when all nodes in the graph are permanently marked. For each node $v \in V$, we compute the difference between $d_h(v)$ and $d_t(v)$, the two distance labels in the shortest path trees of the head and the tail node of a. If the difference $|d_h(v) - d_t(v)|$ for node v is equal to the length ℓ_a of arc a, then we set the flag entry $f_a(r_v)$ to $\texttt{true}$. The running time of the preprocessing is dominated by the time it takes to compute $2m$ times a shortest path tree, which can be done in $\mathcal{O}(m + n \log n)$ time each. For sparse graphs ($m = \mathcal{O}(n)$), such as typical road networks, we get an overall worst-case time complexity of $\mathcal{O}(n^2 \log n)$.

The preprocessing of the geometric containers **[SWW00, WW03]** is done in a similar way by computing two shortest path trees per arc in the graph. Preprocessing our graphs (with up to 24M nodes and 58M arcs) by computing $2m$ times a shortest path tree would take years, but fortunately the arc-flags allow a much faster preprocessing, which is described in the following sections.

3.2. The Preprocessing without All-Pairs Shortest Path. It is not necessary to compute all-pairs shortest paths to set the flags correctly. We can use the following insight: every shortest path from any node s to a region $r_t \in R$ has to enter the region r_t at some arc. If s does not belong to r_t, then there must be an arc (u, v) with $r_u \neq r_t = r_v$; a so-called *boundary arc*. We will see in Lemma 3.1, that in the preprocessing step it is sufficient to take into account only shortest paths to such nodes v which are tail nodes of boundary arcs. Such nodes will be called *boundary nodes*.

LEMMA 3.1 (Boundary nodes). *Consider a graph $G = (V, A)$ and a partition of V in p regions by $r : V \to \{1, \dots, p\}$. If the flag vectors f_a for $a \in A$ are computed by shortest paths to boundary nodes only, then these flag vectors are a superset of any consistent target set V'_a (see Definition 2.1).*

PROOF. Let s and t be arbitrary but fixed nodes which are connected by a shortest path $s = v_1, \dots, v_k = t$. Furthermore, let s and t belong to different regions, i.e. $r_s \neq r_t$. By induction one can easily see that there exists an arc $a = (v_i, v_{i+1})$, $1 \leq i < k$, in this shortest path with $r_{v_i} \neq r_{v_{i+1}} = r_t$. The preprocessing which only considers shortest paths to boundary nodes would have considered the path from s to node v_{i+1} and hence it would have set the flag entry of region r_t on all arcs of the shortest subpath $s, \dots, v_{i+1}$. The flag entry of region r_t of the arcs between v_{i+1} and t are set by definition, because the tail nodes of these arcs belong to the region r_t. Since all flag entries corresponding to the target region r_t are being set for all arcs on the shortest path $s = v_1, \dots, v_k = t$, the modified Dijkstra algorithm finds this shortest path from s to t. □

We can now exploit this property: for a specified region $r' \in R$ and a boundary node b of r' we calculate the set T_b of arcs $a \in A$ with $f_a(r') = \texttt{true}$ and where a is on a shortest path via b to any node in r'. In fact, the reversed arcs corresponding to arcs in the set T_b form a shortest path tree in the reverse graph G_{rev}. A shortest path tree can be computed in time $\mathcal{O}(n \log n)$ on sparse graphs. Therefore, we can compute the flag entries $f_a(r')$ for region r' for all nodes $a \in A$ at once, if we compute a shortest path tree for each boundary node of r'. This can be done in time $\mathcal{O}(kn \log n)$ with $k = |B_{r'}|$, where $B_{r'}$ is the *boundary node set* of r': $B_{r'} = \{v \in r' \mid \exists (u,v) \in A \text{ such that } r_u \neq r_v = r'\}$. The number k of boundary nodes depends on the partition of the nodes. When we search for an appropriate partition for the arc-flags, the following observation helps: the set of boundary arcs of all partitions $r \in R$ represents a multi-way arc separator. Thus, if we want to minimize the number k of boundary nodes (and by this, minimize the preprocessing time), we need to find a minimum multi-way arc separator of the graph. Experiments showed that a minimum multi-way arc separator partition is among the partitions on which the arc-flags achieve the best speedup results, see [**MSS+05**] for details. However, as Section 4 shows, minimizing the number k of boundary arcs (i.e. separator arcs) is not the only objective when searching for a good partition for the arc-flags.

3.3. Preprocessing with Bit-Recycling. A straightforward implementation of the preprocessing phase of our method as it is presented in Section 3.1 leaves much room for improvements. In the previous Section 3.2, we have presented an improved version, where for each single boundary arc a backward Dijkstra search (i.e. a standard Dijkstra search in the reverse graph) has to be conducted to determine the flag vectors of all arcs in our graph. It is a simple observation that a contraction of all nodes of a region to a single super node and then performing one Dijkstra search from this super node does not yield correct arc-flags. Yet, when performing a backward Dijkstra search from two different separator arcs of a common region, the resulting shortest path trees show a strong similarity. More precisely, a large number of arcs that are contained in the first shortest path tree are as well contained in the second one. We call two separator arcs *similar* if they are in a geometrical sense 'closely together' and point in a geometrically similar direction. [1] Especially in the more distant parts of the trees there are only a few differences between the two trees. One can take advantage of the similarity of the shortest path trees of similar arcs for speeding up the preprocessing version in Section 3.2.

Suppose the boundary edges of the region under consideration are given as an ordered list where the ordering is implied by the sequence in which they occur when one walks along the boundary of the region e.g. in clockwise order. If we compute a backward shortest path tree T_i from some edge e_i in this ordering, then the backwards shortest path tree T_{i+1} for the next edge e_{i+1} in the ordered list will in most cases be very similar to the one of e_i, or in other words, e_i and e_{i+1} are similar separator arcs. One can make use of this observation for the computation of T_{i+1} by favouring the propagation of labels along the arcs of T_i.

One way to implement this preprocessing method is to use a customary label-correcting shortest path algorithm with two different label fronts. When computing the shortest path tree for edge e_{i+1}, then in every step the algorithm computes good upper bounds by hastily propagating the currently smallest label of the first label front all the way through the shortest path tree of arc e_i. The second label front unhurriedly moves behind the first one, makes use of the upper bounds to dominate its labels, and guarantees the optimality of the constructed backwards shortest path tree.

Our tests showed that we indeed can save a large portion of the decrease-key operations for the Dijkstra algorithm. Yet, for our road-network instances this technique did not

[1] Lauther [**Lau97**] suggests to use similarities of neighboring nodes for speeding up the preprocessing, but he gives no details on how to make use of this observation.

result in a considerable speedup. The reason is that we are working on sparse, almost planar graphs. In those graphs only a comparably small number of decrease-key operations is necessary during a Dijkstra computation (only for approximately 10% of the labels). However, we expect this bit-recycling method to show its real value not for sparse graphs but rather for dense graphs where many decrease-key operations are performed. Here we would like to point out that our implementation of the arc-flag method does not rely on an embedding of the graph but instead is suitable also for arbitrary (possibly dense) graphs.

3.4. The Preprocessing with Centralized Shortest Path Search. The preprocessing approaches described above required a single Dijkstra run for each of the boundary nodes of every region. As already pointed out in the last subsection, large parts of the computed shortest path trees are almost identical. Hence, it seems to be very promising to "bundle together" the different Dijkstra runs for a given region. Yet, as mentioned before, it is not sufficient to just contract all boundary vertices to a single vertex and run a common Dijkstra from there, since the anticipated graph is now not a tree anymore but rather a more complex graph.

We suggest a different approach, the *centralized shortest path algorithm*. The basic idea is as follows: Instead of starting from just one of the boundary nodes we start the search from all boundary nodes of a region R at once. Let $B = b_1, \ldots, b_{|B|}$ be the set of boundary vertices. Instead of a single distance entry, very vertex v of the graph is assigned a label $L_v : B \rightarrow \mathbb{R}^+$ with $L_v(b_i)$ being the length of the currently shortest path from boundary vertex b_i to vertex v. Furthermore, a heap is utilized containing those vertices that have been visited by the shortest path search and that wait to propagate their labels to their neighbors. Each vertex v in the heap also carries a key $k(v)$ which is used for sorting. Every time a vertex is extracted from the heap, it propagates the complete label. That means that for all arcs $a = (v, w)$ all entries of L_v are checked whether they improve on the corresponding entries of L_w. In accordance to the notation of standard shortest-path search algorithms we will refer to this as relaxation of the arc a.

The success of this methods depends on two factors; the initialization of the labels prior to execution and the organization of the elements in the heap. An obvious initialization assigns infinity to all label entries except for those representing distances of boundary nodes to themselves, which are assigned the value zero. This is the same initialization as used for Dijkstra's algorithm. Another possible initialization is the following: for all boundary vertices a restricted run of Dijkstra's algorithm is performed. We experimented with two forms of restrictions: one limiting the Dijkstra search space, the other aborting Dijkstra's algorithm at some point. The first method, which will be referred to as *limited initialization*, works by limiting the Dijkstra search to the vertices inside the current region. This assigns upper bounds to the label entries of all boundary vertices, if the region is connected. Note that these entries can not be guaranteed to be correct shortest path distances, as a shortest path connecting two boundary vertices might use arcs outside the region. The second method, named *aborted initialization* aborts the Dijkstra run of a boundary vertex when all other boundary vertices have been scanned. This guarantees correct entries for all labels of boundary vertices prior to execution. Additionally, some entries of labels belonging to vertices outside the region are assigned upper bounds on distance entries. Note that those entries can be re-used by the centralized shortest path algorithm. The aborted initialization outperformed the limited initialization in our experiments. Some care has to be taken on the set of vertices inserted into the heap prior to execution of the centralized shortest path algorithm, depending on the initialization. In general, all vertices with labels that can not be guaranteed to contain only correct entries need to be inserted. Actually, the set of vertices that is sufficient for correct termination of the algorithm might be much smaller. Further details can be found in [**Hil07**].

Besides the initialization, the centralized shortest path algorithm highly depends on the organization of the heap. Two different methods of choosing the key value will be presented. The first guarantees a running time bounded by that of $|B|$ runs of Dijkstra's algorithm. The second has a worse theoretical time bound but performs better in practice.

Minimal tentative key. Consider the relaxation of an arc $a = (u, v)$. Let K be the set of values of entries of L_v that have been altered. Now we update the key value of v to $\min(K \cup \{k(v)\})$ (where $k(v)$ is the old key value of v) if v is already contained in the heap and set it to $\min(K)$ if it has to be inserted. We call this method of choosing the key value *minimal tentative key.*

LEMMA 3.2 (Minimal tentative key.). *Utilizing the minimal tentative key for organizing the heap guarantees that*

(1) *the key values of the scanned vertices are non-decreasing*
(2) *every vertex gets scanned at most $|B|$ times*

during a run of the centralized shortest path algorithm.

PROOF. The way the minimal tentative key is chosen guarantees that the key value of a vertex $u \in V$ is always the smallest among those entries that have been recently altered. Therefore, the value of $k(u)$ is the smallest entry responsible for a possible violation of the shortest path optimality criterion on an incident arc $a = (u, v) \in A$. As the priority queue guarantees the key value of a scanned vertex to be minimal over all key values it is an easy insight that the key values are non-decreasing.

Now consider the sequence of scan operations performed on a single vertex $v \in V$ during a whole run of the centralized shortest path algorithm. Obviously, the key values of v when being removed from the queue are non-decreasing. Let $L_v(b) = k(v), b \in B$ be an entry with the same value as the key. For all entries $L_u(b)$ of vertices $u \in V$ being scanned at a later point that violate the shortest path optimality criterion, $L_u(b) \leq L_v(b)$ will hold. Therefore, the entry $L_v(b)$ will not be updated again. So every time a vertex is scanned at least one entry is designated to be permanent. This implies that the number of times a vertex can be scanned is bounded by the number of entries in the label, which is $|B|$. □

Minimum Total Key. Another choice for the key value is to select the minimum over *all* entries in the distance label. This can help to update labels with entries not reflecting shortest path distances quickly because of the following observations. Let $d_b(v)$ denote the shortest path distance from a boundary vertex b to some vertex v. Consider some vertex u with label L_u having an entry $L_u(b) = x > d_b(u)$ and key value $k(u) = L_u(b') \leq L_u(b)$ with $b, b' \in B$. If u is scanned, the false entry $L_u(b)$ will be propagated to all adjacent vertices which will propagate it to their neighbors etc. At some point the entry $L_u(b)$ will get corrected to a value $y \geq d_b(u)$. If the aforementioned minimal tentative key is utilized this results in u being added to the queue with key value $k(u) = y$. With the minimal tentative key value y is guaranteed to be greater than or equal to the current minimum in the queue and therefore u might rest in the queue. However, at every iteration where u is not scanned, labels based on the false distances estimate x might spread further into the graph. *All* of them have to be corrected at a later point in the algorithm.

Now consider the minimum total key choosing the minimum of *all* entries regardless of the subset of entries updated by a relax operation. For relax operations on arcs $a = (u, v) \in A$ where u was never scanned before the minimum total key is equivalent to the minimal tentative key. However, for vertices that have been scanned already, the minimum total key will be the "old" key value. This causes the vertex to be inserted *ahead* of all vertices that have not been scanned until now. Therefore the number of follow-up labels based on a wrong estimate of an entry in u can be assumed to stay small. Note that this ruins the bound on the number of scans performed on a single vertex presented for the minimal tentative key. In fact, vertices might get scanned over and over again resulting in a total count of scanned vertices exceeding $|B|\cdot n$. During our experiments this occurred only for exactly one instance while processing a region of one particular partitioning. However, the overall speed-up of the preprocessing of other regions did compensate for the extra work.

Our experiments show that the minimum total key outperforms the minimal tentative key despite the used initialization. All preprocessing numbers presented in Section 5 have been calculated with the initialization based on abort and the minimum total key. For a detailed discussion see [**Hil07**].

4. Choosing the right partition

The arc-flag acceleration method uses a partition of the graph to precompute information on whether an arc is useful for a shortest path search. Any possible partition can be used for the technique and, as Lemma 2.2 proves, the accelerated Dijkstra algorithm will always return a correct shortest path. However, different partitions do lead to different speedups of the Dijkstra algorithm. The question is, which partitions lead to the best query speedups while minimizing preprocessing effort.

There is a number of objectives which a good partition should fulfill: first, the number of separator arcs should be small, because the preprocessing time directly depends on this number. Second, the size of partitions should be balanced. With almost equally sized regions the 'load' per entry in a flag vector is balanced, i.e. each flag is 'responsible' for the same amount of possible target nodes. Third, the number of almost full flag vectors should be small. For instance, for the partitions that we presented almost one third of the flag vectors have more than 90 `true` entries, see Figure 8. However, a full flag vector means that we almost never can exclude the corresponding arc during an accelerated Dijkstra search. Fourth, the partition should approximate in a geometric sense the consistent target set for each arc as closely as possible, see Figure 3.

Together with Birk Schütz, Dorothea Wagner and Thomas Willhalm different partitions in combination with the arc-flag approach have been studied extensively, see [**MSS⁺06**]. Therefore, in this section, we will only present a summary of these results. Most of these algorithms need a 2D layout of the graph, except for the multi-way arc separator algorithm. The partition algorithms based on a 2D layout can easily be adapted to higher dimensional layouts (time-expanded traffic networks). For the arc-flag approach itself no layout of the graph is necessary as long as a partitioning can be provided for the input graph.

Rectangular partition (grid). Probably the easiest way to partition a graph with a 2D layout is to define the regions using a $x \times y$ grid of the bounding-box. More precisely, we denote with (ℓ, t) the top-left coordinate of the bounding-box of the 2D layout of the graph and with (r, b) the bottom-right one. Furthermore, we define $w = r - \ell$ as the width and $h = t - b$ as the height of the layout. The grid cell or region $G_{i,j}$ with $0 \leq i < x, 0 \leq j < y$ is now defined as the rectangle $\left[\ell + i \cdot \frac{w}{x}; \ell + (i+1) \cdot \frac{w}{x}\right] \times \left[b + j \cdot \frac{h}{y}; b + (j+1) \cdot \frac{h}{y}\right]$. Nodes on a grid line are assigned to one of the neighboring grid cells. Figure 4(a) shows an example

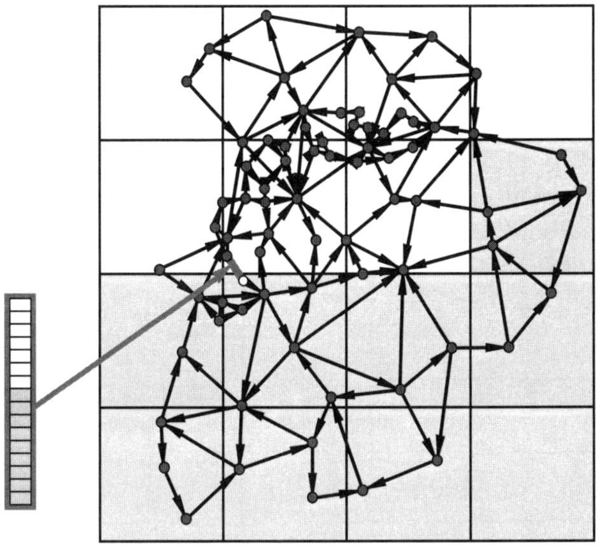

FIGURE 3. The arc-flag method together with a rectangular partitioning. At each arc a, a flag vector f_a is stored such that $f_a(i)$ indicates if a is on a shortest path into region i. The set of regions for which $f_a(i) =$ `true` (grey regions in the figure) approximates in a geometrical sense the consistent target set of arc a, i.e. the set of nodes to which a shortest path starting with a exists.

of a 7×5 grid. The rectangular or grid partition method uses only the bounding-box of the graph. All other properties like the structure of the graph or the density of nodes are ignored and hence it is not surprising that this method is not among the best partitions for our application. In fact, the grid partition always had the worst results in our experiments. Since earlier work on the arc-flag method **[Lau04]** used only grid partitions, we take it as a baseline and compare all other partition algorithms with it.

Quad-trees. A *quad-tree* is a data structure for storing points in the plane. Quad-trees are typically used in algorithmic geometry for range queries since they support fast access to nearest neighbor points. Further applications are in computer graphics, image analysis, and geographic information systems. Quad-trees can be generalized to higher dimensions – for 3D they are called *oct-trees*. Let P be a set of n points in the plane, r_0 its quadratic bounding-box, then the data structure *quad-tree* is recursively defined as follows:

- Root v_0 corresponds to the bounding region r_0.
- Region r_0 and all other regions r_i are recursively divided into four quadrants, while they contain more than one point of P. The four quadratic sub-regions of r_i are sub-nodes of v_i in the quad-tree.

The leaves of a quad-tree form a subdivision of the bounding-box r_0. Even more, the leaves of every sub-tree contain the root from such a subdivision. Since, for our application, we do not want to create a separate region for each node, we use a sub-tree of the quad-tree. More precisely, we define an upper bound $b \in \mathbb{N}$ of points in a region and stop the division if a region contains fewer points than the bound b. The result is a partition of our graph where each region contains at most b nodes. Figure 4(b) shows such a partition with 32 regions. In contrast to the grid partition, this partition reflects the geometry of the graph in a better

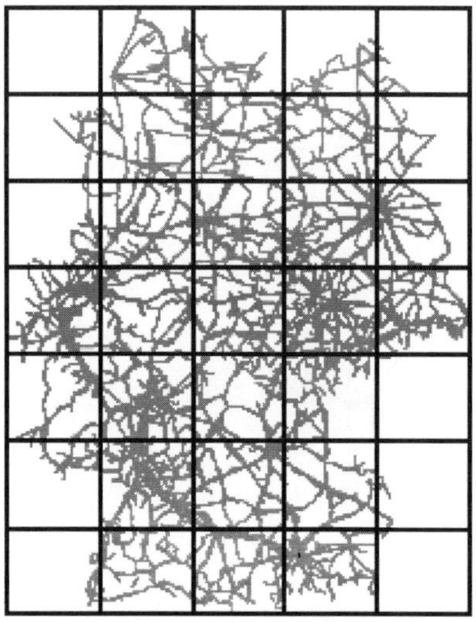

(a) Rectangular partition (35 regions)

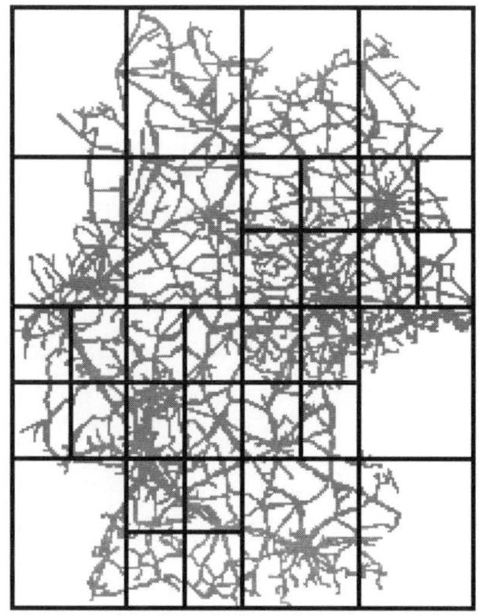

(b) Quad-tree (34 regions)

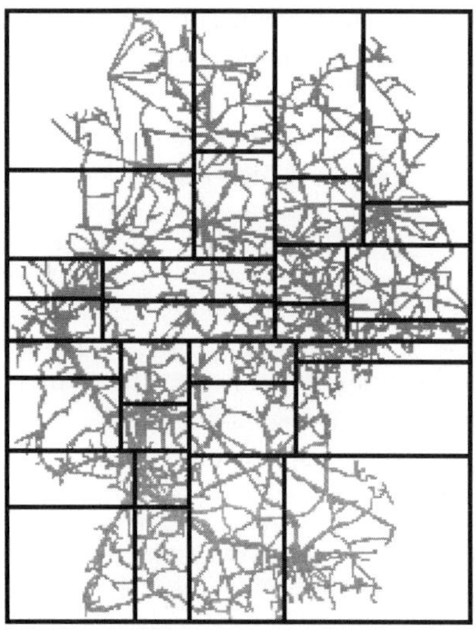

(c) kd-Tree (32 regions)

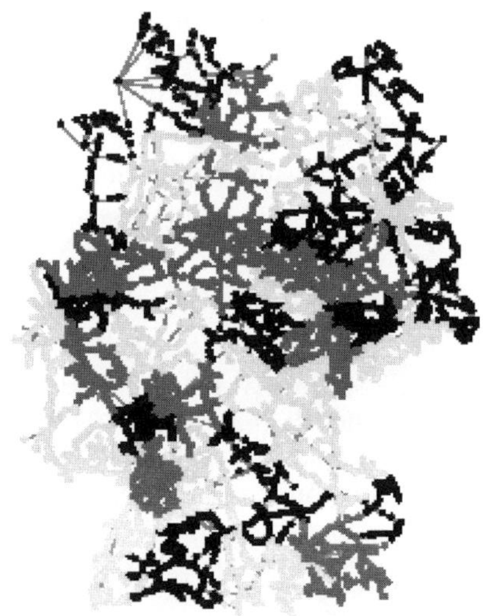

(d) MULTI-WAY ARC SEPARATOR (32 regions).

FIGURE 4. Germany with four different partitions.

way: dense parts will be divided into more regions than sparse parts. The regions generated by this partition have almost balanced size, but the arc separator set can be large.

kd**-Trees.** In the construction of a quad-tree, a region is divided into four equally sized sub-regions. However, equally sized sub-regions do not take the distribution of the points into account. This quad-tree division can therefore be extended to more general subdivision schemes, the so-called *kd-trees*. In the construction of a *kd-tree*, the plane is recursively divided in a similar way as for a quad-tree. In contrast to a quad-tree, the underlying rectangle is decomposed into *two halves* by a straight line parallel to an axis. The axes alternate in the order $x, y, x, y, \ldots$. The positions of the dividing line can depend

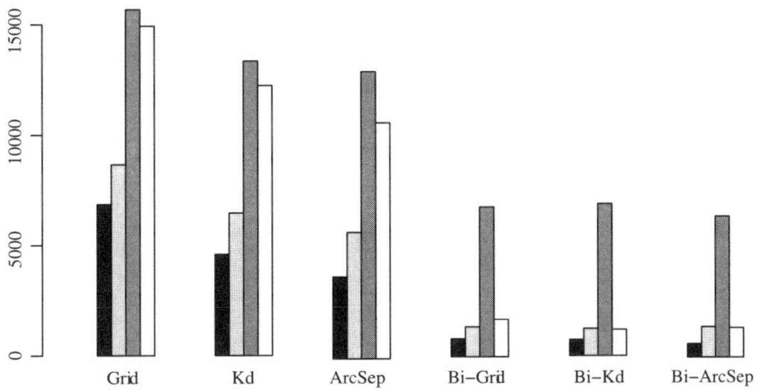

FIGURE 5. Average search spaces of different partitionings on networks of sizes from 362,000 to 1,046,000 nodes.

on the data. Frequently used positions are given by the center of the rectangle (*standard* kd-tree), the *average*, or the *median* of the points inside. If the median of points in general position is used, the number of partitions is always a power of two, and the region sizes are balanced. Figure 4(c) shows a result for the median and 32 regions. In applications with higher dimensions, usually the partition axes are not cycled but the dimension with the largest variance is used. Experiments showed that the kd-tree with median outperforms the other kd-tree alternatives.

Multi-way arc separator. Instead of using geometric information of the graph embedding one can make use of the general graph-theoretic structure of the network itself. In contrast, a geometric subdivision of the graph is very dependant on choosing the right embedding and, especially for the mentioned rectangular partition techniques, one has to choose the partition very carefully to obtain a satisfying speedup of the resulting algorithm. A completely different approach is to ignore all geometric information on the network and simply create a partition of the underlying graph in an appropriate way.

A *multi-way arc separator* is a partitioning of a graph into k (almost equally sized) regions with a small (almost minimal) arc separator, for instance, see [**KK98**]. Efficient implementations of multi-way arc separator algorithms can be obtained free of charge from [**MET95**] or [**PAR06**]. The algorithms implemented in METIS and PARTY are based on multilevel recursive-bisection, multilevel k-way, multi-constraint partition schemes, and other techniques. These partition methods have several advantages for our application: they do not need a layout of the graph and are therefore the most general partition methods among those presented in the present article. The number of arcs in the separator is noticeably smaller than in the other partition methods. The size of the regions is balanced and the number of full flag vectors is among the smallest of all the partitionings we studied (Figure 4(d) shows a partition of a graph generated by METIS).

Figure 4 gives some computational results for queries on the different partition schemes that we described in the previous paragraphs (see [**MSS**$^+$**06**] for further details on these computational results). The size of the preprocessed data per arc is nearly the same for all algorithms; note that exactly the same size could not be realized since, for instance, the number of kd-tree partitions is always a power of two. kd-trees and separator partitioning show equal performance. However, the latter does not require an embedding of the network and produces less separator arcs. Because of these advantages we concentrated our study in the following to arc separator partitionings.

TABLE 1. An overview over different partitionings on the USA instance. For each partitioning the minimal and maximal number of connected components per region as well as the mean are depicted. "PARTY connected" refers to calling PARTY with the "connected" parameter. "Recursive" shows the results of the proposed recursive calls to PARTY.

#regions	method	min	max	mean	$\|B\|$
100	grid	0	305	72.7	49219
	METIS	1	11	1.57	29924
	PARTY	1	1	1	20499
	PARTY connected	1	1	1	32116
	recursive	1	2	1.02	16755
1000	grid	0	141	24.463	157469
	METIS	1	14	1.344	99132
	PARTY	13	59	32.192	522178
	PARTY connected	0	277	19.511	396261
	recursive	1	2	1.006	76741

For selecting the partitioning algorithm (PARTY vs. METIS) we ran extensive tests and finally decided to use PARTY and not METIS in our recent experiments. The newest version of PARTY has several advantages over METIS. First of all METIS is not maintained any more (last update from '95) and we observed instabilities with the provided implementation. Besides this rather technical detail, the partitions obtained by PARTY are more appropriate for our purpose: the number of separator arcs is smaller which reduces the computational effort for pre-computing the arc-flags. More importantly, PARTY allows for retries if non-connected regions are obtained which results in partitionings mainly consisting of connected regions. However, for large numbers of regions, PARTY's performance decreases. Simply applying PARTY recursively, i.e. partitioning the whole graph into a small number of regions and re-partitioning those, clears out these problems (see Table 1).

Hierarchical methods. We would like to finish this section with an outlook on how the presented arc-flag method, together with the multi-way arc separator approach can be further brought forward by employing a hierarchical approach. Let us have a closer look at the search space generated by the arc-flag accelerated Dijkstra search to get an idea of how to compress the arc-flags: as illustrated in Figure 6(a), the accelerated Dijkstra search reduces the search space at the beginning of the search, but once the target region has been reached, almost all nodes and arcs are visited. This is not very surprising, if we consider that usually all arcs of a region were assigned the region-flag of their own region. We could deal with this problem by using a finer partition of the graph but this would lead to larger flag vectors at each arc (requiring more memory and a longer preprocessing). Take the following example: if we use a fine 15×15 grid instead of a coarse 5×5 grid (i.e. each coarse region would be split into 9 additional finer regions), then the preprocessed data will increase from 25 flags (in the coarse case) to 225 flags (in the fine case) per arc. Note that the additional information of the fine grid is mainly needed for arcs close to the target node, e.g. arcs in the target region of the coarse grid. This leads to the idea of splitting each region of the coarse partition into a set of smaller regions (see Figure 6(b)). For each

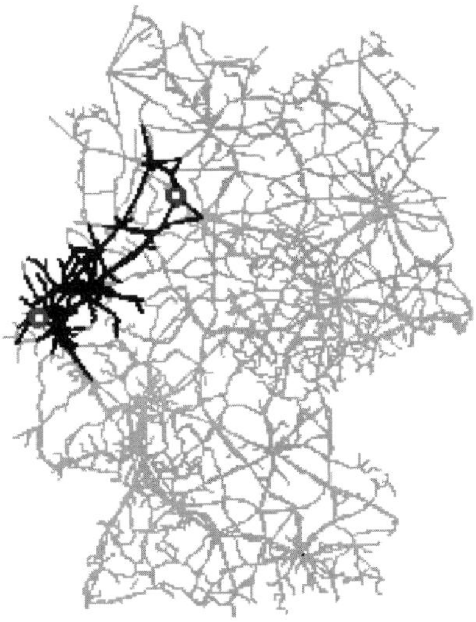

(a) Without two-level arc-flags a search visits almost all arcs in the target region (lower left gray point).

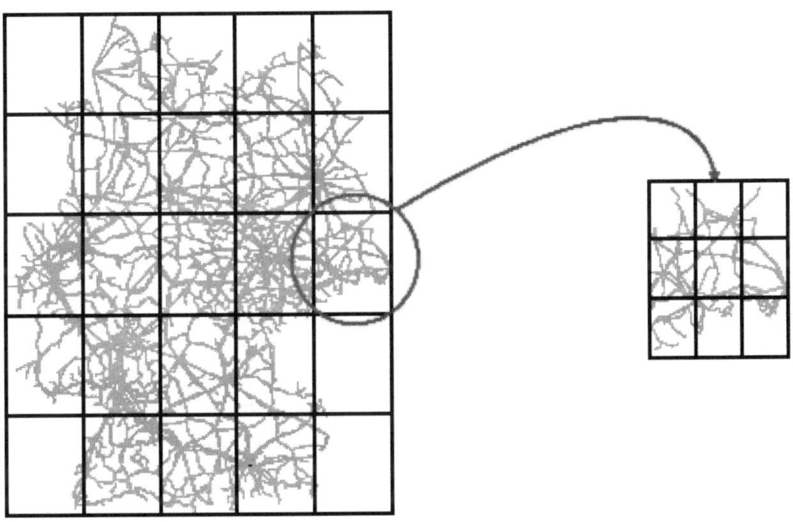

(b) For each arc a, a flag vector is stored for the coarse 5×5 grid and a flag vector for a fine 3×3 grid *in the same coarse region as the arc* a.

FIGURE 6. Two-level partition.

such set of smaller regions (i.e. for each fine partition of a region from the coarse partition) we compute and store additional flag vectors only for those arcs inside the same coarse region. More precisely, we partition the graph induced by the nodes in the (coarse) regions and perform (for each induced graph) another preprocessing where we calculated another flag vector for each arc in the induced graphs. The entries of this additional flag vector per arc are associated with the fine regions of the same coarse region. Hence, each arc gets two flag vectors assigned: one for the coarse partition and one for the fine partition of the coarse region to which the arc belongs. This approach can be iterated to more than 2 levels of finer partitionings.

The advantage of this two-level partition approach is that the preprocessed data is much smaller than for a fine one-level partition. This is because the second flag vector of an arc a is relevant (i.e. computed, stored, and evaluated) only for the coarse region to which the arc a belongs. In the example above, the two-level approach would need only 34 flags per arc (instead of 225). The difference between the search spaces of the arc-flag accelerated Dijkstra search with the one- and with the two-level partition is small. This is the case, because for the one-level partition the entries in flag vectors corresponding to faraway neighboring regions are similar to each other. Therefore, we are not loosing too much information from the one-level flag entries with the two-level partitions. The two-level approach can be viewed as a (lossy) compression of the one-level flag vectors: we accumulate the flag entries for faraway regions. For the two-level partition approach only a slight modification of the search algorithm is required: until the target region is reached, everything remains unaffected, unnecessary arcs are ignored by using the flag vectors of level one. When the algorithm has entered the target region, the second-level flag vector provides further information on whether an arc can be ignored for the search of a shortest path to the target node. Preprocessing the arc-flags for the two-level approach takes slightly longer than for the one-level approach. However, almost the same speedups can be achieved with two-level partitions with only a fraction of the space consumption of the one-level partitions.

5. Experimental Setup and Computational Results

Implementation. Our experiments were implemented in C++ using the GNU g++ compiler version 4.2 with optimization option -O3 on a Linux 2.6 system. Our code is based on the latest Boost Graph Library [**BGL07**] and heavily utilizes the GNU Standard C++ Library. All calculations were performed on a 64 Bit Dual-Core AMD Opteron(tm) Processor 2218 with 2593.6 MHz and 1024 KB of cache as well as 32 GB of main memory.

Table 2 shows our implementation of Dijkstra's Algorithm (strictly speaking a modification of Dijkstra's algorithm). We would like to point out that the acceleration factor of the computational time that we report in this section depends very much on the heap data structure which is used. In previous papers we applied a d-heap implementation ($d = 2$) both for our accelerated and for the standard Dijkstra algorithm implementation. However, in the present implementation we are using a smart queue implementation [**Gol01**] from the Boost Graph Library [**BGL07**] wherever we are in need of a standard shortest path algorithm. In particular, this is used in our reference implementation of Dijkstra's Algorithm and for the initialization of labels prior to the execution of the centralized shortest path algorithm. This modification of Dijkstra's algorithm utilizes a multi-level bucket heap in combination with the caliber heuristic for faster determination of permanent labels (see [**Gol01**]). It is considerably faster than the d-heap implementation that we used in previous publications. Therefore the performance gap in computational time between the accelerated and the standard Dijkstra algorithm has been reduced for graphs of comparable size.

Note however that the centralized shortest path algorithm as well as the bidirectional query used throughout the experiments still utilize a simple binary-heap. Utilizing more sophisticated heap-structures might improve on the running times of both significantly. However, for the centralized shortest path algorithm with minimum total key the key values

TABLE 2. DIMACS benchmark software versus our plain Dijkstra implementation using the Boost Graph Library (BGL).

instance				time [ms]			search space		
name	#nodes	#edges	metric	dimacs	own	×	dimacs	own	×
NY	264346	733846	travel time	29.2	28.9	1.0	133540.0	133525.6	1.0
			distance	24.8	27.2	1.1	133955.9	133958.0	1.0
			unit length	17.8	21.8	1.2	134023.6	134022.0	1.0
BAY	321270	800172	travel time	31.3	38.1	1.2	160744.0	160728.3	1.0
			distance	28.7	36.0	1.3	159490.4	159486.1	1.0
			unit length	21.8	26.2	1.2	156606.8	156613.4	1.0
COL	435666	1057066	travel time	46.1	50.1	1.1	219906.9	219908.5	1.0
			distance	39.5	44.6	1.1	208073.6	208062.3	1.0
			unit length	30.9	35.8	1.2	209462.9	209465.0	1.0
FLA	1070376	2712798	travel time	136.6	130.7	1.0	535082.2	535067.6	1.0
			distance	102.4	115.0	1.1	522623.5	522613.5	1.0
			unit length	99.3	118.8	1.2	522690.5	522733.5	1.0
NW	1207945	2840208	travel time	154.2	187.6	1.2	604655.7	604628.0	1.0
			distance	129.1	162.9	1.3	612029.8	612036.4	1.0
			unit length	107.6	146.2	1.4	591167.8	591161.5	1.0
NE	1524453	3897636	travel time	192.7	282.2	1.5	767389.6	767390.5	1.0
			distance	167.6	211.4	1.3	758473.9	758461.3	1.0
			unit length	150.4	194.4	1.3	756489.6	756471.1	1.0
CAL	1890815	4657742	travel time	241.6	285.7	1.2	896852.8	896873.1	1.0
			distance	220.0	260.0	1.2	955789.7	955774.0	1.0
			unit length	256.1	278.0	1.1	959488.2	959467.9	1.0
LKS	2758119	6885658	travel time	500.4	474.8	0.9	1362438.2	1362443.9	1.0
			distance	338.4	406.3	1.2	1384591.0	1384580.3	1.0
			unit length	325.4	525.6	1.6	1387296.2	1387322.0	1.0
E	3598623	8778114	travel time	821.9	840.2	1.0	1814147.8	1814142.9	1.0
			distance	601.5	639.7	1.1	1833678.1	1833662.9	1.0
			unit length	445.6	608.2	1.4	1791469.0	1791389.7	1.0
W	6262104	15248146	travel time	1110.7	1584.3	1.4	3116286.8	3116280.5	1.0
			distance	951.2	1329.6	1.4	3207533.5	3207549.5	1.0
			unit length	857.0	1458.5	1.7	3225232.8	3225273.9	1.0
CTR	14081816	34292496	travel time	5959.5	6050.8	1.0	7260717.0	7260654.0	1.0
			distance	4229.4	5965.2	1.4	6943986.5	6944001.7	1.0
			unit length	2778.5	5311.4	1.9	6946160.5	6946243.7	1.0
USA	23947347	58333344	travel time	5542.0	7882.3	1.4	12130117.0	12130223.9	1.0
			distance	4392.9	6455.4	1.5	12069795.0	12069747.9	1.0
			unit length	3611.2	5101.1	1.4	12145470.0	12145475.4	1.0

can not be guaranteed to be non-decreasing, thereby excluding many of the faster heap-structures. For the bi-directional search, utilizing the caliber lemma increases the search space compared to standard heap-structures.

It should be noted that the Boost Graph Library (BGL) uses a generic programming framework which makes extensive use of C++ templates. The BGL implementation achieved running times close to the DIMACS reference implementation, see Table 2. The latter is the implementation provided for benchmarking on the challenge homepage[2]. It is remarkable that a flexible and highly reusable generic software architecture approach has no problem to catch up with fine-tuned "handcrafted" implementations. The differences in search space reported in Table 2 are due to the BGL implementation using the caliber heuristic. The running times of our Dijkstra implementation and the competitive

[2] `http://www.dis.uniroma1.it/~challenge9/`

TABLE 3. Summary of time and space requirements for preprocessing with 200 regions. For each metric the time needed for preprocessing and the number of unique flag vectors calculated as well as the resulting extra space is reported.

instance			travel time			distance			unit length		
name	#vertices	#arcs	secs	#vectors	Bytes/arc	secs	#vectors	Bytes/arc	secs	#vectors	Bytes/arc
NY	264346	730100	95.04	46527	5.59	85.27	98435	7.37	79.64	85740	6.94
BAY	321270	794830	82.04	19792	4.62	80.54	49811	5.57	63.37	66829	6.10
COL	435666	1042400	107.68	33849	4.81	98.54	72517	5.74	89.07	100504	6.41
FLA	1070376	2687902	301.2	47471	4.44	298.79	85157	4.79	290.45	50826	4.47
NW	1207945	2820774	347.86	50221	4.45	318.54	118884	5.05	292.78	111530	4.99
NE	1524453	3868020	774.82	91939	4.59	646.42	227458	5.47	641.37	149496	4.97
CAL	1890815	4630444	898.15	59616	4.32	861.6	145434	4.79	732.35	121880	4.66
LKS	2758119	6794808	1903.82	81216	4.30	2006.38	195768	4.72	1597.26	149103	4.55
E	3598623	8708058	2118.18	119175	4.34	1821.48	327547	4.94	1725.62	197961	4.57
W	6262104	15119284	4100.92	133292	4.22	3887.43	312904	4.52	3258.25	221905	4.37
CTR	14081816	33866826	21788.1	291562	4.22	20633.7	826359	4.61	17521.9	575129	4.42
USA	23947347	57708624	36635.8	295660	4.13	35361.2	848633	4.37	28084.9	548603	4.24

TABLE 4. Summary of time and space requirements for preprocessing with 600 regions.

instance			travel time			distance			unit length		
name	#vertices	#arcs	secs	#vectors	Bytes/arc	secs	#vectors	Bytes/arc	secs	#vectors	Bytes/arc
NY	264346	730100	191.19	148479	19.25	173.74	237638	28.41	162.57	217128	26.30
BAY	321270	794830	160.66	75464	11.12	159.9	140127	17.22	137.9	177479	20.75
COL	435666	1042400	209.26	99235	11.14	198.25	173677	16.50	189.88	221005	19.90
FLA	1070376	2687902	629.88	141681	7.95	602.25	255629	11.13	634.06	189961	9.30
NW	1207945	2820774	705.35	163013	8.33	652.37	324411	12.63	655.07	343521	13.13
NE	1524453	3868020	1535.38	308976	9.99	1336.42	669324	16.98	1366.04	508062	13.85
CAL	1890815	4630444	1688.4	196049	7.18	1590.04	453742	11.35	1571.31	428433	10.94
LKS	2758119	6794808	3554.2	350087	7.86	3355.34	785160	12.67	3196.82	622242	10.87
E	3598623	8708058	4088.22	429586	7.70	3612.54	1029551	12.87	3600.47	746840	10.43
W	6262104	15119284	8186.41	450797	6.24	7531.96	1020546	9.06	6952.99	876384	8.35
CTR	14081816	33866826	36901.9	1121594	6.48	34269.9	2903758	10.43	31009.9	2370374	9.25
USA	23947347	57708624	62616.9	1267948	5.65	58851.3	3382666	8.40	51376.2	2405552	7.13

query times achieved by our arc-flag implementation prove the efficiency of generic template code as well as the good performance of the arc-flag method. More details on the implementation can be found in [**Hil07**].

5.1. Instances. We performed computations on the USA road networks from the DIMACS Challenge homepage [**DIM06**] using different length functions on the arcs. In particular we considered distance metric, travel time metric and unit length.

5.2. Results — Preprocessing. Tables 3, 4 and 5 as well as Figure 7 give an overview of the preprocessing. The tables show the total running time of our preprocessing algorithm using the centralized shortest path with minimum total key described in Section 3.4. Furthermore, the number of unique flag-vectors is reported as well as the additional space required per arc. This extra memory results from storing each of the unique flag-vectors once as a bit-set and an index of 4 bytes for each arc to point to its flag-vector.

During the preprocessing we store each flag vector separately for each arc and therefore the space requirements during the preprocessing are much higher, e.g., 200 bits per arc for the instances in Table 3. Additionally, the centralized shortest path algorithm needs to

TABLE 5. Summary of time and space requirements for preprocessing with 1000 regions.

instance			travel time			distance			unit length		
name	#vertices	#arcs	secs	#vectors	Bytes/arc	secs	#vectors	Bytes/arc	secs	#vectors	Bytes/arc
NY	264346	730100	278.22	216396	41.05	255.82	304102	56.07	241.99	282955	52.44
BAY	321270	794830	230.66	117356	22.46	222.9	191698	34.15	214.98	230009	40.17
COL	435666	1042400	296.21	140785	20.88	279.62	225707	31.07	275.34	273818	36.84
FLA	1070376	2687902	910.98	215809	14.04	853.8	376528	21.51	925.99	301501	18.02
NW	1207945	2820774	998.55	250665	15.11	915.44	460623	24.41	932.71	499607	26.14
NE	1524453	3868020	2112.71	466339	19.07	1876.58	937715	34.30	1915.79	756825	28.46
CAL	1890815	4630444	2374.46	310106	12.37	2190.03	671973	22.14	2280.03	666737	22.00
LKS	2758119	6794808	4844.56	577741	14.63	4542.44	1217384	26.40	4482.78	1003382	22.46
E	3598623	8708058	5728.15	676346	13.71	5071.58	1489411	25.38	5104.8	1153177	20.55
W	6262104	15119284	11026.7	714271	9.91	10349.8	1552521	16.84	9869.15	1414805	15.70
CTR	14081816	33866826	56186.1	1830463	10.76	52432.3	4412360	20.29	47186.5	3770229	17.92
USA	23947347	57708624	85170.2	2133396	8.62	78689	5307382	15.50	71134.9	4119036	12.92

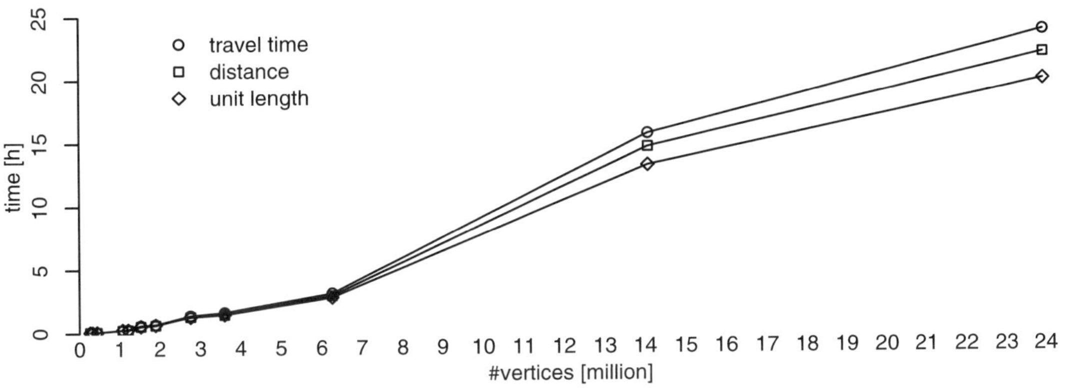

FIGURE 7. Preprocessing times for 1000 regions in hours relative to millions of nodes.

store a large distance matrix, resulting in high memory needs for the preprocessing phase. The smaller number of edges in Tables 3 to 5 result from the removal of parallel edges from the original data presented in Table 2.

An analysis of the calculated flag vectors shows that (depending on the partition) on average more than 40% of the flag vectors have less than 10% `true` entries, see Figure 8. The high amount of almost empty flag vectors justifies the idea for a compression of the vectors. It is important that the decompression algorithm is very fast—otherwise the speedup of the running time will be lost. The two and more level technique described in Section 4 is a suitable lossy compression method for the flag vector entries.

5.3. Results — Queries. Tables 6, 7 and 8 give an overview for random query results for different partitions. For each instance the size of the shortest path, i.e. the number of vertices contained, as well as query times and search space for Dijkstra's algorithm (dijk) and the arc-flag bi-directional query (flag) are reported. Additionally, the resulting speed-up factors for time and search space and the efficiency is given. The efficiency of a shortest path query is the ratio of search space (the number of scanned vertices) and the actual number of vertices in the path; i.e. an optimal query has an efficiency of 100%. The reported numbers are averaged over 1000 queries on each instance and metric. It can be

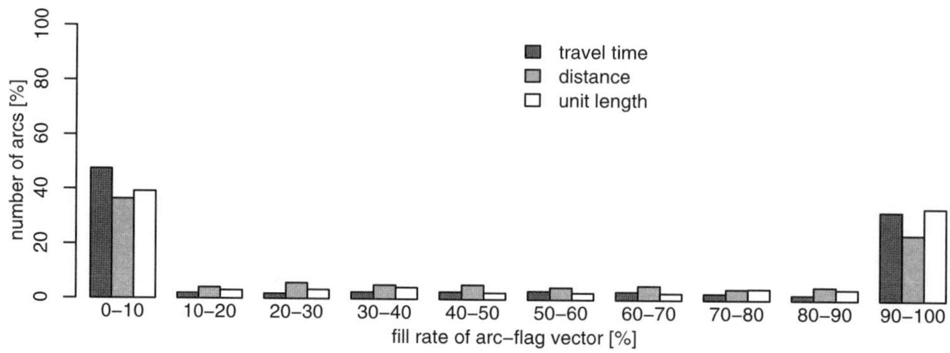

FIGURE 8. Statistics of the fill rate of the flag vectors on instance W with 200 partitions. The y-axis shows the percentage of arcs for which the corresponding flag vector has a certain fill rate, while the x-axis shows the percentages of different fill rates. For instance, an arc a has a flag vector with fill rate 10% if 20 out of 200 flags in the vector have been set to `true`.

observed that the efficiency of queries on the travel time metric is much higher than for the other metrics. For the partitioning into 1000 regions queries on all instances can be calculated with an efficiency of more than 80%, for smaller instances even more than 90% (Table 8).

Random queries are more likely to prioritize pairs of far-away vertices. Therefore experiments focusing on locality have been performed. In Figures 9 and 11 we present box plots[3] for query time and efficiency of local queries, i.e. the query pairs are chosen for different Dijkstra ranks. We define the Dijkstra rank as follows: Suppose we start a run of Dijkstra's algorithm at some vertex v and u is the k-th vertex scanned. Then the Dijkstra rank of u is $\lfloor \log_2(k) \rfloor$. We used the generator supplied with the benchmark suite[4] to generate pairs of local queries. Note that the efficiency improves with the distance of the query pairs. For far-away queries the efficiency is nearly optimal (Figure 9) for both distance and travel time. Figures 12, 13 and 14 present the effect of different partitionings on the local efficiency for the LKS instance. Each point represents the median of the efficiency of 1000 local queries. Larger numbers of regions improve on the efficiency for all metrics, although to different extends. Optimal efficiency for unit length seems hardest to achieve whereas the efficiency for the distance metric is optimal for far away queries. Note the steep increase for the distance metric between, e.g. rank 19 and 20 for 100 partitions. Travel time metric shows the best efficiency and highest improvements even for small numbers of partitions.

In Figure 10 the speed-up factors for query times resulting from partitioning the instances into 600 regions are presented. Again, the arc-flag method performs best on travel time metric, but also achieves considerable speed-ups for the other metrics.

---

[3]Also known as box and whisker plots. The median as well as the quartiles are denoted as small line and box, respectively. The range depicted by the whiskers covers 95% of the data. For ease of readability we omit outliers.

[4]`http://www.dis.uniroma1.it/~challenge9/`

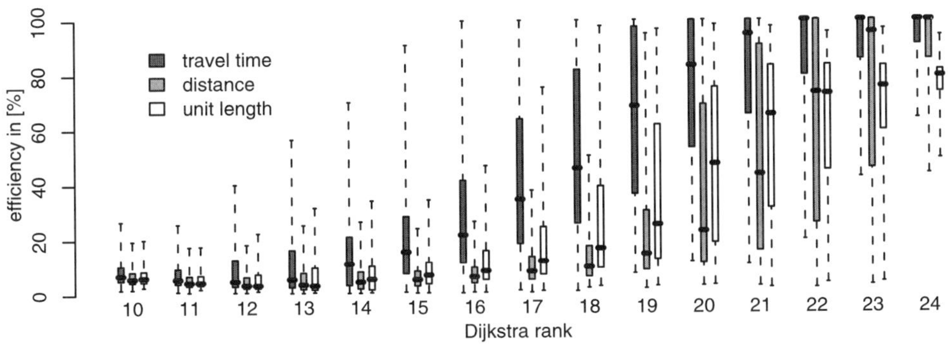

FIGURE 9. Efficiency of Local query results for Dijkstra rank from 10 to 24 on USA instance with 1000 regions.

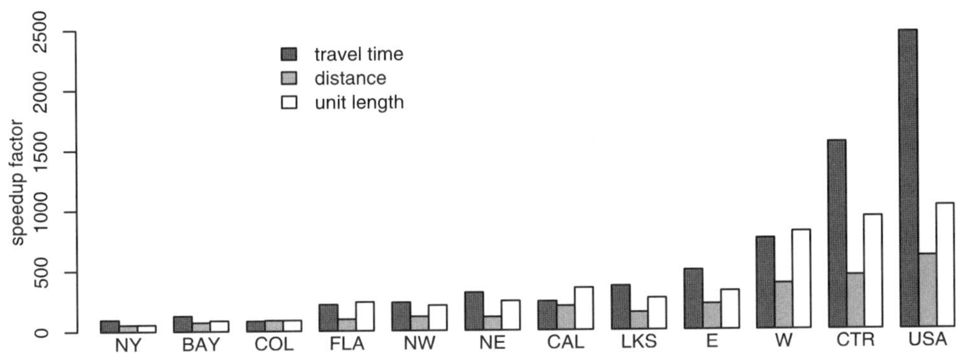

FIGURE 10. Speed up factors for query times using 600 regions.

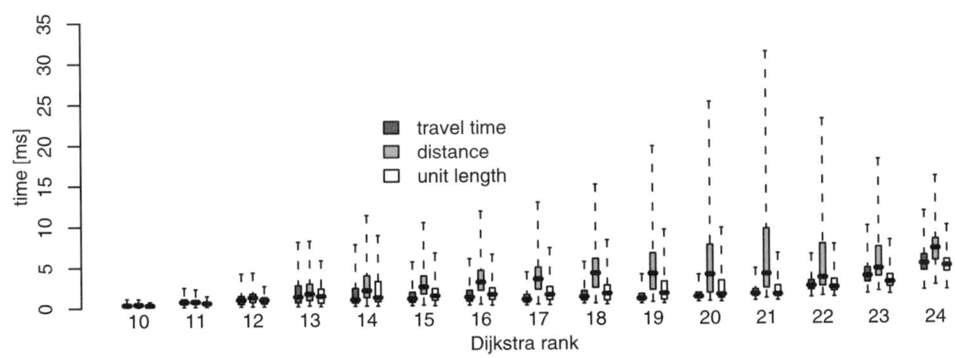

FIGURE 11. Query times in [ms] of Local query results for Dijkstra rank from 10 to 24 on USA instance with 1000 regions.

TABLE 6. Summary of random query results on instances of the USA-road family using 200 regions. For each instance and metric the average size of a shortest path is given, i.e. the number of vertices contained. The search space and the time needed for calculating using our reference Dijkstra as well as the bi-directional accelerated arc-flag query and the resulting factors are depicted. Finally, the resulting efficiency, i.e. the ratio of search space and path size is presented.

instance			query time [ms]			search space			efficiency
name	metric	size	dijk	flag	×	dijk	flag	×	[%]
NY	travel time	346	28.9	0.4	72	133526	442	302	78
	distance	407	27.2	0.7	39	133958	870	154	47
	unit length	267	21.8	0.5	44	134022	593	226	45
BAY	travel time	478	38.1	0.3	127	160728	556	289	86
	distance	467	36.0	0.7	51	159486	769	207	61
	unit length	297	26.2	0.5	52	156613	493	318	60
COL	travel time	743	50.1	0.6	84	219909	918	240	81
	distance	661	44.6	0.8	56	208062	1366	152	48
	unit length	449	35.8	0.5	72	209465	842	249	53
FLA	travel time	831	130.7	0.6	218	535068	1073	499	77
	distance	1019	115.0	1.2	96	522614	1855	282	55
	unit length	555	118.8	0.6	198	522733	776	674	72
NW	travel time	1243	187.6	1.0	188	604628	1506	401	83
	distance	1077	162.9	1.4	116	612036	2234	274	48
	unit length	712	146.2	0.8	183	591162	1355	436	53
NE	travel time	992	282.2	1.0	282	767391	1378	557	72
	distance	1143	211.4	2.6	81	758461	3164	240	36
	unit length	723	194.4	1.3	150	756471	1518	498	48
CAL	travel time	1284	285.7	1.2	238	896873	1709	525	75
	distance	1280	260.0	1.8	144	955774	3000	319	43
	unit length	772	278.0	0.9	309	959468	1507	637	51
LKS	travel time	1985	474.8	1.7	279	1362444	2409	566	82
	distance	2432	406.3	4.0	102	1384580	4663	297	52
	unit length	1300	525.6	1.5	350	1387322	2437	569	53
E	travel time	1784	840.2	2.1	400	1814143	2486	730	72
	distance	2154	639.7	4.8	133	1833663	7107	258	30
	unit length	1279	608.2	1.7	358	1791390	2683	668	48
W	travel time	2916	1584.3	2.6	609	3116281	4197	743	69
	distance	2578	1329.6	6.2	214	3207550	8756	366	29
	unit length	1628	1458.5	2.4	608	3225274	3961	814	41
CTR	travel time	3127	6050.8	5.7	1062	7260654	5667	1281	55
	distance	3471	5965.2	24.5	243	6944002	25400	273	14
	unit length	2039	5311.4	10.8	492	6946244	9361	742	22
USA	travel time	4545	7882.3	4.4	1791	12130224	8171	1485	56
	distance	5181	6455.4	14.2	455	12069748	24545	492	21
	unit length	3000	5101.1	8.5	600	12145475	13958	870	21

6. Discussion and Outlook

We presented two essential improvements of the basic variant of the arc-flag acceleration [**Lau97, Lau04**] for speeding up the p2p shortest path search queries on static road networks. The arc-flag acceleration method is a modification of the standard Dijkstra algorithm and can be used to avoid exploring unnecessary paths during shortest path computations.

Our results show that the arc-flag method together with a multi-way arc separator partitioning yields computational query times between 3.0 and 14.2 milliseconds on the USA instance (24M nodes, 58M arcs), depending on the metric and partitioning. These

TABLE 7. Summary of random query results on instances of the USA-road family using 600 regions.

instance			query time [ms]			search space			efficiency
name	metric	size	dijk	flag	×	dijk	flag	×	[%]
NY	travel time	346	28.9	0.3	96	133526	393	340	88
	distance	407	27.2	0.5	54	133958	582	230	70
	unit length	267	21.8	0.4	55	134022	458	293	58
BAY	travel time	478	38.1	0.3	127	160728	513	313	93
	distance	467	36.0	0.5	72	159486	570	280	82
	unit length	297	26.2	0.3	87	156613	408	384	73
COL	travel time	743	50.1	0.6	84	219909	825	267	90
	distance	661	44.6	0.5	89	208062	900	231	73
	unit length	449	35.8	0.4	89	209465	651	322	69
FLA	travel time	831	130.7	0.6	218	535068	941	569	88
	distance	1019	115.0	1.2	96	522614	1334	392	76
	unit length	555	118.8	0.5	238	522733	681	768	81
NW	travel time	1243	187.6	0.8	234	604628	1354	447	92
	distance	1077	162.9	1.4	116	612036	1503	407	72
	unit length	712	146.2	0.7	209	591162	1020	580	70
NE	travel time	992	282.2	0.9	314	767391	1150	667	86
	distance	1143	211.4	1.9	111	758461	1899	399	60
	unit length	723	194.4	0.8	243	756471	1089	695	66
CAL	travel time	1284	285.7	1.2	238	896873	1456	616	88
	distance	1280	260.0	1.3	200	955774	1869	511	68
	unit length	772	278.0	0.8	348	959468	1157	829	67
LKS	travel time	1985	474.8	1.3	365	1362444	2142	636	93
	distance	2432	406.3	2.8	145	1384580	3289	421	74
	unit length	1300	525.6	2.0	263	1387322	2022	686	64
E	travel time	1784	840.2	1.7	494	1814143	2062	880	87
	distance	2154	639.7	3.0	213	1833663	3919	468	55
	unit length	1279	608.2	1.9	320	1791390	1981	904	65
W	travel time	2916	1584.3	2.1	754	3116281	3472	898	84
	distance	2578	1329.6	3.5	380	3207550	4643	691	56
	unit length	1628	1458.5	1.8	810	3225274	2714	1188	60
CTR	travel time	3127	6050.8	3.9	1551	7260654	4113	1765	76
	distance	3471	5965.2	13.4	445	6944002	11643	596	30
	unit length	2039	5311.4	5.7	932	6946244	4832	1438	42
USA	travel time	4545	7882.3	3.2	2463	12130224	5960	2035	76
	distance	5181	6455.4	10.7	603	12069748	16981	711	31
	unit length	3000	5101.1	5.0	1020	12145475	7711	1575	39

computational query times can be achieved by using only between 4.13 and 15.5 bytes of additional space per arc. The additional space is used for storing the flag vectors and a pointer from each arc to its flag vector. The performance of our method depends on the metric which is used: on the travel time metric the smallest space is required and the fastest query response times can be achieved. The preprocessing of the additional data can be executed within reasonable time, e.g. preprocessing the USA instance takes between 10 and 24 hours. Note that in the static setting assumed by us the preprocessing time should not be an issue since it needs to be done only once.

With regard to routing application such as large web servers (e.g., `maps.google.com`) or small PNDs, the query times of our improved arc-flag method are sufficiently fast: rendering the calculated route in a digital map for its presentation within a graphical user interface takes already much longer than calculating the shortest path itself using our method.

TABLE 8. Summary of random query results on instances of the USA-road family using 1000 regions.

instance			query time [ms]			search space			efficiency
name	metric	size	dijk	flag	×	dijk	flag	×	[%]
NY	travel time	346	28.9	0.2	144	133526	380	351	91
	distance	407	27.2	0.5	54	133958	502	267	81
	unit length	267	21.8	0.3	73	134022	417	321	64
BAY	travel time	478	38.1	0.3	127	160728	507	317	94
	distance	467	36.0	0.5	72	159486	540	295	86
	unit length	297	26.2	0.3	87	156613	390	402	76
COL	travel time	743	50.1	0.4	125	219909	818	269	91
	distance	661	44.6	0.6	74	208062	877	237	75
	unit length	449	35.8	0.4	89	209465	623	336	72
FLA	travel time	831	130.7	0.5	261	535068	906	591	92
	distance	1019	115.0	0.8	144	522614	1218	429	84
	unit length	555	118.8	0.4	297	522733	660	792	84
NW	travel time	1243	187.6	0.7	268	604628	1322	457	94
	distance	1077	162.9	1.0	163	612036	1339	457	80
	unit length	712	146.2	0.6	244	591162	981	603	73
NE	travel time	992	282.2	0.7	403	767391	1107	693	90
	distance	1143	211.4	1.1	192	758461	1638	463	70
	unit length	723	194.4	0.7	278	756471	999	757	72
CAL	travel time	1284	285.7	0.8	357	896873	1399	641	92
	distance	1280	260.0	1.0	260	955774	1643	582	78
	unit length	772	278.0	0.7	397	959468	1074	893	72
LKS	travel time	1985	474.8	1.2	396	1362444	2108	646	94
	distance	2432	406.3	2.2	185	1384580	2998	462	81
	unit length	1300	525.6	1.3	404	1387322	1893	733	69
E	travel time	1784	840.2	1.2	700	1814143	1977	918	90
	distance	2154	639.7	2.5	256	1833663	3181	576	68
	unit length	1279	608.2	1.5	405	1791390	1784	1004	72
W	travel time	2916	1584.3	1.7	932	3116281	3280	950	89
	distance	2578	1329.6	2.3	578	3207550	3830	837	67
	unit length	1628	1458.5	1.6	912	3225274	2468	1307	66
CTR	travel time	3127	6050.8	3.0	2017	7260654	3787	1917	83
	distance	3471	5965.2	8.1	736	6944002	7903	879	44
	unit length	2039	5311.4	4.2	1265	6946244	3887	1787	52
USA	travel time	4545	7882.3	3.0	2627	12130224	5416	2240	84
	distance	5181	6455.4	8.9	725	12069748	12954	932	40
	unit length	3000	5101.1	4.0	1275	12145475	5904	2057	51

For applications on embedded or mobile devices, e.g. PNDs, typically one has to tackle two bottlenecks: the I/O latency and the limited instruction cache of such devices. On the ARM chip family [**ARM08**], a commonly used hardware platform in the embedded or mobile industry, the latter is an even bigger bottleneck. Already implementing a binary heap generates a significant overhead in terms of instruction on an ARM chip. Therefore it is of utmost importance to keep the data structures and procedures on such devices as simple as possible. Although also more complex acceleration methods can be implemented on mobiles devices [**SSV08, GW05a**] it can be expected that a simple method such as arc-flags performs even better on a mobile device. However, it remains a task for future research to confirm this assumption.

The plots showing the results for queries of different Dijkstra ranks demonstrate that there is not much difference in computational effort between near by and far away query pairs. On the travel time metric the average computational effort is nearly constant for all different ranks. This observation can also be made when comparing the average size of

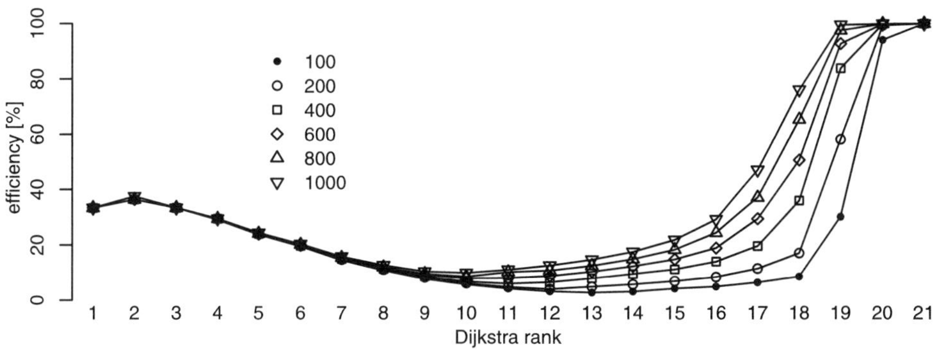

FIGURE 12. Comparison of efficiency of partitionings for local queries on LKS with distance metric. For 1000 queries per Dijkstra rank the median of the efficiency is visualized.

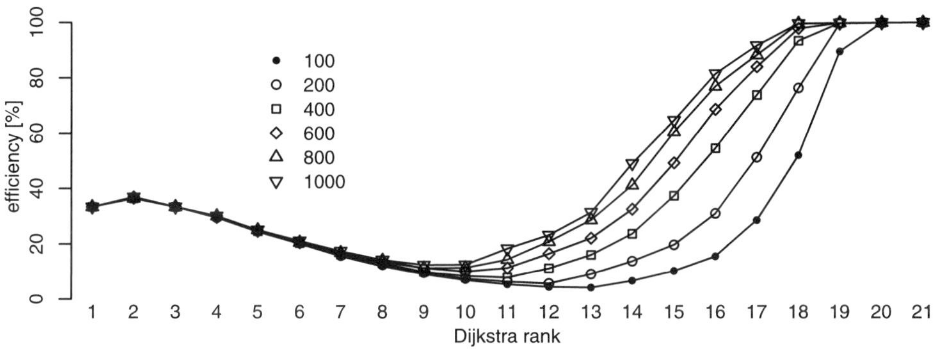

FIGURE 13. Comparison of efficiency of partitionings for local queries on LKS with travel time metric. For 1000 queries per Dijkstra rank the median of the efficiency is visualized.

the search space to the average number of nodes in a path: factors of less than two are achieved on the travel time metric. In all cases, the search space of our arc-flag method is never larger than ten times the actual number of nodes on the shortest paths.

We extended the arc-flag method to general graphs; this is not only relevant for (almost) planar road networks: many of today's problems in traffic optimization do not just consider a usually (almost) planar road network but rather a (time) expanded version of the original network. Typical examples of such applications are time-table graphs for public transport and time expanded graphs for dynamic flows or dynamic routing, see for instance [**GKMS04, Lie06, Sch06**]. There is no reasonable embedding in the plane for such complicated graph structures. Yet, for our acceleration methods we do not need any such embedding.

Furthermore, the arc-flag method provides a (fine-tunable) trade-off between speedup factors and space usage. Depending on the given partitioning, one can regard the arc-flag

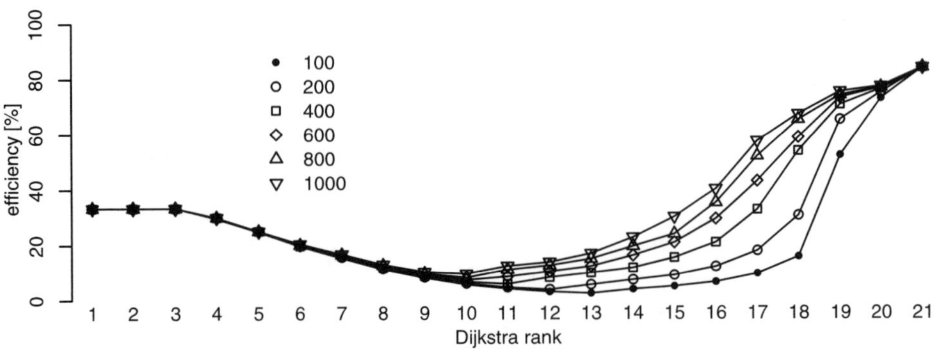

FIGURE 14. Comparison of efficiency of partitionings for local queries on LKS with unit length. For 1000 queries per Dijkstra rank the median of the efficiency is visualized.

acceleration as an interpolation between no precomputed information at all (plain Dijkstra) and full pre-computation (determining all possible shortest paths of the input graph). Whereas the former is achieved by choosing a partitioning of the graph into just one region, the latter means a partitioning of the graph where each node lies in its own region, i.e. it has as many regions as nodes. Thus in theory we can get as close as possible to the ideal shortest path search by increasing the number of regions in the partitioning ('ideal' means that the shortest path algorithm visits only arcs which actually belong to the shortest path itself). Obviously, an increase in the number of regions also entails an increase in preprocessing time and memory consumption.

Again, with regard to possible applications, the scalability and conceptional simplicity of the arc-flag methods makes it very flexible: on large servers with no space restriction one would use arc-flags together with a fine (high resolution) partitioning using many regions while on a PND with hard space and performance restrictions a coarse partitioning already suffices to deliver good speedups with the arc-flags.

References

[ARM08] ARM, *Advanced Risc Machines ltd*, `http://www.arm.com`, February 2008.

[BD08] Reinhard Bauer and Daniel Delling, *SHARC: Fast and Robust Unidirectional Routing*, Proceedings of the 10th Workshop on Algorithm Engineering and Experiments (ALENEX'08), SIAM, 2008, pp. 13–26.

[BDS+08] Reinhard Bauer, Daniel Delling, Peter Sanders, Dennis Schieferdecker, Dominik Schultes, and Dorothea Wagner, *Combining hierarchical and goal-directed speed-up techniques for dijkstra's algorithm*, 2008, Submitted.

[BDW07a] Reinhard Bauer, Daniel Delling, and Dorothea Wagner, *Experimental Study on Speed-Up Techniques for Timetable Information Systems*, Proceedings of the 7th Workshop on Algorithmic Approaches for Transportation Modeling, Optimization, and Systems (ATMOS'07), Internationales Begegnungs- und Forschungszentrum für Informatik (IBFI), Schloss Dagstuhl, Germany, 2007.

[BFM06] Holger Bast, Stefan Funke, and Domagoj Matijevic, *TRANSIT: Ultrafast shortest-path queries with linear-time preprocessing*, DIMACS Implementation Challenge Shortest Paths, 2006.

[BFM+07] Holger Bast, Stefan Funke, Domagoj Matijevic, Peter Sanders, and Dominik Schultes, *In transit to constant time shortest-path queries in road networks*, Proceedings of the 9th Workshop on Algorithm Engineering and Experiments (ALENEX'07), 2007, pp. 46–59.

[BGL07] BGL, *The Boost Graph Library*, `http://www.boost.org/libs/graph/`, March 2007.

[Dij59] Edsger Wybe Dijkstra, *A note on two problems in connexion with graphs*, Numerische Mathematik, vol. 1, Mathematisch Centrum, Amsterdam, The Netherlands, 1959, pp. 269–271.

[DIM06] DIMACS, *9th Implementation Challenge — Shortest Paths*, `http://www.dis.uniroma1.it/~challenge9`, 2006.

[EL99] Reinhard Enders and Ulrich Lauther, *Method and device for computer assisted graph processing*, http://gauss.ffii.org/PatentView/EP1027578, May 1999, Siemens AG.

[FT87] Michael L. Fredman and Robert Endre Tarjan, *Fibonacci heaps and their uses in improved network optimization algorithms*, Journal of the Association for Computing Machinery **34** (1987), no. 3, 596–615.

[GH05] Andrew V. Goldberg and Chris Harrelson, *Computing the shortest path: A^* search meets graph theory*, Proceedings of the 16th Annual ACM-SIAM Symposium on Discrete Algorithms (SODA), Vancouver, BC (Philadelphia, PA, USA) (Adam Buchsbaum, ed.), SIAM, 2005, pp. 156–165.

[GKMS04] Ewgenij Gawrilow, Ekkehard Köhler, Rolf H. Möhring, and Björn Stenzel, *Conflict-free real-time agv routing*, In Proceedings of Operations Research (OR) 2004, 2004.

[GKW06] Andrew V. Goldberg, Haim Kaplan, and Renato Fonseca Werneck, *Reach for A^*: Efficient point-to-point shortest path algorithms*, Proceedings of the 8th Workshop on Algorithm Engineering and Experiments (ALENEX), SIAM, 2006.

[Gol01] Andrew V. Goldberg, *A simple shortest path algorithm with linear expected time*, SIAM J. on Computing **37** (2001), no. 5, 1637–1655.

[Gut04] Ronald J. Gutman, *Reach-based routing: A new approach to shortest path algorithms optimized for road networks*, Proceedings of the 6th Workshop on Algorithm Engineering and Experiments (ALENEX) and the First Workshop on Analytic Algorithmics and Combinatorics (ANALCO), New Orleans, LA, USA (Philadelphia, PA, USA) (Lars Arge, Giuseppe F. Italiano, and Robert Sedgewick, eds.), SIAM, 2004, pp. 100–111.

[GW05a] Andrew V. Goldberg and Renato F. Werneck, *Computing point-to-point shortest paths from external memory*, Proceedings of the 7th Workshop on Algorithm Engineering and Experiments (ALENEX), SIAM, 2005, pp. 26–40.

[Hil07] Moritz Hilger, *Accelerating point-to-point shortest path computations in large scale networks.*, Master's thesis, Technische Universität Berlin, 2007.

[Joh77] Donald B. Johnson, *Efficient algorithms for shortest paths in sparse networks*, Journal of the Association for Computing Machinery **24** (1977), no. 1, 1–13.

[KK98] George Karypis and Vipin Kumar, *A fast and high quality multilevel scheme for partitioning irregular graphs*, SIAM Journal on Scientific Computing **20** (1998), no. 1, 359–392.

[KMS05] Ekkehard Köhler, Rolf H. Möhring, and Heiko Schilling, *Acceleration of shortest path and constrained shortest path computation*, In Proceedings of the 4th International Workshop on Experimental and Efficient Algorithms (WEA) (Heidelberg, Germany) (Sotiris E. Nikoletseas, ed.), Lecture Notes in Computer Science, vol. 3503, Springer, 2005, pp. 126–138.

[KSS+06] Sebastian Knopp, Peter Sanders, Dominik Schultes, Frank Schulz, and Dorothea Wagner, *Fast computation of distance tables using highway hierarchies*, Tech. report, Faculty of Informatics, University of Karlsruhe, 2006.

[Lau97] Ulrich Lauther, *Slow preprocessing of graphs for extremely fast shortest path calculations*, 1997, Lecture at the Workshop on Computational Integer Programming at ZIB (no documentation available).

[Lau04] ______, *An extremely fast, exact algorithm for finding shortest paths in static networks with geographical background*, Geoinformation und Mobilität - von der Forschung zur praktischen Anwendung (Münster, Germany) (Martin Raubal, Adam Sliwinski, and Werner Kuhn, eds.), IfGI prints, vol. 22, Institut für Geoinformatik, Westfälische Wilhelms-Universität, 2004, pp. 219–230.

[Lau06] ______, *An experimental evaluation of point-to-point shortest path calculation on roadnetworks with precalculated edge-flags*, DIMACS Implementation Challenge Shortest Paths, 2006.

[Lie06] Christian Liebchen, *Periodic timetable optimization in public transport*, Ph.D. thesis, Institute of Mathematics, TU Berlin, 2006.

[MET95] METIS, *A family of multilevel partitioning algorithms*, `http://www-users.cs.umn.edu/~karypis/metis/`, 1995.

[MSS+05] Rolf H. Möhring, Heiko Schilling, Birk Schütz, Dorothea Wagner, and Thomas Willhalm, *Partitioning graphs to speed up Dijkstra's algorithm*, Proceedings of the 4th International Workshop on Experimental and Efficient Algorithms (WEA) (Heidelberg, Germany) (Sotiris E. Nikoletseas, ed.), Lecture Notes in Computer Science, vol. 3503, Springer, 2005, pp. 189–202.

[MSS+06] ______, *Partitioning graphs to speed up Dijkstra's algorithm*, ACM Journal of Experimental Algorithms (JEA) **11** (2006), Article No. 2.8.

[PAR06] PARTY, *A Partitioning Library*, `http://wwwcs.uni-paderborn.de/cs/ag-monien/RESEARCH/PART/party.html`, 2006.

[Sch06] Heiko Schilling, *Route assignment problems in large networks*, Ph.D. thesis, Institute of Mathematics, TU Berlin, 2006.

[Sch08] Dominik Schultes, *Route planning in road networks*, Ph.D. thesis, Institute for Theoretical Computer Science, TH Karlsruhe, 2008.

[SS05] Peter Sanders and Dominik Schultes, *Highway hierarchies hasten exact shortest path queries.*, Proceedings of the 13th Annual European Symposium (ESA) (Gerth Stølting Brodal and Stefano Leonardi, eds.), Lecture Notes in Computer Science, vol. 3669, Springer, 2005, pp. 568–579.

[SS06] ______, *Engineering highway hierarchies*, Proceedings of the 14th Annual European Symposium (ESA) (Yossi Azar and Thomas Erlebach, eds.), vol. Lecture Notes in Computer Science 4168, Springer, September 2006, pp. 804–816.

[SS07] Dominik Schultes and Peter Sanders, *Dynamic highway-node routing*, Proceedings of the 6th International Workshop on Experimental Algorithms (WEA), Lecture Notes in Computer Science, vol. 4525, Springer, 2007, pp. 66–79.

[SSV08] Peter Sanders, Dominik Schultes, and Christian Vetter, *Mobile route planning*, 2008, Submitted.

[SWW00] Frank Schulz, Dorothea Wagner, and Karsten Weihe, *Dijkstra's algorithm on-line: An empirical case study from public railroad transport*, ACM Journal of Experimental Algorithms **5** (2000), 12.

[Wil05] Thomas Willhalm, *Engineering shortest paths and layout algorithms for large graphs*, Ph.D. thesis, Faculty of Informatics, University of Karlsruhe, 2005.

[WW03] Dorothea Wagner and Thomas Willhalm, *Geometric speed-up techniques for finding shortest paths in large sparse graphs*, Algorithms - ESA 2003, 11th Annual European Symposium, Budapest, Hungary (Heidelberg, Germany) (Giuseppe Di Battista and Uri Zwick, eds.), Springer-Verlag, Lecture Notes in Computer Science, vol. 2832, 2003, pp. 776–787.

MORITZ HILGER, INSTITUTE OF MATHEMATICS, TU BERLIN, STRASSE DES 17. JUNI 136, D-10623 BERLIN, GERMANY

E-mail address: `Moritz.Hilger@TU-Berlin.DE`

EKKEHARD KÖHLER, MATHEMATICAL INSTITUTE, BTU COTTBUS, POSTFACH 101344, D-03013 COTTBUS, GERMANY

E-mail address: `Ekkehard.Koehler@Math.TU-Cottbus.DE`

ROLF H. MÖHRING, INSTITUTE OF MATHEMATICS, TU BERLIN, STRASSE DES 17. JUNI 136, D-10623 BERLIN, GERMANY

E-mail address: `Rolf.Moehring@TU-Berlin.DE`

HEIKO SCHILLING, INSTITUTE OF MATHEMATICS, TU BERLIN, STRASSE DES 17. JUNI 136, D-10623 BERLIN, GERMANY

E-mail address: `Heiko.Schilling@TU-Berlin.DE`

DIMACS Series in Discrete Mathematics
and Theoretical Computer Science
Volume **74**, 2009

High-Performance Multi-Level Routing

Daniel Delling, Martin Holzer, Kirill Müller, Frank Schulz,
and Dorothea Wagner

ABSTRACT. Shortest-path computation is a frequent task in practice. Owing to ever-growing real-world graphs, there is a constant need for faster algorithms. In the course of time, a large number of techniques to heuristically speed up Dijkstra's shortest-path algorithm have been devised. This work reviews the multi-level technique to answer shortest-path queries exactly [**24, 9**], which makes use of a hierarchical decomposition of the input graph and precomputation of supplementary information. We develop this preprocessing to the maximum and introduce several ideas to enhance this approach considerably, by reorganizing the precomputed data in partial graphs and optimizing them individually.

To answer a given query, certain partial graphs are combined to a search graph, which can be explored by a simple and fast procedure. The concept behind the construction of the search graph is such that query times depend mainly on the number of partial graphs included. This is confirmed by experiments with different road graphs, each containing several million vertices, and time, distance, and unit metrics. Our query algorithm computes the distance between any pair of vertices in no more than 40 μs, however, a lengthy preprocessing is required to achieve this query performance.

1. Introduction

Computation of shortest paths is a central requirement for many applications, such as route planning or network search. Facing real-world data, the need for speed remains unabated: collection of geographic information is enhanced constantly, resulting in increasingly comprehensive road graphs; public-transportation networks often comprise datasets from different means of transportation, such as train, tram, ferry, and even airplane schedules; and the graph representing the WWW is growing faster than ever. There are two basic approaches to tackle this task: relying on approximate algorithms, or devising faster exact ones. We opt for the latter.

2000 *Mathematics Subject Classification.* Primary 68R10; Secondary 90B20.

Key words and phrases. Shortest paths, graph, speed-up technique, decomposition, hierarchy, levels, preprocessing, supersedement, graph compression.

This work was partially supported by the Future and Emerging Technologies Unit of EC (IST priority – 6th FP), under contract no. FP6-021235-2 (project ARRIVAL).

Since its publication in 1959, Dijkstra's famous algorithm for calculation of shortest paths in a directed graph with nonnegative edge weights [**4**] has been subject to many improvements. Due to enormous space requirement (quadratic in the number of vertices), we cannot afford precomputing shortest paths between all pairs of vertices. However, graphs can be preprocessed at an off-line step so that subsequent on-line queries take only a fraction of the time used by Dijkstra's algorithm. One recent speed-up technique [**1, 22**], which also relies on a preprocessing, yields for the European road network that we also use for our experiments a considerable average query time of 4 μs when travel times as edges weights are used, but of comparatively high 38 μs for travel distances.

In this work, we present a further enhancement of the multi-level technique given in [**24**], which is based on a hierarchical decomposition of the input graph and computation of an auxiliary graph containing additional information. The use of this precomputed data allows, at the on-line stage, for reduction in search space and, consequently, query time. We develop the preprocessing to the maximum: our new variant outsources almost all of the effort needed to compute a shortest path to the preprocessing stage. It therefore fits best into an environment where query time is invaluable but long preprocessing times (and a fair amount of precomputed data) can be afforded, such as car navigation systems or web-based route planners. While in [**24, 9**] the multi-level approach was shown to be effective for graphs of up to 100 000 vertices, we are now able to handle much bigger graphs still within reasonable time.

The main differences to the former multi-level technique concern the following issues. During the preprocessing stage, instead of one single multi-level graph we compute a large number of small *partial* graphs. We show that for each possible query, there is a search graph combined of several partial graphs which preserves the distance between the dedicated vertices. This graph is acyclic, and we give a simple linear-time procedure to search it. The advantage of dealing with multiple graphs is that each of them can be optimized individually, which is achieved by two measures: first, omission of edges whose relaxation will never create a shorter path; and second, transformation of the partial graphs into equivalent graphs that preserve all shortest paths but have fewer edges. What is more, we make use of the fact that the preprocessing is parallelizable.

The trade-off between preprocessing effort and query time is adjustable. For fixed parameters, we can provide a guarantee for both the number of edges considered by the search algorithm and the query time. With our implementation, keeping the preprocessed data in secondary storage, we can answer a query through few random accesses to that storage. If the preprocessed data fits entirely into main memory, our query performance is competitive to that of other recent approaches: we obtain query times of less than 40 μs (except for very few outliers) for graphs with up to 24 million vertices, representing the Western European and US road networks. Moreover, our approach yields *equal* performance for all metrics investigated (travel times, travel distances, and unit edge lengths).

In the remainder of this section, we classify our approach in the context of other shortest-path speed-up techniques. The next section briefly reviews the multi-level technique as presented in [**24**], and shows the various refinements made. An experimental study is presented in Section 3, and we conclude in Section 4 addressing some aspects to be explored in the future.

1.1. Related Work. This paper is strongly based on [**18**], which is the master's thesis by one of the authors. It can be seen as a further development in that, e.g., preprocessing time could be improved through refinement of our code. Other pieces of information, however, such as details on some proofs or algorithmic aspects, can, in this work, not be displayed in full length, so we may refer the interested reader to [**18**].

There are a large number of other techniques to speed up single-pair shortest-path algorithms, most of which rely on Dijkstra's algorithm [**4**]. In the following survey, we focus on methods that in a preprocessing step compute some additional information, which is used at the on-line step for answering a shortest-path query. We differentiate between techniques that attach the precomputed data to the graph's vertices or edges, permitting the on-line algorithm to quickly decide which parts of the graph can be pruned [**23, 26, 27, 7, 15, 17, 14, 5**], and such that precompute a hierarchical auxiliary graph, a slender part of which suffices to answer a given shortest-path query [**24, 9, 12, 11, 20, 21**]. We want to briefly review the latter works and point out their relationship to ours.

As mentioned above, the method presented in this work uses the same basic concepts as the one described in [**24, 9**] (which will occasionally be referred to as the *classic* multi-level technique), where the following enhancements are made. The auxiliary data is distributed to many partial graphs, which can afterwards be thinned out and optimized individually. Given start and destination vertices, we combine several partial graphs to obtain an acyclic search graph, which can be explored by the on-line stage in linear time.

The *HiTi model* by Jung and Pramanik [**12**] is similar to the classic multi-level technique, except that it uses edge separators rather than vertex separators. Also with *hierarchical encoded path views*, presented by Jing, Huang, and Rundensteiner [**11**], various partial graphs are computed, which are combined appropriately to form a search graph for a given query. No graph optimization is used, but a compression technique to also keep track of the course of shortest paths is given.

Finally, the *highway hierarchies* technique, introduced by Sanders and Schultes [**20, 21**], computes a hierarchy of coarsenings of the input graph, where the search algorithm proceeds in a bidirectional fashion and needs to consider vertices of only one level of hierarchy at a time. In a further development [**2**], highway hierarchies have been extended to *transit node routing*, which also takes advantage of precomputed all-pair distances of a selection of vertices; the difference to our approach is that these distances are not represented by graphs but matrices, which do not seem to induce as simple means for optimization. The given description of highway node routing is generic enough so that it also covers the basic concept behind our approach. One further difference concerns the way of determining selected vertices; second, transit node routing is designed to yield better speed-ups with the travel time instead of distance metric, which is not true for our technique.

2. Multi-Level Graphs

The formal description of our high-performance multi-level technique (HPML) is divided into two parts. The first one reviews the classic multi-level technique [**24**] and points out the major modifications and enhancements made to it. In the second part, we present the core ingredient to achieve massive reduction in search space and query time, optimization of the partial search graph.

2.1. Enhancing Multi-Level Graphs. A multi-level graph $\mathcal{M}$ extends a weighted digraph $G = (V, E)$ through multiple *levels* of edges, depending on a sequence of vertex subsets. For a pair of vertices $s, t \in V$, a subgraph $\mathcal{M}_{st}$ of $\mathcal{M}$, called *search graph*, with the same s-t distance as in G can be determined efficiently. As the search graph is substantially smaller than G, it allows for answering the given query much faster.

The description of our model is organized as follows. We first fix some notation needed to define multi-level graphs, show how to construct these, and briefly address the issue of parallelizing the preprocessing step. To extract from the whole search graph a search graph $\mathcal{M}_{st}$ sufficent to answer a given query $(s,\ t)$, we have to define an auxiliary datastructure called *component tree*, and give a formal proof that shortest s-t paths in G and $\mathcal{M}_{st}$ are of equal length. The search graph $\mathcal{M}_{st}$ can be transformed into an acyclic graph, which property yields a simple and fast search algorithm compared to Dijkstra's algorithm. Finally, due to construction of our model, short-distance queries have to be treated differently.

2.1.1. *Notation.* To create a multi-level graph, we use a sequence of vertex subsets, denoted by $\mathscr{S} = \langle S_i \rangle$ with $1 \leq i \leq l$. Each S_i is called a *separator set.* The separator sets are decreasing with respect to set inclusion: $V \supset S_1 \supset S_2 \supset \ldots \supset S_l$. Best performance can be achieved when the graph $G - S_i$ falls apart into 'many' components of similar size, while $|S_i|$ is 'small' compared to $|V|$. For the decomposed graph $G - S_i$, we shall use the following definitions.

- By $\mathscr{C}_i$, we denote the set of maximal connected components at level i. A connected component $C \in \mathscr{C}_i$ itself is a weighted graph, whose vertices are referred to by $V(C)$.
- For a vertex $v \in V \setminus S_i$, let $C_i^v \in \mathscr{C}_i$ be the component with $v \in C_i^v$. We call C_i^v the *home component* of v at level i. To simplify notation, we define $C_i^v := \{v\}$ for $i \in \{0, \ldots, l\}$ and $v \in S_i$, and let $S_0 = V$.
- We call a vertex $v \in S_i$ *adjacent* to a component $C \in \mathscr{C}_i$ if there is an edge between v and a vertex in C in either direction. The set of all vertices adjacent to C is denoted by $Adj(C)$. For $v \in S_i$ (i.e., $C_i^v = \{v\}$), we define $Adj(C_i^v) := \{v\}$.
- A component C_i^v together with its adjacent vertices is called the *wrapped component* $G_i^v = G \cap (V(C_i^v) \cup Adj(C_i^v))$.

Figure 1 shows two components (darker shades) and their belonging wrapped components (lighter shades) at levels 1 (smaller components) and 2 (larger components), respectively, as an example for the above definitions. The adjacent vertex sets are $\{v_1, v_2, v_4\}$ and $\{v_3, v_4\}$. Note that vertex v_4 is adjacent to both components, as it is a separator vertex at both levels 1 and 2.

The above definition requires the components C_i^v to be connected; however, we do not rely on this property. If two components C_i^v and C_i^w share the same parent component, we can merge these two components into one new component $C_i^{vw} = C_i^v \cup C_i^w$. The adjacent vertices of the merged component are the vertices that are adjacent to at least one original component. If $Adj(C_i^w) \subseteq Adj(C_i^v)$, merging reduces the total number of components without increasing the number of adjacent vertices for any component. It is advisable to do so, as reducing the total number of components also reduces preprocessing time. For our test instance, merging components leads to a reduction of the total number of components by up to two orders of magnitude.

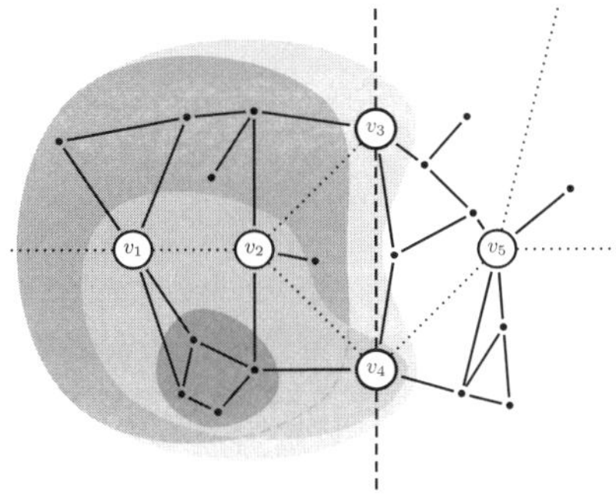

FIGURE 1. Hierarchy due to graph decomposition: components (darker shades) with belonging wrapped components (lighter shades) at levels 1 (smaller components) and 2 (larger components).

2.1.2. *Multi-Level Graph.* Each level of the multi-level graph $\mathcal{M}$ is determined by a set of edges. For each $i \in \{1, \ldots, l\}$, we construct three sets of edges from the following *candidate sets*:

Level edges: $E_i \subseteq S_i \times S_i$.
Upward edges: $U_i \subseteq S_{i-1} \times S_i$.
Downward edges: $D_i \subseteq S_i \times S_{i-1}$.

For a level $i \in \{1, \ldots, l\}$, a candidate edge $(v, w) \in U_i, D_i$ is elected to be an upward or downward edge only if there is a v-w path in G that does not contain any vertices in S_i besides v or w (in other words, both endpoints must be contained in the same wrapped component G_i). The weight of an upward or downward edge is set to the length of a shortest such paths. Note that this equals the v-w distance in the wrapped component G_i; there may exist shorter paths in G that leave G_i.

A level edge at level $i \in \{1, \ldots, l-1\}$ exists if both of its endpoints are contained in the same wrapped component G_{i+1} at level $i+1$. For level l, we simply use all candidate edges: $E_l := S_l \times S_l$. The weight of a level edge matches the distance in G. This constitutes an essential difference to [**24**], where level edges were defined similarly to upward and downward edges. The purpose of this modification is query runtime, allowing to look up distances between vertices in S_i instantly instead of plowing through all level edges, however, at the expense of an increased number of level edges.

Constructing the level edge set naïvely would be too expensive in terms of preprocessing time because determining distances in G may in general require consideration of the whole input graph. We suggest an efficient two-pass construction method. In the first, bottom-up, pass, the upward and downward edge sets are computed: computation of U_i and D_i is performed iteratively using the corresponding edge sets at level $i-1$. The second pass is carried out top-down: to construct E_i, the set of level edges at level i help restrict this computation to a bounded local search; the set E_l is computed directly using U_l.

2.1.3. *Parallelization.* Due to the different levels of hierarchy induced by the vertex subsets $\mathscr{S}$, this construction process does not need to consider the whole input graph G at once. On the contrary, the preprocessing can be split up into tasks so that each one operates on exactly one wrapped component (potentially,

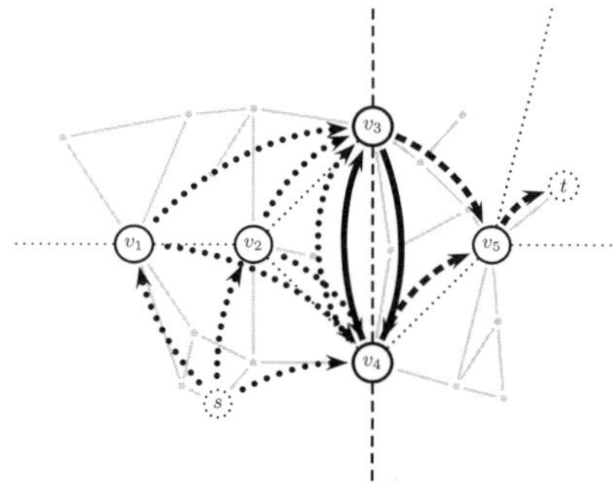

FIGURE 2. The search graph $\mathcal{M}_{st}$: edges from $\mathscr{L}$, $\mathscr{U}_i$, and $\mathscr{D}_i$ are shown as thick lines with solid, dotted, and dashed styles, respectively.

each task could be assigned to a distinct processor, provided that the data flow dependencies between the tasks are obeyed, which would yield speed-up almost linear in the number of processors).

2.1.4. *Component Tree.* The nesting of the separator sets is reflected by the component sets $\mathscr{C}_i$: each component $C_i \in \mathscr{C}_i$ is fully contained in exactly one *parent component* of $\mathscr{C}_{i+1}$, i.e., $C_i \subseteq C'_{i+1}$ for some $C'_{i+1} \in \mathscr{C}_{i+1}$. In addition, we define the *root* or *universe* component $C_{l+1} := G$ that serves as parent for all components in $\mathscr{C}_l$, and a *leaf component* $C_0^v := \{v\}$ for every vertex $v \in V$. For the leaf components, we use C_1^v as parent. The parent relationship naturally induces a tree of components.

2.1.5. *Search Graph.* When for a given pair of vertices $s, t \in V$ simultaneously walking up the component tree from C_0^s and C_0^t towards the root, the paths eventually meet at some component $C_L^s = C_L^t$, the lowest common ancestor of C_0^s and C_0^t. With our notation, the path between C_0^s and C_0^t in the component tree is $(C_0^s, C_1^s, \ldots, C_L^s = C_L^t, \ldots, C_1^t, C_0^t)$. In fact, any s-t path must visit these components in this order.

Now we construct the *search graph* $\mathcal{M}_{st}$, a subgraph of $\mathcal{M}$ with the same s-t distance as in G. The edge set of $\mathcal{M}_{st}$ is the union of the following sets:

$$\begin{aligned}
\mathscr{L} &:= E_{L-1} \cap \left(Adj(C_{L-1}^s) \times Adj(C_{L-1}^t)\right), \\
\mathscr{U}_i &:= U_i \cap \left(Adj(C_{i-1}^s) \times Adj(C_i^s)\right), \text{ and} \\
\mathscr{D}_i &:= D_i \cap \left(Adj(C_i^t) \times Adj(C_{i-1}^t)\right),
\end{aligned}$$

where $i \in \{1, \ldots, L-1\}$.

Figure 2 shows an example, where edges from the sets $\mathscr{L}$, $\mathscr{U}_i$, and $\mathscr{D}_i$ are shown as thick lines with solid, dotted, and dashed styles, respectively. Owing to the altered definition of the level edge set compared to [**24**], we can afford including only a subset of E_{L-1} in the edge set of $\mathcal{M}_{st}$. Note that $\mathscr{L}$ (and therefore also $\mathcal{M}_{st}$) is defined only for $L > 1$. These modifications require a new proof of correctness.

2.1.6. *Correctness.* In the following, we shall prove that $\mathcal{M}_{st}$ can be used for answering the s-t shortest-path query in G. First notice that by definition every edge in $\mathcal{M}_{st}$ has a weight at least as large as the distance between the corresponding vertices in G. Hence, any distance in $\mathcal{M}_{st}$ cannot be smaller than the corresponding

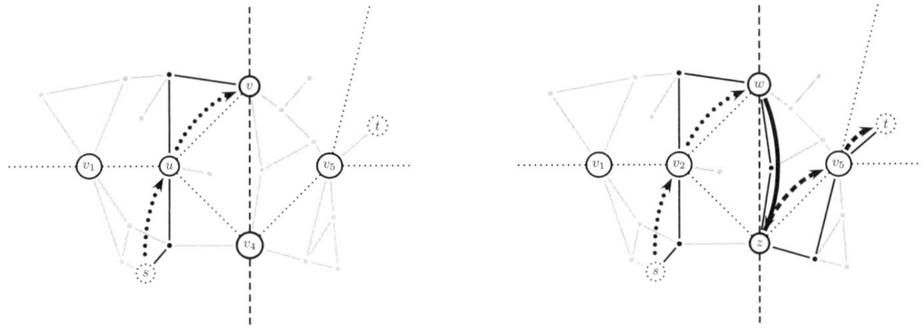

FIGURE 3. Illustration for Lemma 1 (left) and Theorem 2 (right): A shortest path in G (highlighted thin lines) has a corresponding path, of the same length, in $\mathcal{M}_{st}$ (thick lines).

distance in G. It remains to prove that for a shortest s-t path in G there is a path in $\mathcal{M}_{st}$ of equal length.

LEMMA 1. *For $i \in \{0, \ldots, L-1\}$, the distance from s to any vertex $v \in Adj(C_i^s)$ in $\mathcal{M}_{st}$ matches that in G_i^s, the wrapped component around s at level i. Conversely, the distance from any vertex $v \in Adj(C_i^t)$ to t in $\mathcal{M}_{st}$ matches that in G_i^t.*

PROOF. (By induction.) We shall prove only the first part, as the second follows immediately by symmetry. For $i = 0$, the claim is obvious. For $i > 0$, any s-v path in G must contain a vertex in $Adj(C_{i-1}^s)$, let u be the first such vertex. We can split a shortest s-v path at u into two (possibly empty) subpaths. The s-u subpath contains only vertices from G_{i-1}^s and therefore has an equivalent path in $\mathcal{M}_{st}$ by the induction hypothesis. On the other hand, the edge (u, v) is contained in $\mathscr{U}_i$ and its weight corresponds to the length of the u-v subpath. □

THEOREM 2. *If $L > 1$, the s-t distance is equal in the graphs G and $\mathcal{M}_{st}$.*

PROOF. The value L is the level of the lowest common ancestor of C_0^s and C_0^t in the component tree. Therefore, the vertices s and t reside in different home components at level $L-1$, and any s-t path must contain a vertex in S_{L-1}. Let vertex $w \in Adj(C_{L-1}^s)$ and $z \in Adj(C_{L-1}^t)$ be the first or last such vertex, respectively. Again, we split a shortest s-t path at w and z. The s-w and z-t subpaths contain only vertices from G_{L-1}^s and G_{L-1}^t, respectively. According to Lemma 1, these subpaths have equivalent paths in $\mathcal{M}_{st}$. The edge (w, z) is part of $\mathscr{L}$, its weight equals the w-z distance in G. □

2.1.7. *Query.* The search graph $\mathcal{M}_{st}$ can be transformed into an equivalent DAG by creating at most L copies of each vertex. Recall that the edges of $\mathcal{M}_{st}$ are the union of the edge sets $\mathscr{L}$, $\mathscr{U}_i$ and $\mathscr{D}_i$. We call the graph induced by such an edge set a *partial graph*, and distinguish between *level*, *upward*, and *downward parts*, accordingly. A partial graph is a directed bipartite graph, apart from some *twofold* vertices that have both incoming and outgoing edges. We can *unfold* any partial graph into an equivalent directed bipartite graph by creating a copy of each twofold vertex, directing the edges with a twofold vertex as target to the corresponding copies, and adding a zero-cost edge from each original twofold vertex to its copy. In the example in Figure 2, vertex v_4 of the upward part $\mathscr{U}_2$ is twofold, because it has an incoming edge (v_1, v_4) and an outgoing edge (v_4, v_3). Unfolding this partial graph generates a copy v_4' of v_4 with new edges (v_1, v_4'), (v_3, v_4') and (v_4, v_4'), the latter having a length of zero.

After that, each vertex of the unfolded partial graph has either only outgoing or only incoming edges; we distinguish between *source* and *drain* vertices. An unfolded version of the search graph can be created by joining the unfolded partial graphs. The drain vertices of one partial graph match the source vertices of another partial graph. As such, the join can be interpreted as a stacking of partial graphs. All paths traverse this stack in the same direction, thus no cycles exist. Refer to [**18**] for a formal proof.

For a DAG, an s-t shortest-path query can be performed in $O(V + E)$ time. Since the topological structure of our DAG is known in advance, the query algorithm can be reduced to initialization of the vertex distance labels and update of the distance label of each edge's target vertex in the order imposed by the topological structure.

2.1.8. *Nearby Vertices.* Path lookup in $\mathcal{M}_{st}$ works only for $L > 1$, i.e., for source and target vertices from different home components at level 1. For vertices from the same home component $C = C_1^s = C_1^t$, we fall back to Dijkstra's algorithm. However, we avoid leaving the home component in our search. Instead, we use the appropriate edges in $E_1 \cap (Adj(C) \times Adj(C))$ as shortcut for paths that leave C. By keeping the components small, we can state a runtime guarantee for the case of nearby vertices, too.

2.2. Optimizing Partial Graphs. In contrast to the classic variant, where a multi-level graph is stored as a whole, we spread it over a large number of partial graphs (as seen before, any search graph can be constructed through the union of a number of appropriate partial graphs). The foremost advantage is that each of them can be optimized individually using two different techniques: pruning of edges that cannot contribute to a shortest path and conversion of a partial graph into an equivalent one with more vertices but fewer edges.

2.2.1. *Pruning Superseded Edges.* Consider an upward part $\mathscr{U}_i$ at level $i > 1$. This partial graph connects the adjacent vertices of two related components C_{i-1} and C_i at neighboring levels; let G_{i-1} and G_i be the corresponding wrapped components, and consider a fixed edge $(w, v) \in \mathscr{U}_i$. Now, if a shortest w-v path in G_i passes another vertex $z \in Adj(C_{i-1})$ and the w-z subpath does not leave G_{i-1}, then for any s-v path via w there is a path via z that is no longer. This also holds for any search graph that uses $\mathscr{U}_i$: for any s-v path via edge (w, v) there is a path via edge (z, v) that is no longer. That is, we can safely remove the edge (w, v) from the upward part $\mathscr{U}_i$; this edge is called *superseded* by the edge (z, v). An example is shown in Figure 4.

To determine if an edge $(w, v) \in \mathscr{U}_i$ is superseded by another edge, we only need to check local distances between adjacent vertices of C_{i-1}, together with edge weights of the upward part $\mathscr{U}_i$. Edge (w, v) is superseded by edge $(z, v) \in \mathscr{U}_i$, if $c(w, v) = d(w, z) + c(z, v)$ and $d(w, z) > 0$.[1]

The pruning algorithm simply checks each pair of edges in $\mathscr{U}_i$ sharing the same target vertex for supersedement, and removes the superseded ones on the fly. Because supersedement is a strict partial order, the algorithm finds and removes all

[1]Here, c denotes the edge weight function in $\mathscr{U}_i$, and d refers to the local distance in G_{i-1}.

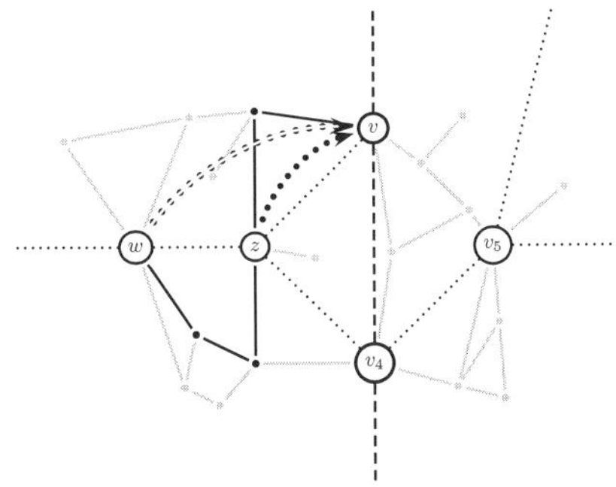

FIGURE 4. Superseded edges. Edge (w, v) (double-dotted line) is superseded by (z, v) (dotted line), because there is a shortest w-v path via z in the input graph (highlighted thin lines).

desired edges (cf. [**18**] for a formal proof). Analogously, we can eliminate superseded edges in downward and level parts. All partial graphs are optimized separately: an edge superseded in one may or may not be superseded in another partial graph.

The pruning algorithm can be further refined by determining *superseded vertices* in a first pass. For upward parts, a vertex w is superseded by another vertex z iff all edges starting at w are superseded by the corresponding edge starting at z. Superseded vertices can be omitted completely in the further processing, because all edges connected to this vertex are superseded. To remove superseded vertices, we iterate over the pairs of source vertices (those with outgoing edges) and check for supersedement by iterating over all outgoing edges. Downward and level parts are handled likewise. For the partial graphs obtained in our experimental evaluation, this first pass takes only a fraction of the time needed for removing all superseded edges, and usually removes many candidate edges, significantly reducing the execution time for the main pruning algorithm.

2.2.2. *Constructing Equivalent Graphs.* A further optimization technique is based on the following idea. The number of edges of a partial graph can be reduced by introducing auxiliary vertices and replacing many original edges with few edges through the new vertices such that distances are preserved. An example is given in Figure 5. Edges highlighted in the left graph are contained in at least one shortest σ-δ path. The edges in the corresponding partial graph are made up from the lengths of these shortest paths. Without optimization this would lead to a complete bipartite graph with 16 edges, while an optimized equivalent graph needs only eleven edges, as shown at the right: the twelve edges from $\{\sigma_1, \ldots, \sigma_4\}$ to $\{\delta_1, \ldots, \delta_3\}$ may be replaced with a star-like graph of seven edges, since the corresponding shortest paths between the vertices in question contain a shared central vertex. In the best case, all underlying shortest paths contain at least one common vertex, and the optimization results in a star-like graph.

We have implemented a simple heuristic that adds a single, so-called *central*, vertex, decides in a greedy manner which of the original vertices to connect to the central vertex, and balances the weight function for the new edges. The goal of the balancing is to remove a maximal number of original edges: if a path via the central vertex is of equal length as the corresponding original edge, the latter can

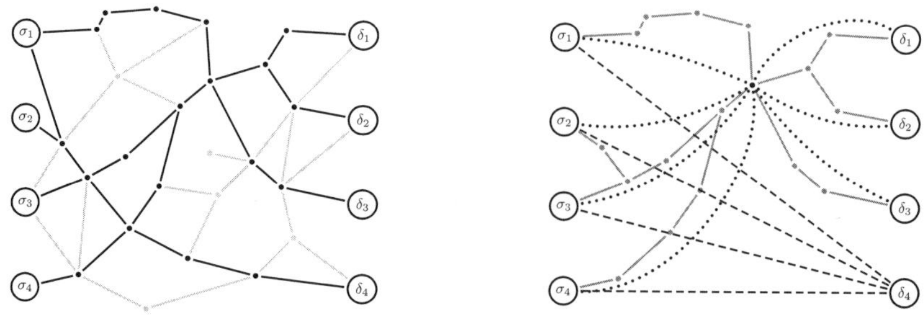

FIGURE 5. Constructing equivalent graphs. Left: sample graph; highlighted edges are contained in a shortest σ-δ path. Right: belonging search graph with edge compression applied; dotted and dashed edges are contained in the graph.

be removed. The balancing must be distance preserving; in particular, no newly introduced path may be shorter than the corresponding original edge.

3. Experiments

As input data to our experimental evaluation, we use road networks of Western Europe, provided by PTV AG for scientific use, and the USA, taken from the TIGER/Line Files [**25**], with around 18 million vertices and 42.6 million edges and 23.9 million vertices and 58.3 million edges, respectively. For both graphs, both distances and travel times are available for each edge; in order to compare our approach to similar ones, we test our graphs also with the unit edge metric.

For the test runs, we used a machine with two AMD Opteron 2218 processors, 32 GB of shared RAM, and 2 × 1 MB of L2 cache, where each processor features two cores clocked at 2.6 GHz; the time measurements given refer to execution on a single core. The program was compiled with the GCC 3.4, using optimization

TABLE 1. DIMACS Challenge benchmarks: query times in *ms* for subgraphs of the US network and different metrics.

	metric	
graph	time	dist
NY	25.4	23.0
BAY	29.9	29.0
COL	43.2	38.8
FLA	119.3	112.0
NW	143.6	143.0
NE	193.2	195.4
CAL	254.5	248.8
LKS	396.6	377.5
E	607.1	558.5
W	1 343.9	1 115.4
CTR	5 892.9	4 778.0
USA	7 741.4	5 908.5

level 3 and the LEDA library (version 5.01). Results on the DIMACS Challenge benchmarks can be found in Table 1.

The subsequent presentation of our results is structured along the two ways of obtaining a *hierarchical decomposition* of the input graphs, planar separators and METIS, the latter being a freely available tool for graph partitioning.

Decomposition. To determine for a given graph sets of selected vertices, the graph is first decomposed in a hierarchical fashion. This process is governed by two parameters, the number of levels and *granularity*, the latter being fixed either through the maximum component size allowed or a maximum number of adjacent vertices per component for each level. Both options have their advantages: limiting the component size generates a balanced decomposition in the sense that each higher-level component contains roughly the same number of lower-level components, while limiting the number of adjacent vertices yields a smaller variance in the search graph sizes and thus allows for predicting the maximum search graph size even before starting the preprocessing. Given an input graph, a hierarchical decomposition is obtained by declaring, iteratively for each level to be generated, vertices selected so long until the granularity criterion for that level is reached. For the next-lower level, repeat this process applied to the decomposition found so far with the specific granularity.

3.1. Planar-Separator Theorem. Decomposing our graphs by means of the planar-separator theorem (PST) [**16, 8**] requires some preparatory step: due to their nature, road graphs are but almost planar since they account for bridges, highway ramps, etc., incurring crossings in those very places. We therefore planarize the input graph first by adding vertices at edge crossings, and eventually have to appropriately retranslate the separation found for the planarized graph into one satisfying the original graph. The planarized input graph is recursively split into two parts so long until the granularity condition holds. Our experiments involve a three-level decomposition with a granularity of 80-40-20 adjacent vertices per component. Note that such a granularity naturally induces an upper bound of $80^2 + 2 \cdot 40 \cdot 80 + 2 \cdot 20 \cdot 40 + 2 \cdot 20 = 14\,440$ edges in the search graph, which can be stated even *before* running our preprocessing.

3.1.1. *Preprocessing.* Decomposition with PST takes about a week for the Europe graph (distance metric),[2] but our implementation caches the course of the computation, and this cache can be reused to create decompositions with any given granularity with just little computational effort. Preprocessing requires about 24 hours on one core and about 8 hours on four cores, resulting in 543 million edges, 288 million of which belong to upward and downward graphs at level 1.

Each edge in a partial graph can be encoded using six bytes (four for the length assigned to it and one each for source and target vertices). Hence, the space needed to store the whole preprocessed data amounts to an overhead of 181 additional bytes per vertex (543 million edges · 6 bytes per edge / 18 million vertices) of the input graph. If we skipped the optimization of the partial graphs, the preprocessing would contain 3 210 million edges; in other words, the optimization step reduces the preprocessing size to 17 %.

For each kind of partial graphs, optimization has a different impact. For all higher-level partial graphs, roughly half of the edges are removed by supersedement.

---

[2]We used METIS to create an initial decomposition into components of about 500 000 vertices each, as our PST implementation is unable to process the input graph as a whole.

However, superseded vertices only exist in level graphs. For most level graphs much smaller equivalent graphs can be found, reducing the total size of the level graphs to 7.5 % in combination with supersedement when the distance metric is used. Note that with time or unit metric, this figure drops to 4.9 % and 5 %, respectively. Our heuristics for computing equivalent graphs has almost no effect for upward and downward graphs. Summarizing, our edge reduction heuristics work very well for level graphs, but yield only moderate results for upward and downward graphs.

Recall that the granularity of the preprocessing naturally induces an upper bound of 14 440 edges in the search graph. However, an analysis of the preprocessed data reveals that due to optimization, the largest search graph contains only 5 262 edges.

3.1.2. *Query Performance.* Unless otherwise stated, we use the road network of Europe, applying the distance metric. We evaluate 10 000 queries picked at random according to an exponential distribution, with a *Dijkstra rank*[3] of between 100 and $|V|$ (we chose an exponential over a uniform distribution as the former captures the intuition that with real-world information systems, short-distance occur more frequently than long-distance queries). Note that only those queries are reported that can be handled by our approach, i.e., s-t queries with s and t in different home components. However, the entry and exit graphs of our technique propose distances from and to all boundary vertices within one component, which could be used as landmark data for low-range queries, i.e., queries within one component. As observed in [**6, 3**], landmark-based routing performs very well for those types of queries (below 1 ms, except for outliers).

We measured the time needed to initialize an array of distance labels and to relax all edges of all partial graphs the search graph consists of. Prior to this, the query algorithm loads the partial graphs from external memory and flushes the L2 cache by performing copy operations on two memory buffers, each as large as the cache. This setup accurately simulates a client-server system that answers random source-target queries and holds all partial graphs in RAM: in general, a partial graph needed for a query is not present in the L2 cache and must be fetched from RAM. (Note that to reduce the impact of outliers, we repeated three times the execution of each query and used the median of the three measurements.)

Figure 6(a) shows the search space size, plotted against the Dijkstra rank, where each dot represents a query. Because our technique is dominated by relaxed edges, search space is measured in these terms rather than by settled vertices (experiments show that the dependency between rank and relaxed edges is indeed linear for Dijkstra's algorithm).

There are three horizontal clouds discernible, at approximately 100, 500, and 2 000 of relaxed edges. They correlate with the number of upward and downward graphs in the search graphs of between 1 and 3, resulting in search graphs constructed from three, five, or seven partial graphs. This observation is supported by Figure 6(b), which depicts the number of partial graphs depending on Dijkstra rank. As expected, with increasing rank the search graphs tend to comprise more partial graphs. Altogether, is seems as if the number of relaxed edges depends more on the number of partial graphs from which the search graph is constructed than on the pure rank of a query.

---

[3]For an s-t query, the Dijkstra rank of vertex v is the number of vertices settled before v is settled.

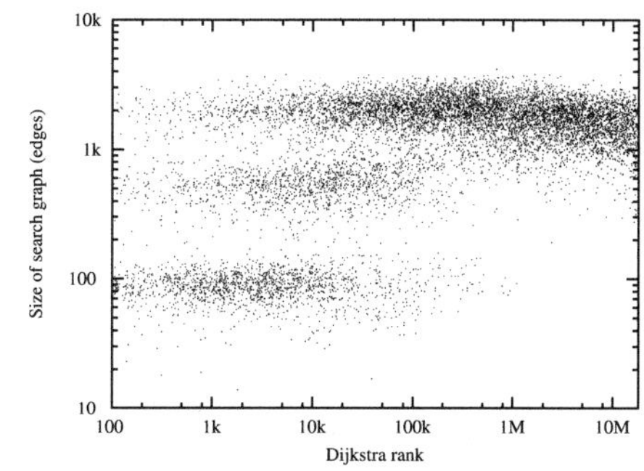

(a) Search space in terms of relaxed edges. Each dot depicts for one query the number of relaxed edges in relation to its Dijkstra rank.

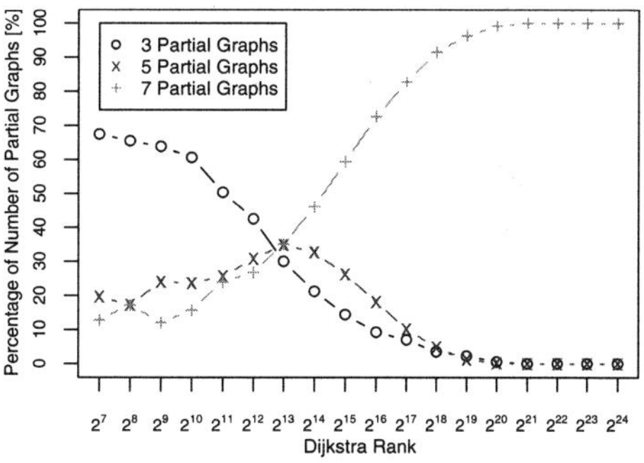

(b) Distribution of number of partial graphs (the search graph is constructed from) in relation to Dijkstra rank.

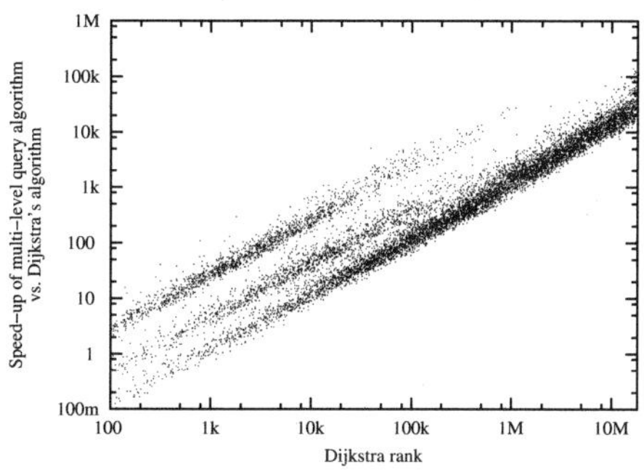

(c) Speed-up of our approach over Dijktra's algorithm in terms of relaxed edges.

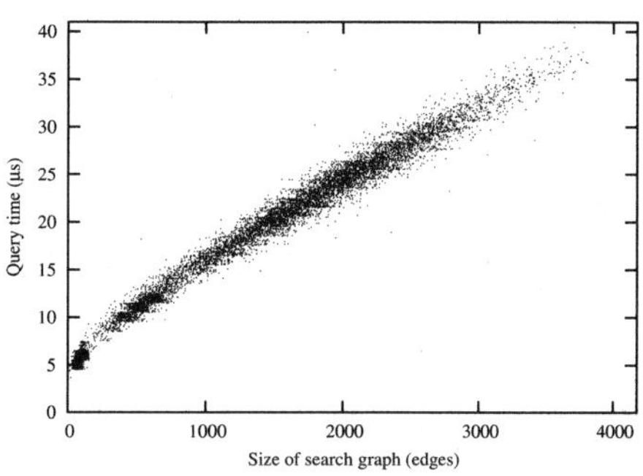

(d) Query time in relation to the search graph size.

FIGURE 6. Results for our approach using the road network of Europe as input. As metric, distances are applied.

In terms of search space, we achieve speed-ups of up to 100 000 for higher Dijkstra ranks (cf. Figure 6(c)). However, a couple of small-rank queries lead to factors of less than 1: such a slow-down occurs when source and target vertices are close to each other in the input graph, but belong to different home components at level 2 or even 3 so that relatively big five- or seven-level search graphs have to explored.

Figure 6(d) depicts search graph size depending on query time: we observe an almost linear relationship. In general, all queries are executed in less than 40 μs. Moreover, for search graphs with 2 000 edges or more, we can state a query runtime of roughly 12 ns per edge.

3.1.3. *Robustness.* Up to now, we have shown that our approach performs very well with the distance metric. In order to prove robustness to the metric applied, we ran a larger series of experiments with both the Europe and the US networks and travel times, distances, and unit lengths. To facilitate comparison of our approach to similar ones, we now employ queries distributed uniformly at random. Table 2 reports average running times as well as the percentages of queries executed.

TABLE 2. Preprocessing and (uniform) random queries performance for different metrics on the European and US network. The size of the preprocessing is given in number of edges in all generated partial graphs. The search space is given in number of relaxed edges within search graph. Note that only those queries are reported which can be performed by our approach. The percentage of executed queries is given in column 6.

		PREPRO	QUERY		
		size	search space	time	executed
graph	metric	[# edges]	[# relaxed edges]	[μs]	[%]
Europe	time	469 M	1 494	18.8	99.90
	dist	543 M	1 617	20.3	99.96
	unit	470 M	1 485	19.3	99.93
USA	time	782 M	1 462	19.3	99.94
	dist	848 M	1 547	20.0	99.94
	unit	774 M	1 441	19.2	99.97

We observe that the metric chosen has almost no impact on query times or preprocessing. Regarding partial-graph optimization, there are no significant differences in the number of additional edges between the various metrics, either. Of all queries, more than 99.9 % were executed; assuming an average time of about

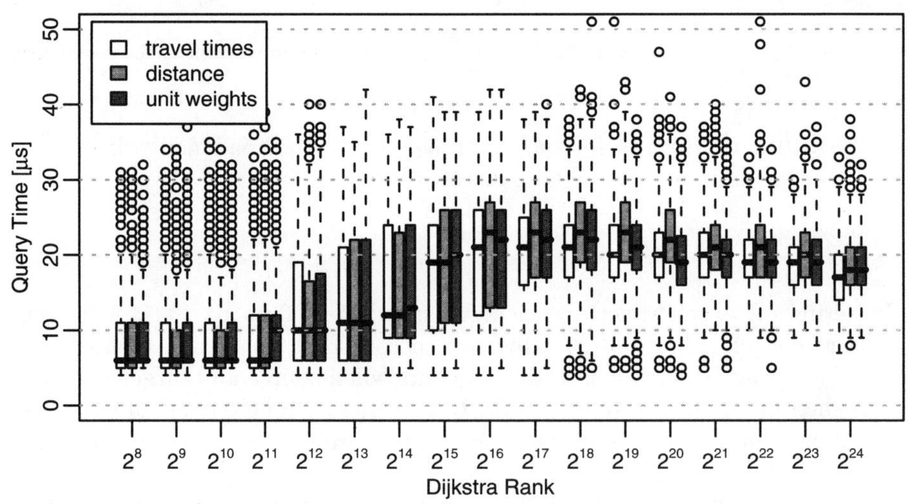

FIGURE 7. Comparison of the query times for different metrics (travel times, distances and unit lengths) using the Dijkstra rank methodology [**20**] on the road network of Europe. The results are represented as box-and-whisker plot [**19**]: each box spreads from the lower to the upper quartile and contains the median, the whiskers extend to the minimum and maximum value omitting outliers, which are plotted individually. Note that only those queries are reported that can be performed by our approach.

TABLE 3. Preprocessing and (uniform) random queries performance for subgraphs of the US networks, taken from the DIMACS homepage. As metric, we apply travel times. Our current implementation cannot handle small graphs. Thus, no results for NY, BAY, and COL are given. The search space is given in number of edges within the partial graphs. In addition to the columns given in Table 2 we also report the ratio of additional edges per node in the original graph (column 4). Note that only those queries are reported which can be performed by HPML. The percentage of executed queries is given in column 7.

	PREPROCESSING			QUERY		
graph	time [min]	size [# edges]	#edges $/\|V\|$	search space	time [μs]	executed [%]
FLA	8	19 M	17.8	331	8.6	90.00
NW	7	22 M	18.2	243	8.9	71.40
NE	444	141 M	92.8	291	8.4	98.60
CAL	324	136 M	72.0	596	11.2	98.40
LKS	320	130 M	47.1	1 071	14.8	99.30
E	39	71 M	19.7	1 824	21.0	99.60
W	89	133 M	21.3	1 440	18.6	99.80
CTR	260	354 M	25.1	1 847	21.6	99.97

1 *ms* for the remaining 0.1 % of (low-range) queries, the average query times as of Table 2 would increase by 1 *µs*. Hence, we wind up with an overall average time of less than 22 *µs* for all inputs.

In order to check whether this robustness with respect to metrics holds for all types of queries (low-, mid-, and long-range), Figure 7 shows the performance of our approach using the Dijkstra rank methodology [**20**]. As input we use the European instance. Strikingly, the performance of our approach is (almost) independent of the applied metric for all types of queries.

All of the above-said is confirmed by experiments with different subgraphs of the US network and distance metric, the results being summarized in Table 3. However, on smaller graphs the number of executed queries drops to values of 71.4%. Preprocessing times differ greatly from that for the whole graph as now only two instead of three levels are used.

3.1.4. *Comparison.* Table 4 contrast the results of our approach to the most prominent speed-up techniques presented at the DIMACS workshop applied to all graphs and metrics, with queries distributed uniformly at random.

The results show clearly that in terms of preprocessing time, HPML cannot compete with any other technique. Comparing query times with the time metric, our approach as well as the TNR-variants all yield values of less than 20 µs, where the latter techniques still outperform ours. The very strength of our approach unfolds when the distance metric is used: HPML query times do not change significantly, while with transit node routing they increase by factors of up to 26. Using the unit metric, all of these approaches yield similar performance. To sum up, our findings corroborate the robustness of our approach regarding edge metric, which does not hold for transit node routing.

TABLE 4. Performance of the most prominent speed-up techniques in comparison to our high-performance multi-level (HPML) approach. More precisely, we report preprocessing and query times for highway hierarchies star (HH*) [**3**], REAL [**6**], grid-based transit node routing (grid-TNR) [**1**], and transit node routing based on highway hierarchies (HH-TNR) [**22**].

metric	technique	Europe PREPRO [h:mm]	Europe QUERY [μs]	USA PREPRO [h:mm]	USA QUERY [μs]
time	HH*	0:22	550	0:28	600
	REAL	2:20	1110	2:01	1050
	HH-TNR	1:15	4.3	1:25	3.3
	Grid-TNR	58:00	13	7:00	17.8
	HPML	$\approx$ 24:00	18.8	$\approx$ 36:00	19.3
dist	HH*	0:49	1950	0:59	1740
	REAL	1:30	1160	2:18	1800
	HH-TNR	2:42	37.6	3:37	86.1
	Grid-TNR	29:00	56	9:00	69.4
	HPML	$\approx$ 24:00	20.3	$\approx$ 36:00	20.0
unit	HH*	0:27	990	0:32	890
	REAL	3:49	1140	2:27	1160
	HH-TNR	0:53	13.1	3:59	19.8
	Grid-TNR	17:00	12	9:00	30.3
	HPML	$\approx$ 24:00	19.3	$\approx$ 36:00	19.2

3.2. METIS. As an alternative to compute graph decompositions we also use the METIS collection [**13**]. These tools allow to divide graphs into a given number of *partitions* of roughly equal size, where the edge cut, i.e., the number of edges with source and target vertex located in different partitions, is minimized. Since our preprocessing technique requires a selection of vertices instead of edges, we subsequently compute a greedy vertex cover on the edge cut obtained from METIS. This procedure can be carried out recursively to obtain a hierarchical decomposition.

METIS runs amazingly fast for our test instances: decomposition into any number of components requires less than one minute compared to one week for PST; thus, a hierarchical decomposition can be obtained in about ten minutes. As a further advantage, the input graph does not have to be planar. On the other hand, METIS produces separators that are about 20 % larger than those generated by PST; similar results were reported in [**8**]. While a decomposition with limited component sizes can be obtained quite naturally using METIS, we cannot easily create a decomposition with limited maximum number of adjacent vertices per component; to achieve the latter, we have to perform recursive two-way partitioning, as described for PST.

For our evaluation, we settled for a decomposition into three levels, as preliminary experiments showed that two levels would consume too much space, while employing a fourth level would not pay off. Further tests suggested granularities of 120 000-4 000-360 of adjacent vertices for the Europe and 120 000-3 300-300 for the US network. To compare METIS with PST, we used the very same granularity also for PST: compared to the granularity used in Section 3.1, the preprocessing time for the Europe graph and distance metric is slightly smaller, whereas search graph size doubles on average.

Unfortunately, the overall results were not as promising as with adjacent-vertex granularities. When METIS is used, the size of preprocessed data grows by 50 % and the average search graph size doubles. For all preprocessings, the maximal search graph was more than ten times larger than the average; this is unfavorable if a tight guarantee for query times is required. Summing up, the quality of the decomposition is of utmost importance for the efficiency of our speed-up technique.

4. Conclusion

We have shown how to enhance the classic multi-level approach [**24**] in such a way that an even greater deal of the effort to compute a shortest path can be shifted to the preprocessing stage. The main developments concern distribution of the multi-level graph to many partial graphs. These permit to be sparsified fairly easily, which leads to massive reduction of the total amount of precomputed data and hence of query times.

In an experimental study with road graphs, where an extensive preprocessing as well as larger amounts of additional data could be afforded, our approach proved highly effective: speed-ups achieved over Dijkstra's algorithm reach factors of up to around 3 000. Furthermore, our approach has shown to be robust to different edge metrics with respect to both preprocessing and query performance: except for very few outliers, query time does not exceed 40 μs.

For future work, we see the following points of improvement: concerning our implementation, use of custom-tailored data structures as well as of locality, as exploited in [**6, 3**]; at an algorithmic level, development of alternative heuristics to construct equivalent graphs, and combination with other speed-up techniques (in [**10, 5**], certain combinations were shown to perform better than the individual techniques). Another interesting question would be that of dynamization: due to the hierarchical nature of our approach, we believe that only small parts of the preprocessed data need to be updated upon an edge change in the input graph. Finally, storage and retrieval of the course of a shortest path is a major requirement for practical applications, which could be solved through similar concepts as in [**11**].

References

1. Holger Bast, Stefan Funke, and Domagoj Matijevic, *TRANSIT: Ultrafast Shortest-Path Queries with Linear-Time Preprocessing*, 9th DIMACS Challenge on Shortest Paths, November 2006, An updated version of the paper appears in this book.
2. Holger Bast, Stefan Funke, Domagoj Matijevic, Peter Sanders, and Dominik Schultes, *In transit to constant time shortest-path queries in road networks*, 9th Workshop on Algorithm Engineering and Experiments (ALENEX), 2007, pp. 46–59.
3. Daniel Delling, Peter Sanders, Dominik Schultes, and Dorothea Wagner, *Highway Hierarchies Star*, 9th DIMACS Challenge on Shortest Paths, November 2006, An updated version of the paper appears in this book.
4. Edsger W. Dijkstra, *A note on two problems in connexion with graphs*, Numerische Mathematik **1** (1959), 269–271.
5. Andrew Goldberg, Haim Kaplan, and Renato Werneck, *Reach for A*: Efficient point-to-point shortest path algorithms*, Proc. Algorithm Engineering and Experiments, SIAM, 2006, pp. 129–143.

6. Andrew V. Goldberg, Haim Kaplan, and Renato F. Werneck, *Better Landmarks within Reach*, 9th DIMACS Challenge on Shortest Paths, November 2006, An updated version of the paper appears in this book.
7. Ronald J. Gutman, *Reach-based routing: A new approach to shortest path algorithms optimized for road networks.*, Proc. Algorithm Engineering and Experiments, SIAM, 2004, pp. 100–111.
8. Martin Holzer, Grigorios Prasinos, Frank Schulz, Dorothea Wagner, and Christos Zaroliagis, *Engineering planar separator algorithms*, Proc. European Symposium on Algorithms, LNCS, vol. 3669, Springer, 2005, pp. 628–639.
9. Martin Holzer, Frank Schulz, and Dorothea Wagner, *Engineering multi-level overlay graphs for shortest-path queries*, Proc. Algorithm Engineering and Experiments, SIAM, 2006, pp. 156–170.
10. Martin Holzer, Frank Schulz, and Thomas Willhalm, *Combining speed-up techniques for shortest-path computations*, Proc. Workshop on Experimental and Efficient Algorithms, LNCS, vol. 3059, Springer, 2004, pp. 269–284.
11. Ning Jing, Yun-Wu Huang, and Elke A. Rundensteiner, *Hierarchical encoded path views for path query processing: An optimal model and its performance evaluation*, IEEE Trans. Knowledge and Data Engineering **10** (1998), no. 3, 409–432.
12. Sungwon Jung and Sakti Pramanik, *HiTi graph model of topographical roadmaps in navigation systems*, Proc. Data Engineering, IEEE Computer Society, 1996, pp. 76–84.
13. George Karypis, *METIS*, 1998, `http://www-users.cs.umn.edu/~karypis/metis`.
14. Ekkehard Köhler, Rolf H. Möhring, and Heiko Schilling, *Acceleration of shortest path and constrained shortest path computation.*, Proc. Workshop on Experimental and Efficient Algorithms, Springer, 2005, pp. 126–138.
15. Ulrich Lauther, *An extremely fast, exact algorithm for finding shortest paths in static networks with geographical background*, Geoinformation und Mobilität — von der Forschung zur praktischen Anwendung, vol. 22, IfGI prints, Institut für Geoinformatik, Münster, 2004, pp. 219–230.
16. Richard J. Lipton and Robert Endre Tarjan, *A separator theorem for planar graphs*, SIAM Journal on Applied Mathematics **36** (1979), no. 2, 177–189.
17. Rolf H. Möhring, Heiko Schilling, Birk Schütz, Dorothea Wagner, and Thomas Willhalm, *Partitioning graphs to speed up Dijkstra's algorithm.*, Proc. Workshop on Experimental and Efficient Algorithms, LNCS, vol. 3503, Springer, 2005, pp. 189–202.
18. Kirill Müller, *Design and implementation of an efficient hierarchical speed-up technique for computation of exact shortest paths in graphs*, Master's thesis, Department of Informatics, University of Karlsruhe, Germany, June 2006, online available at `http://i11www.iti.uka.de/teaching/theses/files/da-kmueller-06.pdf`.
19. R Development Core Team, *R: A Language and Environment for Statistical Computing*, `http://www.r-project.org`, 2004.
20. Peter Sanders and Dominik Schultes, *Highway hierarchies hasten exact shortest path queries*, Proc. European Symposium on Algorithms, LNCS, vol. 3669, Springer, 2005, pp. 568–579.
21. ———, *Engineering highway hierarchies*, Proc. 14th European Symposium on Algorithms, LNCS, vol. 4168, Springer, 2006, pp. 804–816.
22. ———, *Robust, Almost Constant Time Shortest-Path Queries on Road Networks*, 9th DIMACS Challenge on Shortest Paths, November 2006, An updated version of the paper appears in this book.
23. Frank Schulz, Dorothea Wagner, and Karsten Weihe, *Dijkstra's algorithm on-line: An empirical case study from public railroad transport*, Proc. Workshop on Algorithm Engineering, Springer, 1999, Also published in: ACM Journal of Experimental Algorithms **5** (2000), pp. 110–123.
24. Frank Schulz, Dorothea Wagner, and Christos Zaroliagis, *Using multi-level graphs for timetable information in railway systems*, Proc. Algorithm Engineering and Experiments, LNCS, vol. 2409, Springer, 2002, pp. 43–59.
25. U.S. Census Bureau, Washington, DC, *UA Census 2000 TIGER/Line Files*, `http://www.census.gov/geo/www/tiger/tigerua/ua_tgr2k.html`, 2002.
26. Dorothea Wagner and Thomas Willhalm, *Geometric speed-up techniques for finding shortest paths in large sparse graphs*, Proc. European Symposium on Algorithms, LNCS, vol. 2832, Springer, 2003, pp. 776–787.

27. Dorothea Wagner, Thomas Willhalm, and Christos Zaroliagis, *Geometric containers for efficient shortest-path computation*, ACM Journal of Experimental Algorithmics **10** (2005), 1–30.

Daniel Delling, Universität Karlsruhe (TH), Fakultät für Informatik, Postfach 69 80, 76128 Karlsruhe, Germany
E-mail address: delling@ira.uka.de

Martin Holzer, Universität Karlsruhe (TH), Fakultät für Informatik, Postfach 69 80, 76128 Karlsruhe, Germany
E-mail address: mholzer@ira.uka.de

Kirill Müller, Universität Karlsruhe (TH), Fakultät für Informatik, Postfach 69 80, 76128 Karlsruhe, Germany
E-mail address: mail@kirill-mueller.de

Frank Schulz, PTV Planung Transport Verkehr AG, Stumpfstrasse 1, 76131 Karlsruhe, Germany
E-mail address: frank.schulz@ptv.de

Dorothea Wagner, Universität Karlsruhe (TH), Fakultät für Informatik, Postfach 69 80, 76128 Karlsruhe, Germany
E-mail address: wagner@ira.uka.de

DIMACS Series in Discrete Mathematics
and Theoretical Computer Science
Volume **74**, 2009

Reach for A*: Shortest Path Algorithms with Preprocessing

Andrew V. Goldberg, Haim Kaplan, and Renato F. Werneck

ABSTRACT. We study the point-to-point shortest path problem with preprocessing. Given an input graph, we preprocess it so as to be able to answer a series of source-to-destination queries efficiently. Our work is motivated by an algorithm of Gutman [ALENEX'04], based on the notion of *reach*, which measures how important each vertex is with respect to shortest paths. We present a simplified version of his algorithm that does not require explicit lower bounds during queries. We also show how the addition of shortcuts to the graph greatly improves the performance of both preprocessing and queries. Finally, we combine a reach-based algorithm with landmark-based A* search to obtain a wide range of space-time trade-offs. For our motivating application, driving directions for road networks, the resulting algorithm is very efficient and practical. The road networks of the USA and Western Europe have roughly 20 million vertices, but on average our algorithm must visit fewer than a thousand to find the distance between two points. Our algorithm also works reasonably well on 2-dimensional grid graphs with random arc weights.

1. Introduction

We study the *point-to-point shortest path problem* (P2P): given a directed graph $G = (V, A)$ with nonnegative arc lengths and two vertices, the source s and the destination t, find a shortest path from s to t. Although there is a single input graph, typically there are many source/destination queries. We therefore allow preprocessing of the input graph, but limit the size of the precomputed data to a (moderate) constant times the graph size. Preprocessing time is limited by practical considerations. For example, in our motivating application, driving directions on large road networks, quadratic-time algorithms are impractical. We are interested in exact shortest paths only.

Finding shortest paths is a fundamental problem. The single-source problem with nonnegative arc lengths has been studied most extensively [**3, 5, 10, 11, 15, 16, 17, 18, 22, 27, 36, 47, 51**]. Near-optimal algorithms are known both in theory, with near-linear time bounds, and in practice, with running times within a small constant factor of the breadth-first search time.

The P2P problem with no preprocessing has been addressed, for example, in [**26, 39, 45, 52**]. While no nontrivial theoretical results are known for the general P2P problem, there has been work on the special case of undirected planar graphs with slightly super-linear preprocessing space. The best bound in this context is due to Fakcharoenphol and Rao [**14**]. Algorithms for approximate shortest paths that use preprocessing have also been studied; see e.g., [**4, 28, 48**].

Previous work on exact P2P algorithms with preprocessing includes, e.g., [**19, 23, 24, 30, 33, 37, 40, 43, 44, 50**]. We focus our discussion here on the most relevant recent developments in preprocessing-based algorithms for road networks. Such methods have two components: a *preprocessing algorithm*, which computes auxiliary data, and a *query algorithm*, which computes an answer for a given s-t pair.

Gutman [**24**] introduced the notion of *vertex reach*. Informally, the reach of a vertex v is large if v is close to the middle of some long shortest path and small otherwise. Gutman proposes a simple modification of Dijkstra's algorithm that can prune an s-t search based on (upper bounds on) vertex reaches and (lower bounds on) vertex distances from s and to t. He uses Euclidean distances as lower bounds, and observes that the efficiency can be improved if reaches are combined with Euclidean-based A* search [**25, 38**], which uses lower bounds on the distance to the destination to direct the search towards it.

Goldberg and Harrelson [**19**] (see also [**23**]) have shown that the performance of A* search (without reaches) can be significantly improved if Euclidean lower bounds are replaced by landmark-based lower bounds. These bounds are obtained by storing (in the preprocessing step) the distances between every vertex and a small set of special vertices, the landmarks. During queries, one can use this information, together with the triangle inequality, to obtain lower bounds on the distance between any two vertices in the graph. This leads to the ALT (A* search, landmarks, and triangle inequality) method for the point-to-point problem.

Sanders and Schultes [**40, 41**] use the notion of *highway hierarchies* to design efficient algorithms for road networks. The preprocessing algorithm builds a hierarchy of increasingly sparse *highway networks*; queries start at the original graph and gradually move to upper levels of the hierarchy, greatly reducing the search space. To magnify the natural hierarchy of road networks, the algorithm adds *shortcuts* to the graph: additional edges with the same lengths as the original shortest paths between their endpoints.

In this paper, our first major contribution is to show that reaches can be used to prune the search even when explicit lower bounds (such as those obtained by Euclidean bounds or landmarks) are not available. By making the search bidirectional, we can use the bounds implicit in the search itself to prune it.

The second major contribution of our paper is to explain how shortcuts can be used in the context of reach-based algorithms. This significantly improves both preprocessing and query efficiency. The resulting algorithm is called RE. Although the preprocessing algorithm needs to be modified in order to generate shortcuts, the query algorithm remains the same regardless of whether shortcuts are present or not.

Our third major contribution is to show that the ALT method can be combined with reach-based pruning in a natural way, leading to an algorithm we call REAL. We also show that by maintaining landmark data only for high-reach vertices, one can greatly reduce the memory requirements of REAL. Furthermore, if we use some of the saved space for more landmarks, we can win in both space and time.

In addition, we introduce several improvements to preprocessing and query algorithms for landmark- and reach-based methods.

We evaluate the efficiency of our algorithms through experiments, mostly on road networks with three length functions (travel times, travel distances, and unit lengths). Our experiments show practical results for all three metrics. The road networks of Western Europe or the United States, each with roughly 20 million vertices, can be preprocessed in an hour or less. The average query with our fastest algorithm takes roughly one millisecond on a standard workstation and scans fewer than 1000 vertices.

We have also obtained good results for 2-dimensional grids with random arc lengths. Although not as good as for road networks, the results prove that our techniques have more general applicability. To show the limitations of these techniques, we also experimented with grids of higher dimension and with random graphs. On these inputs, our heuristics do not achieve significant performance gains.

This paper is organized as follows. Section 2 reviews Dijkstra's algorithm and some variants, and establishes the notation used throughout the paper. Section 3 formalizes the definition of *reach* and explains how it can be used to prune a point-to-point shortest path search. Section 4 deals with reach computation: how reaches (or upper bounds on reaches) can be computed in reasonable time during the preprocessing stage of our algorithm. Section 5 reviews the ALT algorithm and shows how it can be combined with reach-based pruning. Section 6 presents our experimental results. Final remarks are made in Section 7, including a brief comparison with recent work presented at the 9th DIMACS Implementation Challenge [**8**].

2. Preliminaries

The input to the preprocessing stage of a P2P shortest path algorithm is a directed graph $G = (V, A)$ with n vertices and m arcs, and nonnegative lengths $\ell(a)$ for every arc a. Besides the source s and the sink t, the query stage takes as input the data produced by the preprocessing stage, which includes the graph itself (potentially modified) and auxiliary information. The goal is to find a shortest path from s to t. We denote by $\text{dist}(v, w)$ the shortest-path distance from vertex v to vertex w with respect to ℓ. In general, $\text{dist}(v, w) \neq \text{dist}(w, v)$.

The *labeling method* for the shortest path problem [**31, 32**] finds shortest paths from the source to all vertices in the graph. It works as follows (see e.g., [**46**]). It maintains for every vertex v its distance label $d(v)$, parent $p(v)$, and status $S(v) \in \{\texttt{unreached}, \texttt{labeled}, \texttt{scanned}\}$. Initially $d(v) = \infty$, $p(v) = nil$, and $S(v) =$ `unreached` for every vertex v. The method starts by setting $d(s) = 0$ and $S(s) =$ `labeled`. While there are labeled vertices, it picks a labeled vertex v and *scans* it by *relaxing* all arcs out of v and setting $S(v) =$ `scanned`. To relax an arc (v, w), one checks if $d(w) > d(v) + \ell(v, w)$ and, if true, sets $d(w) = d(v) + \ell(v, w)$, $p(w) = v$, and $S(w) =$ `labeled`.

If the length function is nonnegative, the labeling method terminates with correct shortest path distances and a shortest path tree. Its efficiency depends on the rule to choose a vertex to scan next. We say that $d(v)$ is *exact* if it is equal to the distance from s to v. If one always selects a vertex v such that, at selection time, $d(v)$ is exact, then each vertex is scanned at most once. Dijkstra [**11**] (and independently Dantzig [**5**]) observed that if ℓ is nonnegative and v is a labeled vertex with the smallest distance label, then $d(v)$ is exact. The labeling method with the minimum label selection rule is known as *Dijkstra's algorithm*.

For the P2P case, note that when the algorithm is about to scan the sink t, we know that $d(t)$ is exact and the s-t path defined by the parent pointers is a shortest path. We can terminate the algorithm at this point. Intuitively, Dijkstra's algorithm searches a ball with s in the center and t on the boundary.

One can also run Dijkstra's algorithm on the *reverse graph* (the graph with every arc reversed) from the sink. The reverse of the t-s path found is a shortest s-t path in the original graph.

The *bidirectional algorithm* [**5, 13, 38**] alternates between running the forward and reverse versions of Dijkstra's algorithm, each maintaining its own set of distance labels. We denote by $d_f(v)$ the distance label of a vertex v maintained by the forward version, and by $d_r(v)$ the distance label of a vertex v maintained by the reverse version. (We will still use $d(v)$ when the direction would not matter or is clear from the context.) During initialization, the forward search scans s and the reverse search scans t. The algorithm also maintains the length of the shortest path seen so far, μ, and the corresponding path. Initially, $\mu = \infty$. When an arc (v, w) is relaxed by the forward search and w has already been scanned by the reverse search, we know the shortest s-v and w-t paths have lengths $d_f(v)$ and $d_r(w)$, respectively. If $\mu > d_f(v) + \ell(v, w) + d_r(w)$, we have found a path that is shorter than those seen before, so we update μ and its path accordingly. We perform similar updates during the reverse search. Note that, to maintain the current best path, it is enough to remember the arc (v, w) whose relaxation gave the current value of μ, as the forward search maintains the s-v path of length $d_f(v)$ and the reverse search maintains the w-t path of length $d_r(w)$. We can alternate between the searches in any order. In our experiments, we strictly alternate between the searches, balancing the work between them.

Intuitively, the bidirectional algorithm searches two touching balls centered at s and t: we can stop when the search in one direction selects a vertex already scanned in the other. A slightly tighter criterion is to terminate the search when $\text{top}_f + \text{top}_r \geq \mu$, where top_f and top_r denote the top keys (values) in the forward and reverse priority queues, respectively (i.e., the smallest labels of unscanned vertices in each direction). To see why this is a valid criterion, suppose there exists an s-t path P whose length is less than μ. Then there must exist an arc (v, w) on this path such that $\text{dist}(s, v) < \text{top}_f$ and $\text{dist}(w, t) < \text{top}_r$, which means that v and w must have been scanned already. Suppose, without loss of generality, that v was scanned first; then, when scanning w, path P would have been detected. Since it was not, P cannot exist and μ must indeed be the length of the shortest s-t path.

3. Reach-Based Pruning

Given a path P from s to t and a vertex v on P, the *reach of v with respect to P* is the minimum of the length of the prefix of P (the subpath from s to v) and the length of the suffix of P (the subpath from v to t). See Figure 1. The *reach* of v, $r(v)$, is the maximum, over all **shortest** paths P through v, of the reach of v with respect to P. Throughout this paper, we assume that shortest paths are unique. If the input has ties, these can be broken in several ways; we describe our tie-breaking procedure in Section 4.6.

The intuition behind the notion of reach is simple. A vertex has high reach only if it is close to the middle of some very long shortest path. On road networks,

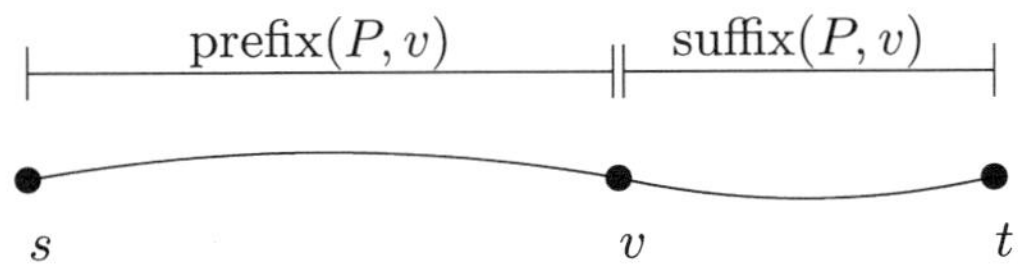

FIGURE 1. The reach of v with respect to the shortest path P between s and t is the minimum between the lengths of its prefix and its suffix (with respect to v).

high-reach vertices roughly correspond to highways, whereas low-reach vertices correspond to local intersections. Take a vertex v representing an intersection between two local roads in a small town. There are shortest paths containing v that start close to v and end somewhere far away, and shortest paths containing v that start far away and end close to v. Usually, however, it is not the case that there is a *shortest* path that starts far away from v, passes through v, and ends far away from v.

The knowledge that a particular vertex v has low reach can be used to prune it during searches. Once we are far away from the source s and the target t, there is no point in visiting v: we know it cannot be on the shortest path from s to t. The remainder of this section will make these observations more formal.

For large graphs, computing exact reaches is impractical with current algorithms. Instead, we efficiently compute *upper bounds* on the reach of every vertex, which is enough for our purposes. Section 4 will explain in detail how this can be done. For now, assume that we have valid reach upper bounds; how they were obtained is immaterial.

We denote an upper bound on $r(v)$ by $\overline{r}(v)$. Let $\underline{\text{dist}}(v, w)$ denote a lower bound on the distance from v to w. The following fact allows us to use reaches for pruning an s-t Dijkstra's search:

> Suppose $\overline{r}(v) < \underline{\text{dist}}(s, v)$ and $\overline{r}(v) < \underline{\text{dist}}(v, t)$. Then v is not on a shortest path from s to t, and therefore Dijkstra's algorithm does not need to label or scan v.

This holds for the bidirectional algorithm as well.

Note that upper bounds on reaches are not enough: we still need lower bounds on distances from s (which the search itself can provide) and to t. Gutman [**24**] proposed using Euclidean lower bounds to find lower bounds on distances to t. Unfortunately, this only works when vertex coordinates are available, which is not always the case. Even when they are available, as in the case of road networks, the lower bounds are not particularly tight, especially when length functions other than travel distances are used.

We propose a simpler (and more effective) strategy: make the search bidirectional, and extract implicit lower bounds from the bidirectional search itself. During an execution of a bidirectional version of Dijkstra's algorithm, consider the search in the forward direction, and let γ be the smallest distance label of a labeled vertex in the reverse direction (i.e., the topmost key in the reverse heap). If a vertex v has not been scanned in the reverse direction, then γ is a lower bound on the distance from v to the target t. The same idea applies to the reverse search: we use the topmost key in the forward heap as a lower bound on the distance from the source to any vertex not yet scanned in the forward direction.

When we are about to scan v, we know that $d_f(v)$ is the distance from the source to v. So we can prune the search at v if v has not been scanned in the reverse direction, $\bar{r}(v) < d_f(v)$, and $\bar{r}(v) < \gamma$. When using these bounds, the stopping condition is the same as for the standard bidirectional algorithm (without pruning). As in the original case, we can alternate between the searches in any way. We call the resulting procedure the *bidirectional bound* algorithm. See Figure 2.

FIGURE 2. Pruning using implicit bounds. Assume v is about to be scanned in the forward direction, has not yet been scanned in the reverse direction, and that the smallest distance label of a vertex not yet scanned in the reverse direction is γ. Then v can be pruned if $\bar{r}(v) < d_f(v)$ and $\bar{r}(v) < \gamma$.

A variant of this method is the *self-bounding* algorithm, which can prune a vertex based on its own distance label, regardless of the other direction. Assume we are about to scan a vertex v in the forward direction (the procedure in the reverse direction is similar). If $\bar{r}(v) < d_f(v)$, we prune the vertex. Note that if the distance from v to t is at most $\bar{r}(v)$, the vertex will still be scanned in the reverse direction, given the appropriate stopping condition. It is easy to see that the following stopping condition works.

> Stop the search in a given direction when either there are no labeled vertices or the minimum distance label of labeled vertices for the corresponding search is at least half the length of the shortest path seen so far.

The self-bounding algorithm can safely ignore the lower bound to the destination because it leaves to the other search to visit vertices that are closer to it. Note, however, that when scanning an arc (v, w), even if we end up pruning w, the self-bounding algorithm must check if w has been scanned in the opposite direction; if so, it must check if the candidate path using (v, w) is the shortest path seen so far.

The natural *distance-balanced* algorithm falls into both of the above categories. It balances the radii of the forward search and the reverse search by scanning in each iteration the labeled vertex with minimum distance label, considering both directions. The distance label of this vertex is also a lower bound on the distance to the target, as the search in the opposite direction has not selected the vertex yet. This algorithm, which we call RE, is the one we tested in our experiments, given its simplicity and the fact that it can be naturally combined with A* search, as Section 5.2 will show. Although it could be implemented with only one priority queue, we use two for consistency with the other algorithms we implemented.

3.1. Early Pruning. Our implementation checks whether a vertex w can be pruned not only when scanning it, but also when considering whether to insert it into the heap or not. We call this *early pruning.* When processing an arc (v, w) in the forward search, we are actually evaluating a path from s to t that passes through v and w. The distance from s to w on this path is $d_f(v) + \ell(v, w)$; a lower bound on the distance from w to t is γ (the topmost value in the reverse heap), assuming that w has not been scanned yet. If $\bar{r}(w) < \min\{d_f(v) + \ell(v, w), \gamma\}$, we can prune the search at w. The same procedure applies to the reverse search.

Besides avoiding heap insertions, early pruning can actually reduce the number of arcs we have to scan. Suppose that the arcs in the adjacency list of v are sorted in decreasing order of upper bounds on reaches. In other words, if (v, w) and (v, x) are such that $\bar{r}(w) < \bar{r}(x)$, then (v, x) appears before (v, w) in the adjacency list of v. If, when scanning (v, w), we determine that $\bar{r}(w) < \gamma$ and $\bar{r}(w) < d_f(v)$, then we can not only prune (v, w), but implicitly prune all remaining arcs in the adjacency list of v. Note that there will be cases where $\bar{r}(w)$ will be greater than $d_f(v)$ but smaller than $d_f(v) + \ell(v, w)$; we can still prune w, but we must continue traversing the adjacency list of v. To make this optimization possible, we can sort the adjacency lists when reading the graph for queries (as we did in our experiments) or during the preprocessing step.

To ensure that implicit pruning is indeed correct, there is still a special case we must address. As described, the routine assumes, when processing (v, w), that γ is a valid lower bound on the distance to t from every neighbor z of v that appears after (v, w) in the adjacency list. This may not be true if z has already been scanned: the algorithm will miss the arc (v, z) when scanning v, and it can conceivably be part of the shortest path. Fortunately, this arc will have already been scanned from z during the reverse search. At that point, however, v had not yet been scanned in the forward direction, and therefore the algorithm will not have checked if (v, z) belongs to the shortest path. This can remedied by performing an additional check when scanning v itself: if v is labeled in the other direction, the algorithm checks whether $d_f(v) + d_r(v) \leq \mu$ and updates the shortest path seen so far accordingly.

3.2. Improving Locality. When reaches are available, a typical point-to-point query spends most of its time scanning high-reach vertices. Except at the very beginning of the search, low-reach vertices are pruned. During repeated searches, most vertices visited have high reach. This suggests an obvious optimization: during preprocessing, reorder the vertices such that high-reach vertices are close together in memory to improve cache locality.

The simplest way to achieve this would be to sort vertices in non-increasing order of reach. This, however, will destroy the locality of the input: in many applications (including road networks), the original vertex order has high locality.

Instead, we adopt the following approach to order the vertices. We partition the vertices into two equal-sized sets: the first contains the $n/2$ vertices with highest reach, and the other contains the remaining vertices. We keep the original relative ordering in each part, then recursively process the first part. Besides improving locality, this reordering also facilitates other optimizations, such as reach-aware landmarks (described in Section 5.4).

4. Reach Computation

Having seen how reach upper bounds are used to prune the search, we now turn to the problem of computing these bounds. The standard algorithm for computing exact reaches builds shortest path trees from each vertex in the graph. The shortest path tree rooted at vertex r compactly represents all shortest paths that start at r.[1] The reach of a vertex v restricted to these paths is given by the minimum between its *depth* (the distance from r) and its *height* (the distance to its farthest descendent in the tree). The reach of v with respect to the entire graph is the maximum reach of v with respect to all shortest path trees.

Building n shortest path trees is too expensive for large road networks. Fortunately, as already mentioned, it is enough to compute upper bounds on reaches. Gutman [**24**] suggested an algorithm for this purpose that works in rounds. Each round tries to find upper bounds for reaches that are smaller than a threshold ϵ by growing *partial shortest path trees* of depth greater than 2ϵ. Intuitively, a high-reach vertex (i.e., a vertex with reach at least ϵ) must be close to the middle of some shortest s-t path of length slightly bigger than 2ϵ. This path will be among those considered when a partial shortest path tree is grown from s. At the end of a round, vertices with bounded reach are removed from the graph, the threshold ϵ is increased (by a multiplicative factor), and the procedure is repeated in the resulting subgraph, now with a larger threshold. This process continues until all reaches have been bounded.

Since the threshold increases substantially as the algorithm progresses, so does the depth of the shortest path trees obtained. Therefore, the efficiency of the algorithm depends crucially on how fast the graph shrinks due to the elimination of low-reach vertices. Intuitively, one would like the number of vertices in the partial trees to remain approximately constant from one round to the next. Unfortunately, when Gutman's algorithm is applied to road networks, the trees increase in size, rendering the algorithm impractical for large graphs.

4.1. Our Approach. Our algorithm is based on the same basic approach (growing partial trees) as Gutman's, but we suggest several improvements that lead to significant speedups.

Our most important improvement is adding *shortcut arcs* (or simply *shortcuts*) to the graph during the preprocessing procedure. A shortcut (v, w) is a new arc with length equal to that of an existing path between v and w. If we break ties appropriately (giving preference to shortcuts), we may decrease the reach of internal vertices on the original v-w path. This speeds up both preprocessing (because more vertices are eliminated after each round) and queries (because vertices with lower reach are more likely to be pruned). The resulting algorithm becomes practical for large road networks, such as those of the USA and Western Europe.

Our method for generating shortcuts is based on the one suggested by Sanders and Schultes [**40, 41**] in the context of highway hierarchies. The idea is to bypass a low-degree vertex v by adding for each pair of arcs $[(u, v), (v, w)]$ an arc (u, w) of length $\ell(u, v) + \ell(v, w)$ and deleting v and all arcs adjacent to it. To avoid introducing too many arcs, we prefer to bypass vertices of low degree. Although this shortcut strategy is local, its repeated application may introduce shortcuts

[1]Note that r can represent either a reach value (as a function) or a tree root. We rely on context to resolve the ambiguity.

representing very long paths. This happens on road networks, where these paths often correspond to portions of highways between two important exits.

A second important novelty of our preprocessing algorithm is that it computes upper bounds on the reaches of *arcs*, not vertices. Let P be the shortest path from s to t, and assume it contains an arc (v, w). The *reach of* (v, w) *with respect to* P is the minimum between the distance from s to w and the distance from v to t. The *reach of* (v, w) (with respect to the entire graph), denoted by $r(v, w)$, is the maximum, over all shortest paths P containing (v, w), of the reach of (v, w) with respect to P. The main advantage of computing arc reaches is that it allows for more efficient shortcutting: a high-reach vertex can be bypassed as soon as enough low-reach arcs incident to it are eliminated.

We now outline the preprocessing algorithm in more detail. Like Gutman's algorithm, it works in *rounds* (or *levels*). At level i, it tries to bound all arc reaches below some threshold ϵ_i, which grows exponentially with i. Level i starts by removing from the graph every arc whose reach was bounded in the previous level. It then eliminates (bypasses) some low-degree vertices by adding shortcuts between its original neighbors. Finally, it grows partial shortest path trees from all remaining vertices, and uses these trees to find upper bounds on the reaches of arcs whose reaches are less then ϵ_i. The algorithm proceeds until all arcs have been eliminated, i.e., until all reaches have been bounded.[2]

At this point, arc reaches are converted to vertex reaches, which are used during queries. We could employ arc reaches during queries, but vertex reaches require less space and are easier to use.

Although the partial shortest path trees grown in a given level contain only vertices and arcs that have not been eliminated from the graph, our goal is to find reach upper bounds that are valid for the original graph. We must therefore take into consideration the arcs that have already been deleted, either in previous iterations or when introducing shortcuts in the current iteration. We do that implicitly, by associating *penalties* with each vertex v in the current graph, as Section 4.2 will explain in detail. Intuitively, penalties are used to artificially extend the lengths of all shortest paths that start or end at v in the original graph, implicitly accounting for the fact that these paths could be extended to vertices that have been previously eliminated by the preprocessing routine.

Unfortunately, although they help find valid upper bounds, penalties often lead to overly conservative reach estimates. We may end up with upper bounds that are significantly above the actual reach values, which makes pruning less effective during queries. This is especially true for high-reach vertices, the last to have their reaches bounded. This is unfortunate, since they are arguably the most important vertices in the graph: queries prune most low-reach vertices and spend most of their time traversing high-reach ones.

To minimize this issue, the preprocessing algorithm also has a *refinement phase*. After all reach upper bounds are obtained, we compute exact vertex reaches on the subgraph induced by the δ vertices with highest reach upper bound, where δ is a user-defined parameter (set to $2\lceil\sqrt{n}\rceil$ in most of our experiments). The remaining vertices are considered only implicitly, as penalties.

[2]We reiterate that these arcs are "eliminated" during preprocessing only; all original arcs and shortcuts will be present in the final graph, on which queries will be performed.

The remainder of this section describes in detail each component of the preprocessing algorithm: the partial-trees procedure (Section 4.2), shortcut generation (Section 4.3), and the refinement step (Section 4.4). Other implementation details, including parameter choices, are discussed in Section 4.5.

4.2. Approximate Reaches: Growing Partial Trees. We now describe the main routine executed in each iteration of our preprocessing algorithm. Given a graph $G = (V, A)$ and the threshold ϵ, our goal is to find valid reach upper bounds for arcs in A whose actual reaches are smaller than ϵ. (This discussion deals with a single iteration, and therefore assumes that ϵ is fixed.) For the remaining arcs, the upper bound is ∞. While the algorithm is allowed to report false negatives (i.e., it may find an upper bound of ∞ for arcs whose actual reach is less than ϵ), it must never report a false positive.

Fix an arc (v, w). To prove that $r(v, w) < \epsilon$, we must consider all shortest paths that contain (v, w). Fortunately, we do not have to evaluate all such paths explicitly: it suffices to process only *minimal paths.* Let $P_{st} = (s, s', \ldots, v, w, \ldots, t', t)$ be the shortest path between s and t, and assume that (v, w) has reach at least ϵ with respect to this path. Path P_{st} is *ϵ-minimal* with respect to (v, w) if and only if the reaches of (v, w) with respect to $P_{s't}$ and $P_{st'}$ are both smaller than ϵ.

The algorithm works by growing a *partial tree* T_r from each vertex $r \in V$. It runs Dijkstra's algorithm from r, but stops as soon as it can prove that all minimal paths starting at r are part of the tree. In order to determine when to stop growing the tree, we need the notion of *inner vertices.* Let v be a vertex in this tree, and let x be the first vertex (besides r) on the path from r to v. We say that v is an *inner vertex* if either $v = r$ or $d(x, v) < \epsilon$, where $d(x, v)$ denotes the distance (in the tree) between x and v. Note that, when v is not an inner vertex, no path P_{rt} starting at r will be ϵ-minimal with respect to a tree arc (u, v): if the reach of (u, v) is greater than ϵ with respect to P_{rt}, it will also be greater than ϵ with respect to P_{xt}.

The tree arcs whose heads are inner vertices are those whose reaches we will try to bound; we call them *inner arcs.* The partial tree T_r must be large enough to include all of them, as well as enough descendants to bound their height accurately. More precisely, we must make sure that every inner vertex v has one of two properties: (1) v has no labeled (unscanned) descendent; or (2) v has at least one scanned descendent whose distance from $p(v)$ (the parent of v in T_r) is ϵ or greater. When these conditions are satisfied, we do not need to grow the tree any further because all ϵ-minimal paths starting at r are already taken into account. In practice, however, this condition often leads to very large partial trees. As Section 4.2.1 will explain, we use a relaxed version of the second condition that only guarantees that the distance (in the partial tree) to every *labeled* vertex from its closest inner ancestor is greater than ϵ. When processing the partial tree, we consider both scanned and labeled vertices (which will be leaves) as belonging to it. This ensures that there are no false positives, but may generate false negatives.

Once the tree is built, processing it is straightforward. For each vertex v, we know its *depth*, i.e., the distance from the root to v. In $O(|T_r|)$ time, one can also compute the *height* of every inner vertex v, defined as the distance within the tree from v to its farthest descendent (either scanned or labeled). The reach of a tree arc (v, w) with respect to the partial tree is the minimum between the depth of w and the height of v. The reach bound for (v, w) with respect to the entire graph is the maximum over all such reaches, considering all partial trees that contain (v, w)

and have w as an inner vertex. If this maximum is at least ϵ, we declare the reach bound to be ∞.

As already observed, the notion of running several rounds of partial-tree computation to find reach upper bounds is due to Gutman [**24**]. Our algorithm improves on his in three important ways. First, we add shortcuts between two rounds of partial-trees computation, which causes the graph to shrink much faster (on road networks). Second, we compute arc reaches instead of vertex reaches, which decreases the degrees of high-reach vertices more quickly during preprocessing and allows them to be bypassed (which reduces their reach). Finally, we only grow partial shortest path trees from vertices that have not been eliminated yet (as the next subsection will explain). Gutman's algorithm, in contrast, also grows trees from eliminated vertices with "live" neighbors, which is significantly slower. Without these modifications, the preprocessing algorithm would be impractical for very large graphs, and query performance would be significantly worse.

4.2.1. *Dealing with penalties.* As described, the algorithm assumes that partial trees are grown from every vertex in the graph. We would like, however, to run the partial-trees routine even after some of the vertices have been eliminated (because they were bypassed or their reach was bounded in a previous iteration), growing partial trees from the remaining vertices only. Eliminated vertices must be taken into account, however, since they may belong to the shortest paths that determine the reaches of the remaining vertices.

We use the notion of *penalties* to account for the eliminated vertices. If v is a vertex that remains in the graph, its *in-penalty* is the maximum over the reaches of all arcs (u, v) that have already been eliminated. Similarly, the *out-penalty* of v is the maximum reach of all arcs (v, w) that have already been eliminated. The intuition behind penalties is simple. Suppose that the reach of an arc (x, y) (still in the graph) is determined by the shortest path between s and t. Some arcs on this path may have already been eliminated from the graph, because they have small reach. But consider the largest subpath s'-t' of s-t that remains in the graph and contains (x, y): if we implicitly "extend" it on both ends (by adding *in-penalty*(s') to the prefix and *out-penalty*(t') to the suffix), the reach of (x, y) with respect to this subpath will be at least as large as the original reach.

To consider penalties when processing partial trees, it suffices to modify some of the definitions used by the procedure. The (redefined) depth of a vertex v within a tree T_r, denoted by $depth_r(v)$, is the distance from r to v plus the in-penalty of r. Similarly, the height of a vertex is redefined to take out-penalties into account. We implicitly attach a *pseudo-leaf* v' to each vertex v in T_r and set the length of the arc between v and v' to be the out-penalty of v. Heights are computed not with respect to T_r, but with respect to the pseudo-tree obtained when the pseudo-leaves are taken into account. As before, the reach of a tree arc (v, w) with respect to T_r is the minimum between the (modified) depth of w and the (modified) height of v.

Next we describe two simple modifications to the way partial trees are grown. They use penalty information to reduce the number of inner vertices in each tree, and with them the number of scanned vertices. This makes the algorithm faster without changing the reach bounds it obtains.

First, when deciding whether v is an inner vertex with respect to a root r, we compute $depth_r(v)$ taking in-penalties into account, i.e., as $d(r, v) + \textit{in-penalty}(r)$. To be considered an inner vertex, v must satisfy one of the following conditions:

(1) $v = r$ or (2) $p(v) = r$ or (3) $depth_x(v) < \epsilon$ (recall that x is the second vertex on the path from r to v). There is one exception to condition (3): if $depth_r(p(v)) <$ *in-penalty*$(p(v))$, v will not be considered an inner vertex (because its modified depth will be even higher in the tree rooted at $p(v)$), and neither will its descendents. Note that, together, these definitions imply that the parent of an inner vertex must also be an inner vertex.

The second modification is in the stopping criterion. We grow the tree until none of the labeled (unscanned) vertices is *relevant.* All inner vertices are relevant. To decide whether an outer vertex is relevant, we keep track of the *extension* of each vertex w, denoted by $ext_r(w)$: if u is the last inner vertex on the path from r to w, $ext_r(w)$ is defined as $d(p(u), w)$. An outer vertex v is relevant if its parent is relevant and $ext_r(p(v)) + out\text{-}penalty(p(v)) \leq \epsilon$. This definition ensures that, when v is not relevant, every inner ancestor of v will have height greater than ϵ even if we stop growing the tree at $p(v)$.

Our implementation introduces a third modification: we relax the notion of relevance to allow the algorithm to stop sooner. To be relevant, in addition to the conditions above, a vertex v must be such that $ext_r(v) + out\text{-}penalty(v) \leq \rho\epsilon$, with $\rho \geq 1$ (we used $\rho = 1.1$ in our experiments). Even if we end up not scanning v because of this rule, the algorithm remains correct because, as a labeled vertex, v is still guaranteed to appear in the final partial tree. Unlike the first two modifications, this relaxed definition may lead to worse reach bounds, since the parent u of v in the partial tree when it stops growing may not be v's actual parent (which may remain unscanned). As a result, the reaches of u and its ancestors may appear to be artificially high; in particular, the iteration may end up assigning infinite bounds to vertices whose actual reaches are less than ϵ. In practice, we observed that the heuristic stopping criterion makes the algorithm significantly faster and has little effect on the quality of the bounds.

4.2.2. *Improvable arcs.* When an iteration of the reach algorithm starts, we assume that all the arcs that remain in the graph have reach estimate $\bar{r}(v, w) = 0$. We grow partial shortest path trees one at a time. For each inner arc (v, w) in a tree, we check if its reach with respect to the tree is greater than $\bar{r}(v, w)$. If so, we update $\bar{r}(v, w)$ to its reach value in the current tree (or to ∞, if the reach in the current tree is greater than ϵ).

The fact that the reach estimate can only increase as the algorithm progresses (within a single iteration) can be used to speed up the computation. Assume we have already grown a few partial shortest path trees and the current reach estimate of (v, w) is $\bar{r}(v, w)$. When growing a new partial shortest path tree T_r, suppose (v, w) is again an inner arc. Without any further processing, we know that the reach of (v, w) with respect to this tree will be no greater than $depth_r(w)$. If this value is smaller than (or equal to) $\bar{r}(v, w)$, we can safely say that T_r is irrelevant for computing the reach of (v, w). To formalize this notion, we say that a vertex v is *improvable* with respect to T_r if $depth_r(v) > \bar{r}(p(v), v)$.

We can now redefine the *extension* $ext_r(w)$ of a vertex w as $d(u, w)$, where u is the last inner vertex on the path from r to w that is improvable. Compared with the previous definition, the extension of a vertex can only be larger (or remain the same). As a result, fewer vertices will be considered relevant, thus allowing the search to stop sooner.

4.2.3. *Dealing with long arcs.* The partial-trees algorithm aims at analyzing ϵ-minimal paths. Typically, these paths have length close to 2ϵ, which is the usual depth of the trees examined. In fact, if all arc lengths are much smaller than ϵ, it is not hard to see that all partial trees will have comparable depth.

Some road networks, however, contain arcs that are significantly longer than average. An obvious example is an arc representing a ferry route. Partial shortest path trees that include these arcs tend to be much deeper. Since for road networks the total number of vertices in a tree is roughly quadratic in the depth, this can significantly slow down the preprocessing procedure.

To speed up the algorithm, we used a hard bound on the total depth of the tree. We set it to 4ϵ, which is large enough so that most trees are unaffected, but small enough to prune exceptional cases significantly. Unfortunately, we cannot simply stop growing the tree when it reaches this value and process the result. Because the standard stopping criteria will not be observed, this might lead to false positives (i.e., some reaches will be underestimated).

Consider what happens, for instance, when we grow a partial tree from a root r, and let (r, v) be a very long arc (e.g., $\ell(r, v) > 4\epsilon$) incident to r. Suppose that the shortest path from r to some vertex x consists of the arcs (r, v), (v, w), and (w, x), with $\ell(v, w) < \epsilon$ and $\ell(w, x) \gg \epsilon$. Clearly, $r(v, w)$ is greater than ϵ. An algorithm that simply stops growing partial trees when their depth reaches 4ϵ will not detect this path, however. When growing the tree from r, it would stop after its depth reached 4ϵ. At this point, v will be labeled (with parent r), but not scanned, and vertex w may not have been visited at all. When growing a tree from v, we would only see the path (v, w, x), where the reach of (v, w) is smaller than ϵ.

To avoid this situation, we must use a modified notion of in-penalty when growing partial trees. For every vertex v, we determine the length $\ell(u, v)$ of its longest remaining incoming arc. If it is ϵ or larger, we set $\textit{in-penalty}'(v) \leftarrow \max\{\textit{in-penalty}(v), \ell(u, v)\}$. Intuitively, whenever there is a long arc incident to v, we ensure that the in-penalty of v is large enough to "catch" all necessary ϵ-minimal paths. In the example above, even though we would abort the search too soon when growing a tree from r, we would still determine that (v, w) has high reach when growing the tree from v itself.

Of course, the downside of this approach is that it may lead to more false negatives, since in some cases arc (u, v) will not be on the shortest paths between u and the inner vertices of the partial tree rooted at v. Very long arcs are relatively rare, however, and the considerable speed-up allowed by this technique makes it worthwhile in practice.

4.2.4. *Converting arc reaches to vertex reaches.* Before the refinement step, we convert the upper bounds on arc reaches into upper bounds on vertex reaches. Consider a vertex v, and let P be the shortest path that determines $r(v)$. Assume that $r(v) > 0$, i.e., that v is not an endpoint of P. Let (u, v) and (v, w) be the arcs of P entering and leaving v. Clearly, the reaches of these arcs with respect to P are at least $r(v)$; conversely, $r(v) \leq \min\{r(u, v), r(v, w)\}$.

Unfortunately, we do not know which neighbors of v are the ones that determine the reach (i.e., which ones are u and w). But it is easy to verify that $r(v) \leq \min\{\max_x\{r(x, v)\}, \max_y\{r(v, y)\}\}$. In other words, a valid upper bound for $r(v)$ is the minimum over the highest incoming arc reach and the highest outgoing arc reach.

Often, however, both maxima are achieved at the same neighbor $x = y$. Although the upper bound in this case is still valid, it may be much higher than necessary, since we know that $u \neq w$ on the path that determines the highest reach. We can exclude this case as follows:

(1) Find x', the incoming neighbor of v that maximizes $r(x', v)$, then find y', the outgoing neighbor of v that is *different from* x' and maximizes $r(v, y')$. Let δ' be the minimum of $r(x', v)$ and $r(v, y')$.
(2) Find the outgoing neighbor y'' of v such that $r(v, y'')$ is maximized, and the incoming neighbor x'' of v that is different from y'' and maximizes $r(x'', v)$. Let δ'' be the minimum of $r(x'', v)$ and $r(v, y'')$.

Note that δ' and δ'' may be different from each other in a directed graph. A valid upper bound on $r(v)$ is the maximum of δ' and δ''. As a special case, if v has only one neighbor (even if it is both incoming and outgoing), the reach of v will be zero.

4.3. Adding Shortcuts. To bypass a vertex v, we first examine all pairs $[(u, v), (v, w)]$ of incoming/outgoing arcs with $u \neq w$. For each pair, if the arc (u, w) is not in the graph, we add an arc (u, w) of length $\ell(u, v)+\ell(v, w)$. Otherwise, we set $\ell(u, w) \leftarrow \min\{\ell(u, w), \ell(u, v) + \ell(v, w)\}$. Finally, we delete v and all arcs adjacent to it.

In principle, any vertex in the graph could be subject to this procedure. Bypassing high-degree vertices, however, would significantly increase the number of arcs in the graph. To avoid an excessive expansion, we consider the ratio c_v between the number of new arcs added and the number of arcs deleted by the procedure above. A vertex is deemed *bypassable* if $c_v \leq c$, where c is a user-defined parameter (we follow the notation proposed by Sanders and Schultes [**41**]). Higher values of c will cause the graph to shrink faster during preprocessing, but may increase the final number of arcs substantially. For road networks, we used 0.5 for the first level, 1.0 for the second, and 1.5 for the remaining levels. This prevents the algorithm from adding too many shortcuts at the beginning of the preprocessing algorithm (when the graph is larger but shrinks faster) and ensures that the graph will shrink fast enough as the algorithm progresses. For grids, which do not have a natural hierarchy, we used higher values of c from the beginning.

We impose some additional constraints on a vertex v to deem it bypassable (besides having a low value of c_v). First, we require both its in-degree and its out-degree to be bounded by a constant (5 in our experiments). This guarantees that the total number of arcs added by the algorithm will be linear in n. We also consider two additional measures (besides c_v) related to v: the length of the longest shortcut arc introduced when v is bypassed, and the largest reach of an arc adjacent to v (when v is about to be removed). The maximum between these two values is the *cost* of v, and it must be bounded by $\epsilon_i/2$ during iteration i for the vertex to be considered bypassable (recall that ϵ_i is the threshold for bounding reaches at iteration i). As explained in Section 4.2.3, long arcs and large penalties can decrease the quality of the reach upper bounds provided by the preprocessing algorithm. Imposing these additional bypassability criteria prevents such long arcs and large penalties from appearing too soon.

On any given graph, many vertices may be bypassable. When a vertex is bypassed, the fact that we remove existing arcs and add new ones may affect the bypassability of its neighbors. Therefore, the order in which the vertices are processed matters. Vertices with low expansion (c_v) and low cost are preferred, since they are the least likely to affect the bypassability of their neighbors. When deciding which vertex v to bypass next, we take the vertex that minimizes the product of these two measures (expansion and cost). When a vertex is bypassed, its neighbors must have their priorities updated. We use a priority queue to efficiently determine which vertex to bypass next.

4.3.1. *Computing reaches of deleted arcs.* As already mentioned, when a vertex v is bypassed, it is deleted from the graph alongside all arcs currently incident to it. At this moment, we must find a valid upper bound on the reaches of these deleted arcs. Consider an incoming arc (u, v). We know that any shortest path P containing this arc will *not* continue beyond v using one of the existing outgoing arcs (since we break ties by preferring shortcuts, such path would contain one of the newly inserted shortcuts instead). Therefore, P will either stop at v or proceed through a previously deleted arc. In the latter case, *out-penalty*(v) bounds the length of the suffix of P that starts from v. It follows that we can safely set the upper bound $\bar{r}(u, v)$ on the reach of (u, v) to $\ell(u, v)$ + *out-penalty*(v). The same argument applies to an outgoing arc (v, w): we set $\bar{r}(v, w) \leftarrow \ell(v, w)$+*in-penalty*$(v)$.

The penalties associated with the neighbors of v must also be updated to take the reaches of the newly eliminated arcs into account.

4.4. Exact Reaches: Refinement Step. The refinement step computes exact vertex reaches on the subgraph induced by the δ vertices with highest reach, where δ is a user-defined parameter. Vertices and arcs not in this induced subgraph are considered implicitly, as penalties associated with every remaining vertex in the graph. To simplify notation, in this section we denote by n the number of vertices of the induced subgraph in which we compute exact reaches.

As already mentioned, exact reaches can be computed by growing shortest path trees from each of the n vertices in the graph, then computing the depth and height of each vertex within these trees. We developed an algorithm that has the same worst-case complexity, but can be significantly faster in practice on road networks. Even though it is still prohibitively expensive for large road networks, it is very useful when applied in the refinement step on a much smaller subgraph.

Our method follows the same principle as the basic algorithm: build a shortest path tree from each vertex in the graph and compute reaches within these trees. Our improvement consists of building parts of these trees *implicitly* by reusing previously found subtrees.

The algorithm works as follows. First, it partitions the vertices of the graph into k subsets, for a given parameter k (usually around $\sqrt{n}$). The algorithm will work with any such partition, but some are better than others, as we shall see. We call each subset a *region* of the graph. The *frontier* of a region A, denoted by $f(A)$, is the set of vertices $v \in A$ such that there exists at least one arc (v, w) with $w \notin A$. The remainder of the region consists of *internal vertices.*

Given any set $S \subseteq V$, we say that a vertex v is *stable* with respect to S if it has the same parent in all shortest path trees rooted at vertices in S; otherwise, we call it *unstable.* For any region A, the following holds:

LEMMA 4.1. *If $v \in V \setminus A$ is stable with respect to $f(A)$, then v is stable with respect to A.*

PROOF. Let $p_r(v)$ denote the parent of v in T_r (the shortest path tree rooted at some vertex r) and $p_{f(A)}(v)$ denote the common parent of v in all shortest path trees rooted at $f(A)$. Suppose the lemma is not true, i.e., that there exist a vertex $r \in A$ and a vertex $v \in V \setminus A$ such that $p_r(v) \neq p_{f(A)}(v)$. Consider the path P from r to v in T_r. Because $r \in A$ and $v \notin A$, at least one vertex in P must belong to $f(A)$. Let s be the last such vertex, i.e., the one closest to r. The subpath of P from s to v is itself a shortest path, and therefore it must appear in the shortest path tree rooted at s (we assume all shortest paths are unique). But recall that v has $p_r(v)$ as its parent on this path, which contradicts our initial assumption that the parent of v in all trees rooted at vertices of $f(A)$ is $p_{f(A)}(v) \neq p_r(v)$. □

A vertex $v \in V$ is considered *tainted* with respect to $f(A)$ if at least one of the following conditions holds: (1) v is unstable with respect to $f(A)$; (2) v has an unstable descendent in at least one of the $|f(A)|$ trees; or (3) $v \in A$. If none of these conditions holds, v is *untainted.* An untainted vertex v will be the root of the exact same subtree in every shortest path tree rooted at $f(A)$. The lemma above also ensures that it would be the root of the same subtree if we grew a shortest path tree from any internal vertex of A as well.

Our algorithm takes advantage of this. In its first stage, it grows full shortest path trees from every vertex in $f(A)$. For these trees, it computes the height and the depth of every vertex, as usual. The second stage of the algorithm grows *truncated trees* from every internal vertex of A (i.e., every vertex in $A \setminus f(A)$). These truncated trees contain only tainted vertices; no untainted vertex is ever visited. Even so, it is still possible to compute the reach of the tainted vertices as if we had grown the entire tree. The depth can be computed as before. For the height, we need to consider vertices that were not visited.

This is done as follows. Consider a maximal untainted subtree rooted at a vertex w. The height of this tree can be easily precomputed. Because w is untainted, its parent $p(w)$ will be the same (tainted) vertex in every shortest path tree rooted at A. Therefore, w imposes an *implicit penalty* on $p(w)$ equal to w's own height plus the length of the arc $(p(w), w)$. The *extended penalty* of a tainted vertex is defined as the maximum between its own out-penalty and the implicit penalties associated with its untainted children. Note that the extended penalty needs to be computed only once, after all trees rooted at the frontier are built. When trees are built from internal vertices, the height of each vertex visited (which must be tainted) is computed as usual, using extended penalties instead of out-penalties.

Our description so far allows us to correctly compute the reach of each tainted vertex, but we also need to determine the reach of the untainted vertices. Although the height of an untainted vertex v is the same across all trees, the depth varies, and so does the reach. Fortunately, we do not need to know the reach of v within each tree; it suffices to know the *maximum* reach. Since the height is constant, the maximum is realized in the tree that maximizes the depth of v. If, when growing full and truncated shortest path trees, we remember the maximum depth of each tainted vertex, we can later compute (in linear time) the maximum depth of all untainted vertices. Since their heights are known, their reaches (with respect to the trees rooted at A) can be easily determined.

4.4.1. *Regions.* The algorithm above is correct regardless of how the regions are chosen. In particular, if each region has exactly one vertex, we have the standard algorithm. Of course, there are better choices of regions. There are two main goals to achieve: (1) the size of the frontier should be small compared to the size of the entire region and (2) the number of tainted vertices should be minimized. On road networks, these two goals are conflicting. In general, a larger region will have a smaller fraction of its vertices in the frontier. However, it also increases the probability of an external vertex being tainted. A good compromise is to choose regions with roughly $\sqrt{n}$ vertices.

To create a partition with at least k sets, we pick k vertices at random to be centers and determine their Voronoi regions (we use $k = 2\lceil\sqrt{n}\rceil$). Recall that the Voronoi region associated with a center v is the set of vertices w that are closer to v than to any other center (ties are broken arbitrarily). One can compute the Voronoi diagram of a graph with a multiple-source version of Dijkstra's algorithm. If there are unreachable vertices, they are assigned to regions by themselves.

Although the Voronoi diagram is a simple way of defining the regions, it is certainly not the best conceivable partition. A topic for future research is to compute regions quickly with relatively smaller frontiers.

4.5. Other Details. As already mentioned, the reach threshold grows exponentially as the preprocessing algorithm progresses: we set $\epsilon_{i+1} \leftarrow 3\epsilon_i$ for each round i. It remains to determine ϵ_1, the threshold during the first round. Ideally, our choice should be such that the first iteration takes roughly as much time as each of the remaining iterations. A large value will make the first iteration comparatively slow and will not give the algorithm the chance to add shortcuts when needed. In contrast, a very small value will introduce penalties too early, which will decrease the accuracy of the reach upper bounds computed in subsequent iterations.

The value of ϵ_1 depends on an integer input parameter k_1 (we used $k_1 = 1000$ in our experiments). We pick $\lfloor n/k_1 \rfloor$ root vertices at random. For each root, we grow a partial shortest path tree with exactly k_1 vertices and take note of its radius (the distance label of the last scanned vertex). We set ϵ_1 to be half the minimum among all such radii. This ensures that, during the first iteration of the preprocessing algorithm, not many partial shortest path trees have more than k_1 vertices. Our choice of k_1 is fairly robust for road networks: small changes do not have much effect on the performance of either preprocessing or queries.

4.6. Correctness. We argued that the transformation of arc reaches into vertex reaches and the refinement step are correct. For completeness, we now present a (somewhat tedious but straightforward) proof that the main stage of our algorithm, which grows partial trees and adds shortcuts to the graph, is correct: it finds valid upper bounds on reaches. We focus on the basic version of the partial-trees algorithm, which takes penalties into account when processing the tree but not when growing it. The three improvements described in Section 4.2.1 are not taken into account, and neither are the special stopping criteria introduced in Section 4.2.3 to handle very long arcs. We have already argued why these accelerations do not affect correctness.

Before we proceed to the proof, we must explain in more detail an important element of the partial-trees algorithm. Recall that we assume that ties are broken so that a shortcut is always preferred to the arcs it replaces; furthermore, we assume that all original shortest paths are unique. We deal with these issues by working with *canonical paths*. A canonical path is a shortest path with additional properties, including uniqueness. We require the following properties, which are sufficient and easy to work with, but may not be necessary.

(1) A canonical path is a simple shortest path.
(2) For every pair (s,t), there is a unique canonical path between s and t.
(3) A subpath of a canonical path is a canonical path.
(4) There is an implementation of Dijkstra's algorithm that always finds canonical paths.
(5) (Non-shortcut property.) A path Q is *not* a canonical path if Q contains a subpath P with more than one arc such that the graph contains a shortcut arc for P.

Regarding Property 1, note that, if the graph has no cycles of zero-length arcs, all shortest paths are simple. Property 5 ensures that adding shortcut arcs decreases vertex reaches.

We implement canonical paths as follows. For each original arc a, we generate a length perturbation $\ell'(a)$. When computing the length of a path, we separately sum lengths and perturbations along the path, and use the perturbation to break ties in path lengths.

First suppose there are no shortcut arcs. If the perturbations are chosen uniformly at random from a big enough range of integers, with high probability all shortest paths (with respect to length and perturbations) are canonical paths. In our implementation, perturbations were picked uniformly at random from the range $[1, 65535]$. Shortcut arcs are added after the perturbations are introduced. The length and the perturbation of a shortcut arc are equal to the sum of the corresponding values for the arcs on the path that we shortcut. To break ties in a graph with shortcuts, we use the number of hops of the paths (fewer hops are better) after considering perturbations. Note that Dijkstra's algorithm can easily maintain the number of hops of candidate paths. It is not hard to see that the shortest paths that win the tie breaking are canonical.

Our way of dealing with canonical paths is conceptually simple but has two disadvantages: the memory overhead for storing perturbations during queries (which is minor) and a small probability of failure due to ties in sums of perturbations. Regarding the latter issue, we have never observed the algorithm working incorrectly during our extensive experiments. In general, our method can use any tie breaking rule that gives preference to paths with shortcuts and for which a (suitably modified) Dijkstra's algorithm finds the corresponding canonical paths. It is possible that other tie-breaking approaches, such as that of [**9**], will work. However, this is not completely obvious.

Our preprocessing algorithm computes upper bounds on reaches with respect to the set of canonical paths as defined above using tie-breaking by perturbations and hops. During queries, however, we can completely ignore both tie-breaking rules, and simply prune by reach. If we only prune a vertex v if $r(v) < \text{dist}(s,v)$ and $r(v) < \text{dist}(v,t)$ (as our algorithms do), the canonical path will not be pruned. The algorithm may not necessarily return the canonical path as its answer, but it is guaranteed to find a path of the same length.

Once we have the notion of canonical shortest paths, we can prove that the upper bounds on arc reaches computed during the first stage of the preprocessing algorithm are indeed valid. The proof is by a straightforward induction with a somewhat tedious case analysis.

THEOREM 4.2. *For every arc (v, w) we compute an upper bound $\overline{r}(v, w)$ on the reach of (v, w) in the graph obtained from the original graph G by adding all shortcuts.*

PROOF. We break the process into its elementary steps, where an elementary step is either bypassing a vertex or a round of partial tree computations (after which we delete every arc whose reach is finitely bounded). Let G_i be the original graph with all shortcuts added up to the end of step i, and let G'_i be the graph that the algorithm is left with after step i. We prove by induction on the steps that the following invariants hold.

(1) For every arc (v, w) that we have deleted until (and including) step i (i.e., $(v, w) \notin G'_i$), $\overline{r}(v, w)$ upper bounds the reach of (v, w) in G_i.
(2) For every vertex $v \in G'_i$, *out-penalty*(v) upper bounds the reach in G_i of any arc (v, w) that we have already deleted.
(3) For every vertex $v \in G'_i$, *in-penalty*(v) upper bounds the reach in G_i of any arc (u, v) that we have already deleted.

Note that Invariants (2) and (3) immediately follow from Invariant (1) because of the way we update the penalties.

We outline a proof of the induction step. The basis is trivial. Assuming these invariants are true at the end of step $i - 1$, we must prove that they also hold at the end of step i.

Since adding shortcuts can only reduce reaches, Invariant (1) continues to hold for all arcs deleted prior to step i. We will establish Invariant (1) for every arc that we delete during step i.

Assume step i is bypassing a vertex v. Let (v, w) (or (w, v)) be an arc that we delete when bypassing v. Then, from the correctness of the penalties after step $i - 1$ and the discussion in Section 4.3, it follows that $\overline{r}(v, w)$ indeed bounds the reach of (v, w) in the graph G_i, which includes also all shortcut arcs added when we bypassed v.

Assume step i is a partial tree computation. Let ϵ be the threshold of the current level, and let (v, w) be an arc whose reach we bound at this step ($\overline{r}(v, w) < \epsilon$). Assume (to get a contradiction) that the reach of (v, w) in G_i is $r(v, w) > \overline{r}(v, w)$. Note that the partial tree computation is done in the graph G'_{i-1} and, since we do not bypass vertices in this step, $G_i = G_{i-1}$.

Assume first that $r(v, w) \geq \epsilon$, and let $P = (s, \ldots, v, w, \ldots, t)$ be an ϵ-minimal path with respect to (v, w) in G_i. For a vertex $x \in P$, let $p(x)$ be the vertex preceding x on P and let $f(x)$ be the vertex following x on P. Let $P' = (z, \ldots, v, w, \ldots, y)$ be the maximal subpath of P in G'_{i-1} containing (v, w). Since if $z \neq s$ the reach of the arc $(p(z), z)$ in G_{i-1} is at least $\min\{\text{dist}(s, z), \text{dist}(p(z), t)\}$, it follows from the correctness of the penalties after step $i - 1$ that *in-penalty*$(z) + \text{dist}(z, w) \geq \epsilon$. Let z' be the last vertex on the prefix of P' up to and including v such that *in-penalty*$(z') + \text{dist}(z', w) \geq \epsilon$. In the partial tree rooted at z', w is inner. (Note that this is true also if $z' = v$.)

If $y \neq t$ then the reach of $(y, f(y))$ is at least $\min\{\text{dist}(y, t), \text{dist}(s, f(y))\}$. So (regardless of whether y is t or not) by the correctness of the penalties after step $t - 1$ we get that $\text{dist}(v, y) +$ *out-penalty*$(y) \geq \epsilon$. Let y' be the last vertex after v on P' such that $\text{dist}(v, y') +$ *out-penalty*$(y') < \epsilon$ (or v if there is no such vertex). Then y' must be scanned, and $f(y')$ must be a labeled child of y', during the shortest path computation from z'. Therefore, in the partial tree rooted at z' the depth of

w and the height of v are both at least ϵ. So in this tree the arc (v, w) should have gotten a bound of ∞ on its reach, which is a contradiction.

If $r(v, w) < \epsilon$, let $P = (s, \ldots, v, w, \ldots, t)$ be an $r(v, w)$-minimal path. A similar argument shows that there is a partial tree in which the bound on the reach of (v, w) is at least $r(v, w)$, which gives a contradiction. □

4.7. Cardinality Reach and Highway Hierarchies. Having fully described our reach-based algorithm (RE), we now discuss its relationship to the HH algorithm of Sanders and Schultes [**40, 41**]. We show similarities between the algorithms other than those pointed out in [**40, 41**].

We introduce a variant of reach that we call *c-reach* (cardinality reach). Given a vertex v on a shortest path P, grow equal-cardinality balls centered at the endpoints of P until v belongs to one of the balls. Let $c_P(v)$ be the cardinality of each of the balls at this point. The c-reach of v, $c(v)$, is the maximum, over all shortest paths P, of $c_P(v)$. Note that if we replace cardinality with radius, we get the definition of reach. To use c-reach for pruning the search, we need the following values. For a vertex v and a nonnegative integer i, let $\rho(v, i)$ be the radius of the smallest ball centered at v that contains i vertices. Consider a search for the shortest path from s to t and a vertex v. We do not need to scan v if $\rho(s, c(v)) < \text{dist}(s, v)$ and $\rho(t, c(v)) < \text{dist}(v, t)$. A direct implementation of this pruning method would require maintaining $n - 1$ values of ρ for every vertex.

The main idea behind HH preprocessing is to use the partial-trees algorithm for c-reaches instead of reaches. Given a threshold h, the algorithm identifies vertices that have c-reach below h (local vertices). Consider a bidirectional search. During the search from s, once the search radius advances past $\rho(s, h)$, one can prune local vertices in this search. One can do similar pruning for the reverse search. This idea is applied recursively to the graph with low c-reach vertices deleted. This gives a hierarchy of vertices, in which each vertex needs to store a ρ-value for each level of the hierarchy it is present at. The preprocessing phase of HH also adds shortcuts and uses other heuristics to reduce the graph size at each iteration.

An important property of the HH query algorithm, which makes it similar to the self-bounding algorithm discussed in Section 3, is that the search in a given direction never goes to a lower level of the hierarchy. Our self-bounding algorithm can be seen as having a "continuous hierarchy" of reaches: once a search leaves a reach level, it never comes back to it.

5. Reaches and A* Search

This section reviews A* search and the ALT algorithm, and shows how the latter can be combined with reach pruning in a natural way.

5.1. A* Search. Suppose we need to find shortest paths in a graph G with distance function ℓ. A *potential function* maps vertices to reals. Given a potential function π, the *reduced cost* of an arc is defined as $\ell_\pi(v, w) = \ell(v, w) - \pi(v) + \pi(w)$. Suppose we replace ℓ by ℓ_π. Then for any two vertices x and y, the length of every x-y path (including the shortest) changes by the same amount, $\pi(y) - \pi(x)$. Thus the problem of finding shortest paths in G is equivalent to the problem of finding shortest paths in the transformed graph.

Now suppose we are interested in finding the shortest path from s to t. Let π_f be a (perhaps domain-specific) potential function such that $\pi_f(v)$ gives an estimate on the distance from v to t. In the context of this paper, A^* *search* [**12, 25**] is an algorithm that works like Dijkstra's algorithm, except that at each step it selects a labeled vertex v with the smallest *key*, defined as $k_f(v) = d_f(v) + \pi_f(v)$, to scan next. It is easy to see that A* search is equivalent to Dijkstra's algorithm on the graph with length function ℓ_{π_f}. If π_f is such that ℓ_π is nonnegative for all arcs (i.e., if π_f is *feasible*), the algorithm will find correct shortest paths. We refer to the class of A* search algorithms that use a feasible function π_f with $\pi_f(t) = 0$ as *lower-bounding algorithms*, since $\pi_f(v)$ is guaranteed to be a lower bound on the distance from any vertex v to t.

We combine A* search and bidirectional search as follows. Let π_f be the potential function used in the forward search and let π_r be the one used in the reverse search. Since the latter works in the reverse graph, each original arc (v, w) appears as (w, v), and its reduced cost w.r.t. π_r is $\ell_{\pi_r}(w, v) = \ell(v, w) - \pi_r(w) + \pi_r(v)$, where $\ell(v, w)$ is the length in the original graph. We say that π_f and π_r are *consistent* if, for all arcs (v, w), $\ell_{\pi_f}(v, w)$ in the forward graph is equal to $\ell_{\pi_r}(w, v)$ in the reverse graph. This is equivalent to $\pi_f + \pi_r =$ constant.

If π_f and π_r are not consistent, the forward and reverse searches use different length functions. When the searches meet, we have no guarantee that the shortest path has been found. Assume π_f and π_r are feasible (but not necessarily consistent) and give lower bounds to the sink and from the source, respectively. Ikeda et al. [**26**] suggest using an *average function*, defined as $p_f(v) = \frac{\pi_f(v) - \pi_r(v)}{2}$ for the forward computation and $p_r(v) = \frac{\pi_r(v) - \pi_f(v)}{2} = -p_f(v)$ for the reverse one. To make the algorithm more intuitive, we add $\pi_r(t)/2$ to the forward function (which ensures that $p_f(t) = 0$) and $\pi_f(s)/2$ to the reverse function (making it zero at s). Because the added terms are constant, the functions remain consistent. Although p_f and p_r usually do not give lower bounds as good as the original ones, they are feasible and consistent.

Recall that the standard bidirectional algorithm can stop as soon as $\text{top}_f + \text{top}_r \geq \mu$, where top_f is the top key in the forward heap, top_r is the top key in the reverse heap, and μ is the length of the shortest path found so far. For bidirectional A* with consistent lower bounds, we can use a similar stopping criterion, but in the transformed graph.

Let v_f and v_r be the top vertices in each heap (with keys top_f and top_r, respectively). The standard stopping criterion states that we can stop as soon as $\text{dist}(s, v_f) + \text{dist}(v_r, t) \geq$ [best path seen so far]. In the transformed graph, the distance between s and v_f corresponds to $d_f(v_f) + p_f(v_f) - p_f(s) = k_f(v_f) - p_f(s) = \text{top}_f - p_f(s)$. Similarly, the distance between v_r and t in the transformed graph is $d_r(v_r) + p_r(v_r) - p_r(t) = k_r(v_r) - p_r(t) = \text{top}_r - p_r(t)$. Finally, the length of the shortest path seen so far is $\mu - p_f(s) + p_f(t)$. The stopping criterion then becomes:

$$[\text{top}_f - p_f(s)] + [\text{top}_r - p_r(t)] \geq \mu - p_f(s) + p_f(t).$$

Since we fixed $p_f(t) = 0$, this translates into

$$\text{top}_f + \text{top}_r \geq \mu + p_r(t).$$

5.2. The ALT Algorithm. The ALT algorithm [**19, 23**] is an A*-based method that uses landmarks and triangle inequality to compute feasible lower bounds. We select a small subset of vertices (a constant number) as *landmarks* and, for each vertex in the graph, precompute distances to and from every landmark. Consider a landmark L. If we are using distances *to* L, then, by the triangle inequality, $\text{dist}(v, L) - \text{dist}(w, L) \leq \text{dist}(v, w)$; if we use distances *from* L, then $\text{dist}(L, w) - \text{dist}(L, v) \leq \text{dist}(v, w)$. To get the tightest lower bound, one can take the maximum of these bounds, over all *active* landmarks (i.e., those in use for the current search). These bounds are feasible, and we use the average function to ensure consistency. Intuitively, the best lower bounds on $\text{dist}(v, w)$ are given by landmarks that appear "before" v or "after" w. The version of ALT algorithm that we use balances the work of the forward search and the reverse search (see Section 2). This version had better overall performance than other variants [**21**].

5.2.1. *Active Landmarks.* Each s-t query starts with just two active landmarks, those that give the best lower bounds on the distance from s to t (one using distances *to* the landmark, the other using distances *from* the landmark). Periodically, the algorithm checks whether adding another landmark to the set of active ones would be advantageous. We call this a *checkpoint.*

Of course, checkpoints are expensive, so we avoid running them too often. We allow each search (forward and reverse) to have at most ten checkpoints. In addition, we require that at least $8k$ vertices be scanned between checkpoints in the same direction, where k is the total number landmarks available. For details, see [**23**].

5.2.2. *Landmark Generation.* In our experiments, we use the *avoid* algorithm to select landmarks. Among the landmark selection algorithms studied in [**23**], *avoid* is the second best in terms of solution quality. It is only surpassed by *maxcover*, which is essentially *avoid* followed by a simple local search and is about five times slower. We have discovered a minor issue with our previous implementation of *avoid* that caused it to obtain slightly worse landmarks than it should. The fixed version of *avoid* provides slightly better landmarks, and is almost as good as *maxcover.*

The *avoid* method works by adding one landmark at a time. In each iteration, it tries to pick a landmark in a region that is still not well-covered by existing landmarks. To do so, it first builds a complete shortest path tree T_r rooted at a vertex r picked in a randomized fashion. More precisely, r is picked with probability proportional to the square of its distance to the closest existing landmark (if there is no landmark yet, r is picked uniformly at random). The algorithm then assigns a weight to each vertex v, defined as the difference between the distance from r to v and the lower bound on that distance given by the existing landmarks (if there are no landmarks yet, the weight is constant). Therefore, vertices with weak lower bounds will have larger weights. The algorithm determines the vertex p that, among all vertices that have no landmark as a descendent, maximizes the sum of the weights of its descendents in T_r (we call this sum the *size* of p). A leaf of the subtree of T_r rooted at p is then selected as the new landmark. More precisely, we follow a path from p to a leaf by picking in each step a child with maximum size.

Computing the lower bound on the distance from the root r to a particular vertex takes time proportional to the number of landmarks already selected, which makes *avoid* quadratic in the number k of landmarks. This is not a major issue when only 16 landmarks are selected (as in most of our experiments), but the algorithm does get measurably slower with 64 or more landmarks. We propose a simple modification to the algorithm to make it linear in the number of landmarks.

When processing the shortest path tree rooted at r, we still define the weight of v to be the difference between the distance from r to v and the current lower bound on this distance, but only if v belongs to a set of n/k *relevant vertices.* For all other vertices, we define the weight as zero. The set of relevant vertices is picked uniformly at random (unless reaches are available, as Section 5.4 will explain). For the range of values of k we tested (up to 64), the landmarks produced by this method were not measurably worse than those obtained when n vertices are considered relevant, as in [**21**].

As observed in [**23**], we create a separate file for each landmark. For each vertex (in order), it contains the distances *to* and *from* the landmark stored contiguously. To save space, we use a simple compression scheme that exploits the fact that, on road networks, distances to and from the same landmark are usually very similar, and that vertices with similar identifiers are usually close together in the graph. Although each (uncompressed) distance could be stored as a 32-bit value, there are long runs in which the first 16 bits in each distance are identical. These 16 bits are represented only once, together with an 8-bit count on the number of times it is repeated (runs with more than 255 elements are split); each distance in the run is then represented by its 16 least significant bits. This approach, which is a run-length encoding scheme,[3] leads to compression ratios that are close to 50%. We observe that compression is used only in disk. Once in main memory, landmark distances are kept uncompressed as 32-bit integers. We still keep a separate array for each landmark, with "from" and "to" distances stored contiguously.

5.3. Combining Reaches and Landmarks. Reach-based pruning can be easily combined with A* search: the basic idea is to just run A* search and prune vertices (or arcs) based on reach conditions. Specifically, when A* search is about to scan a vertex v, we extract from the key of v the length of the shortest path from the source to v. Furthermore, $\pi_f(v)$ is a lower bound on the distance from v to the destination. If the reach of v is smaller than both $d_f(v)$ and $\pi_f(v)$, we prune the search at v.

Note that we must actually use landmark-based lower bounds for pruning. Implicit bounds cannot be used with A* search because the search grows balls with respect to reduced costs, which have little correlation with the original lengths.

The reason why reach-based pruning works is that, although A* search uses transformed lengths, the shortest paths remain invariant. This applies to bidirectional search as well. We use $d_f(v)$ and $\pi_f(v)$ to prune in the forward direction, and $d_r(v)$ and $\pi_r(v)$ to prune in the reverse direction. Pruning by reach does not affect the stopping condition of the algorithm: the usual condition for A* search can still be applied. We call our implementation of the bidirectional A* search algorithm with landmarks and reach-based pruning REAL. As we did for ALT, our implementation of REAL balances the work of the forward and reverse searches.

Note that REAL has two preprocessing algorithms: the one used by RE (which computes shortcuts and reaches) and the one used by ALT (which chooses landmarks and computes distances between them and all vertices). These two procedures can be independent from each other: since shortcuts do not change distances, landmarks can be generated regardless of what shortcuts are added. Furthermore, queries are still independent of the preprocessing algorithm: they only take as input the graph

[3]See e.g., http://www.data-compression.info/Algorithms/RLE/.

with shortcuts, the reach values, and the distances to and from landmarks. How this data was obtained is immaterial. We will see in Section 5.4, however, that it might be useful to take reaches into account when generating landmarks.

5.3.1. *Optimizations.* Our implementation of REAL actually uses early pruning, as in RE. When scanning arc (v, w) in the forward direction, we prune w if $\overline{r}(w) < \min\{d_f(v) + \ell(v, w), \pi_f(w)\}$. (We restrict our discussion to the forward search; the reverse one is similar.) Note that this computation requires knowledge of $\pi_f(w)$, a lower bound on the distance from w to the target. Computing it requires retrieving from memory the distances between w and all active landmarks, and computing the triangle inequality in each case. Although this takes constant time, it is relatively expensive in practice. Before actually computing it, we use $\pi_f(v) - \ell(v, w)$ as a lower bound on the distance from w to t; since v is the vertex being scanned, $\pi_f(v)$ is readily available. If (1) $\overline{r}(w) < d_f(v) + \ell(v, w)$ and (2) $\overline{r}(w) < \pi_f(v) - \ell(v, w)$, we can prune w. Only if condition (2) fails and condition (1) succeeds do we need to compute $\pi_f(w)$.

Also as in RE, we can use arc sorting to prune some arcs implicitly. We sort the arcs (v, w) in the adjacency list of v by $\overline{r}(w) + \ell(v, w)$ (in non-increasing order). Suppose that, while scanning v, we find an arc (v, w) such that (1′) $\overline{r}(w) + \ell(v, w) < d_f(v)$ and (2′) $\overline{r}(w) + \ell(v, w) < \pi_f(v)$. This implies that conditions (1) and (2) above are true for w and therefore (v, w)—and all arcs that succeed it—can be pruned. Note that conditions (2) and (2′) are equivalent, but condition (1) may succeed while (1′) fails. In this case we still prune the arc, but keep traversing the adjacency list.

We also try to prune a vertex after we remove it from the heap (and before we scan it). This is still useful because the lower bound on the distance to the target may have improved since the vertex was inserted into the heap, due to the activation of new landmarks.

5.4. Reach-Aware Landmarks. Although landmark generation and reach computation can be completely independent, we can use landmarks more efficiently when reaches are available. We can reduce the memory requirements of the algorithm by storing landmark distances only for high-reach vertices. As we shall see, however, this results in some degradation of query performance. If we add more landmarks, we get a wide range of trade-offs between query performance and memory requirement. We call the resulting method, a variant of REAL, the *partial landmark algorithm.*

Queries for the partial landmark algorithm work as follows. Let R be the reach threshold: we store landmark distances for all vertices with high reach, i.e., with reach at least R. We start a query by running the bidirectional Dijkstra algorithm with reach pruning (but without A* search), until either the algorithm terminates or both balls searched have radius R. In the latter case, we know that, from this point on, we need to examine only vertices with reach R or more. We switch to A* search (still with pruning by reach) by removing all labeled vertices from the heaps and reinserting them using new keys that incorporate lower bound values.

Recall that, for every vertex v it visits, A* search may need lower bounds on the distance from v to t (in the forward search) or from s to v (in the reverse search). They are computed with the triangle inequality, which requires distances between these vertices (v, s, and t) and the landmarks. These are guaranteed to be available for v, which has high reach, but not for s or t, which are arbitrary vertices. We therefore need to specify how to compute a lower bound on the distance between a low-reach vertex (s or t) and a high-reach one (v).

Suppose s has low reach (t is treated symmetrically). Let s', the *proxy* for s, be the high-reach vertex that is closest to s.[4] One can compute proxies during preprocessing and store them, or compute them during the initialization phase of the query algorithm; we choose the latter approach. Two executions of a multiple-source version of Dijkstra's algorithm (one in the forward graph and one in the reverse graph) suffice to compute both the proxies and the appropriate distances.

As already mentioned, when processing a high-reach vertex v, the A* search needs lower bounds on $\text{dist}(s, v)$ and $\text{dist}(v, t)$. Given a landmark L, we can obtain these bounds using either distances *from* L or distances *to* L. With the help of proxies, these bounds can be easily computed.

A lower bound on $\text{dist}(s, v)$ using distances *to* L is given by

$$\text{dist}(s, v) \geq \text{dist}(s', L) - \text{dist}(v, L) - \text{dist}(s', s). \tag{5.1}$$

Using distances *from* L, the lower bound can be computed as follows:

$$\text{dist}(s, v) \geq \text{dist}(L, v) - \text{dist}(L, s') - \text{dist}(s', s). \tag{5.2}$$

Lower bounds on distances from v to the target t can be computed similarly. Using distances *from* landmarks, the following relation applies:

$$\text{dist}(v, t) \geq \text{dist}(L, t') - \text{dist}(L, v) - \text{dist}(t, t'). \tag{5.3}$$

With distances *to* landmarks, the appropriate expression is

$$\text{dist}(v, t) \geq \text{dist}(v, L) - \text{dist}(t', L) - \text{dist}(t, t'). \tag{5.4}$$

Note that distances between L and v, s' and t' are computed during the preprocessing stage, since all three vertices have high reach. As already mentioned, the distances between every vertex and its proxy (or, more precisely, proxies) are computed in the initialization phase of the query algorithm.

The quality of the lower bounds obtained by the partial landmark algorithm depends not only on the number of landmarks available, but also on the value of R. In general, the higher the reach threshold, the farther the proxy s' will be from s (and t' from t), thus decreasing the accuracy of the lower bounds. If all landmark distances are available, R will be zero, and the algorithm will behave exactly as the standard REAL method. By decreasing the number of distances per landmark (representing distances only to higher-reach vertices), R will increase; a trade-off between memory usage and query efficiency is thus established.

It turns out, however, that we can often improve both memory usage and query times. Starting from the base algorithm, we can increase the number of landmarks while decreasing the number of distances per landmark so that the total memory usage is lower. On large road networks, we can find parameter values for which both memory use and query times decrease.

We note that, in our experiments, we do not pick the reach threshold R explicitly. Instead, we actually pick how many vertex distances we want to store per landmark; the value of R will be fully determined by this choice.

---

[4]Each vertex s actually has two proxies: the high-reach vertex s' that minimizes $\text{dist}(s', s)$ and the high-reach vertex s'' that minimizes $\text{dist}(s, s'')$. We will assume they are the same to simplify the discussion, but they need not be.

5.4.1. *Landmark Generation.* As observed in Section 5.2.2, we use a fast version of the *avoid* method to generate landmarks. We also use it to generate partial landmarks, with a small modification. Instead of picking the set of relevant vertices uniformly at random, we pick the first n/k vertices in the vertex list. Because vertices are approximately sorted in decreasing order of reach (as seen in Section 3.2), only vertices with high reach are taken into account, which greatly simplifies the implementation when only partial landmark data is stored. In terms of solution quality, these two approaches are roughly equivalent.

6. Experimental Results

We implemented our algorithms in C++ and compiled them with Microsoft Visual C++ 2005. All tests were performed on a dual-processor, 2.4 GHz AMD Opteron machine running Microsoft Windows Server 2003 with 16 GB of RAM, 32 KB instruction and 32 KB data level 1 cache per processor, and 2 MB of level 2 cache. Our code is single-threaded and runs on a single processor at a time (but is not pinned to a particular processor).

Our implementation uses two kinds of priority queues. We use 4-heaps when the number of elements in a priority queue is small, such as during queries and when building partial trees during preprocessing. For single-source shortest path computations on the entire graph, we use multi-level buckets [**17, 22**]. Multi-level buckes are faster in general, but 4-heaps are competitive when the number of elements is small, and have a smaller memory overhead.

One of the main goals of our experimental analysis is to measure the performance of our algorithm on the road networks of the USA and Europe, made available for the 9th DIMACS Implementation Challenge. The experiments, reported in Section 6.1, include both random and local queries. We also investigated how much various ingredients of our algorithms contribute to the overall performance. In particular, we study the preprocessing/query and the time/space trade-offs.

Section 6.2 considers how our algorithms behave on other graph classes. In particular, we present results on grid graphs in 2 and 3 dimensions with random arc weights. We also conducted preliminary experiments on higher-dimensional grids and random graphs. Since our algorithm performed poorly on these graphs, we omit the detailed results.

We ran the ALT (with 16 landmarks), RE and REAL-(i, j) algorithms, where REAL-(i, j) uses i landmarks and maintains landmark data for n/j highest-reach vertices. We call the parameter j the *sparsity* of the landmarks. For instance, REAL-(64,16) maintains 64 landmarks, but with distances only to $n/16$ vertices with high reaches; REAL-(16,1) uses 16 landmarks, each with distances to all vertices in the graph. All landmarks (for ALT and all variants of REAL) were generated with the *avoid* method. On grid graphs, we also ran our own implementation of the bidirectional version of Dijkstra's algorithm, denoted by BD. For machine calibration purposes, we ran the DIMACS Challenge implementation of the P2P version of Dijkstra's algorithm, denoted by D, on the largest road networks.

For the graphs of the USA and Europe with travel times as arc lengths, additional data from previous works was available at the time of writing. In particular, we report the results obtained by the highway hierarchy-based algorithm of Sanders and Schultes, as reported in [**41**]. Their sequential code was run on a dual-core 2.0

GHz AMD Opteron machine, which is comparable to our machine. In fact, because of a different memory architecture, D runs about 2% faster on their machine. Their algorithm has two versions [**41**]: HH-mem, entirely based on highway hierarchies, and HH, which replaces high levels of the hierarchy by a table with distances between all pairs of vertices in the corresponding graph.

6.1. Road Networks. The graphs representing the USA (Tiger/Line) [**49**] and (Western) Europe [**35**] road networks belong to the 9th Implementation DIMACS Challenge [**8**] data set. The USA is symmetric and has 23 947 347 vertices (road intersections) and 58 333 444 arcs (directed road segments); Europe is directed, with 18 010 173 vertices and 42 560 279 arcs. To evaluate the performance on smaller graphs, we tested the subgraphs of USA described in Table 1. All graphs are strongly connected. We used two natural metrics as length functions: travel times and travel distances. In addition, to test the robustness of our algorithm, we considered a third metric (uniform) in which all arcs have unit length. For all metrics, lengths were represented as 32-bit integers.

TABLE 1. USA road networks from the TIGER/Line collection.

NAME	DESCRIPTION	VERTICES	ARCS	LAT. (N)	LONG. (W)
USA	—	23 947 347	58 333 344	—	—
CTR	Central USA	14 081 816	34 292 496	[25.0; 50.0]	[79.0; 100.0]
W	Western USA	6 262 104	15 248 146	[27.0; 50.0]	[100.0; 130.0]
E	Eastern USA	3 598 623	8 778 114	[24.0; 50.0]	[$-\infty$; 79.0]
LKS	Great Lakes	2 758 119	6 885 658	[41.0; 50.0]	[74.0; 93.0]
CAL	California and Nevada	1 890 815	4 657 742	[32.5; 42.0]	[114.0; 125.0]
NE	Northeast USA	1 524 453	3 897 636	[39.5, 43.0]	[$-\infty$; 76.0]
NW	Northwest USA	1 207 945	2 840 208	[42.0; 50.0]	[116.0; 126.0]
FLA	Florida	1 070 376	2 712 798	[24.0; 31.0]	[79; 87.5]
COL	Colorado	435 666	1 057 066	[37.0; 41.0]	[102.0; 109.0]
BAY	Bay Area	321 270	800 172	[37.0; 39.0]	[121; 123]
NY	New York City	264 346	733 846	[40.3; 41.3]	[73.5; 74.5]

6.1.1. *Random queries.* Our first experiment consists of running our algorithms on 1000 random queries. In each case, the source and the target are picked independently and uniformly at random from the set of all available vertices. Table 2 presents results for the USA graph, and Table 3 shows the corresponding results for the European road network. In each case, all three metrics (length functions) are considered.

For each algorithm, we report the average query time (in milliseconds), the average number of scanned vertices and, when available, the maximum number of scanned vertices (over the queries in the set). Also shown are the total preprocessing time (in minutes) and the total disk space (in megabytes) required by the preprocessed data. For D, this is the space required to store only the graph itself; for ALT, this includes the graph and landmark data; for RE, it includes the graph with shortcuts (which has roughly two-thirds more arcs) and an array of vertex reaches; finally, the data for REAL includes the graph with shortcuts, the array of reaches, and landmark data. All files are stored in binary format, with 32-bit lengths, distances, and vertex identifiers.

TABLE 2. Random queries on the USA graph.

METRIC	METHOD	PR. TIME (min)	DISK SPACE (MB)	QUERY AVG SC.	QUERY MAX SC.	QUERY TIME (ms)
TIMES	ALT	17.6	2563	187968	2183718	295.44
	RE	27.9	893	2405	4813	1.77
	REAL-(16,1)	45.5	3032	592	2668	0.80
	REAL-(64,16)	113.9	1579	538	2534	0.86
	HH	18	1686	1076	—	0.88
	HH-mem	65	919	2217	—	1.60
	D	—	536	11808864	—	5440.49
DISTANCES	ALT	15.2	2417	276195	2910133	410.73
	RE	46.4	918	7311	13886	5.78
	REAL-(16,1)	61.5	2923	905	5510	1.41
	REAL-(64,16)	120.5	1575	670	3499	1.22
	D	—	536	11782104	—	4576.02
UNIFORM	ALT	14.0	1992	240801	3922923	312.97
	RE	28.8	865	3496	6830	2.58
	REAL-(16,1)	42.7	2321	790	2968	1.09
	REAL-(64,16)	96.6	1315	573	3067	0.95
	D	—	536	11724870	—	3636.92

TABLE 3. Random queries on Europe.

METRIC	METHOD	PR. TIME (min)	DISK SPACE (MB)	QUERY AVG SC.	QUERY MAX SC.	QUERY TIME (ms)
TIMES	ALT	12.5	1597	82348	993015	120.09
	RE	45.1	648	4371	8486	3.06
	REAL-(16,1)	57.7	1869	714	3387	0.89
	REAL-(64,16)	102.6	1037	610	2998	0.91
	HH	15	1570	884	—	0.8
	HH-mem	55	692	1976	—	1.4
	D	—	393	8984289	—	4365.81
DISTANCES	ALT	9.6	1622	240750	3306755	321.11
	RE	31.5	681	7259	13059	5.20
	REAL-(16,1)	41.1	1938	855	4867	1.20
	REAL-(64,16)	76.0	1084	562	2596	0.91
	D	—	393	8991955	—	2934.24
UNIFORM	ALT	10.6	1488	140291	2137518	188.84
	RE	37.5	625	4109	10574	2.84
	REAL-(16,1)	48.1	1720	1048	5888	1.28
	REAL-(64,16)	88.7	964	644	3631	0.94
	D	—	393	9054599	—	3000.50

Note that for ALT, RE and REAL, query performance is worst when lengths are travel distances, but not drastically so. This can be explained by the fact that the natural hierarchy of road networks, which is exploited by both ALT and RE, is more pronounced with travel times (since highways usually have higher speed limits). With travel distances, a local road running alongside a major freeway may seem more attractive. Even with unit lengths the hierarchy is more pronounced than with travel distances, as the number of vertices scanned by RE shows. Preprocessing, however, can be slower—probably because the partial trees become too dense with our default choice of parameters. We made no attempt to tune the algorithm for unit lengths.

Although the two graphs have similar sizes, the algorithms behave differently when lengths are travel times. ALT is relatively more efficient on Europe than on USA, while RE is faster on the USA graph. By combining ALT and RE, REAL is more robust than either method, and has similar performance on both graphs. In fact, even if we consider all six combinations of graphs and metrics, REAL-(64,16) was remarkably consistent: the average number of scans ranged from 538 to 670.

For a fixed graph, REAL-(16,1) and REAL-(64,16) have almost identical query performance, the latter being slightly better on most metrics. Given that REAL-(64,16) requires about half as much disk space, it has the edge for these queries. However, preprocessing for REAL-(64,16) takes roughly twice as long. As already observed, RE is less robust than REAL: the query times of RE on USA, for example, vary from 1.77 ms for travel times to 5.78 ms for travel distances, while the corresponding numbers for REAL-(64,16) are 0.86 and 1.22 ms.

We can compare RE and REAL to HH-mem and HH for the travel-time metric, for which the data is available. RE performs worse than HH-mem, but the difference is small, particularly on the USA graph. For queries, the performance of REAL-(64,16) is similar to that of HH, and REAL-(64,16) uses a little less space. Preprocessing for HH, however, is faster, especially on Europe.

6.1.2. *Graph size dependence.* To test how the performance of our algorithms depends on graph size, we ran 1000 random queries on subgraphs of the USA road network. The results are reported in Tables 4 and 5. Although query times tend to increase as the graph grows, they are not strictly monotone: the graph structure affects the performance of the algorithm. It is clear, however, that reach-based algorithms have better asymptotic performance than ALT. As the graph size increases by two orders of magnitude, so does the time for ALT queries. Running times of RE and REAL change by a factor of six or less for travel times, and more for travel distances. The speed advantage of REAL-(64,16) relative to REAL-(16,1) holds only for large graphs: for smaller ones, the loss of precision due to proxies (defined in Section 5.4) is relatively higher, since the average shortest path is shorter.

Regarding preprocessing time, with a fixed number of landmarks ALT is roughly linear in the graph size. For 16 landmarks, the preprocessing of ALT is faster than the preprocessing of RE, and the ratio of the two does not change much as the graph size increases. With 64 landmarks, the times for landmark selection and reach computation are roughly the same: preprocessing takes about twice as long for REAL-(64,16) as for RE.

6.1.3. *Local queries.* Up to this point, we have reported data only on random queries. A more realistic assumption for road networks is that most queries will be more local. To define the notion of local queries formally, we use the concept of *Dijkstra rank.* Suppose we run Dijkstra's algorithm from s, and let v be the k-th vertex it scans. Then the Dijkstra rank of v with respect to s is $\lfloor \log_2 k \rfloor$. To generate a local query with rank r, we pick s uniformly at random from V, and pick t uniformly at random from positions $[2^r, 2^{r+1})$ in the scanning order. Note that our definition of Dijkstra rank differs slightly from the one proposed by Sanders and Schultes [**40**], but it has a similar purpose. For each of the large graphs (Europe and

TABLE 4. Data for subgraphs of USA with travel times.

GRAPH	METHOD	PREP. TIME (min)	DISK SPACE (MB)	QUERY AVG SC.	QUERY MAX SC.	QUERY TIME (ms)
NY	ALT	0.1	25	2735	23739	1.73
	RE	0.6	11	1104	2634	0.56
	REAL-(16,1)	0.7	30	226	1601	0.22
	REAL-(64,16)	1.1	17	383	1099	0.38
BAY	ALT	0.1	31	3251	31953	2.00
	RE	0.3	12	770	2114	0.42
	REAL-(16,1)	0.4	37	183	921	0.20
	REAL-(64,16)	0.9	20	238	939	0.25
COL	ALT	0.2	45	6533	70857	4.80
	RE	0.4	16	756	2503	0.39
	REAL-(16,1)	0.5	53	172	780	0.19
	REAL-(64,16)	1.2	28	240	1228	0.25
FLA	ALT	0.5	106	8195	126608	5.77
	RE	0.9	39	814	1751	0.47
	REAL-(16,1)	1.4	125	197	1067	0.19
	REAL-(64,16)	3.1	68	210	1131	0.22
NW	ALT	0.6	125	13106	166870	9.91
	RE	1.0	43	1091	3551	0.66
	REAL-(16,1)	1.6	149	211	1115	0.27
	REAL-(64,16)	3.8	78	230	1379	0.28
NE	ALT	0.7	152	12161	148647	10.91
	RE	1.8	58	1474	3466	0.86
	REAL-(16,1)	2.6	182	298	1527	0.33
	REAL-(64,16)	5.5	98	324	1234	0.42
CAL	ALT	1.0	198	22418	227812	23.28
	RE	1.9	70	1220	3700	0.73
	REAL-(16,1)	2.9	234	296	1581	0.36
	REAL-(64,16)	6.4	123	362	2123	0.44
LKS	ALT	1.5	292	21616	218084	22.78
	RE	3.2	105	1687	4015	1.12
	REAL-(16,1)	4.8	347	331	1669	0.39
	REAL-(64,16)	10.0	183	368	1800	0.45
E	ALT	2.1	375	26712	516585	29.62
	RE	3.9	132	1562	3410	1.03
	REAL-(16,1)	5.9	444	322	1575	0.42
	REAL-(64,16)	13.5	232	322	1749	0.41
W	ALT	4.1	683	75494	888950	97.61
	RE	6.4	233	1675	4223	1.03
	REAL-(16,1)	10.4	802	405	2028	0.56
	REAL-(64,16)	24.8	414	400	1782	0.58
CTR	ALT	13.4	1613	119303	1335378	234.89
	RE	25.3	522	2277	5237	2.14
	REAL-(16,1)	38.7	1880	545	2559	1.00
	REAL-(64,16)	86.9	937	453	2227	0.97
USA	ALT	17.6	2563	187968	2183718	295.44
	RE	27.9	893	2405	4813	1.77
	REAL-(16,1)	45.5	3032	592	2668	0.80
	REAL-(64,16)	113.9	1579	538	2534	0.86

TABLE 5. Data for subgraphs of USA with travel distances.

GRAPH	METHOD	PREP. TIME (min)	DISK SPACE (MB)	QUERY AVG SC.	QUERY MAX SC.	QUERY TIME (ms)
NY	ALT	0.1	24	3083	35210	2.08
	RE	0.7	11	1622	3346	0.92
	REAL-(16,1)	0.8	29	222	1321	0.25
	REAL-(64,16)	1.2	17	473	1498	0.44
BAY	ALT	0.1	29	4875	80310	3.47
	RE	0.3	12	968	2252	0.48
	REAL-(16,1)	0.5	35	192	965	0.20
	REAL-(64,16)	0.9	20	264	1268	0.25
COL	ALT	0.2	42	5774	78677	3.52
	RE	0.4	16	1087	3405	0.58
	REAL-(16,1)	0.6	51	178	1181	0.19
	REAL-(64,16)	1.2	28	267	1446	0.27
FLA	ALT	0.5	101	12394	153621	9.95
	RE	1.1	39	1196	2776	0.62
	REAL-(16,1)	1.6	121	239	1162	0.28
	REAL-(64,16)	3.2	66	246	1128	0.31
NW	ALT	0.5	117	13994	131191	10.89
	RE	1.1	43	1391	4015	0.83
	REAL-(16,1)	1.6	142	219	1143	0.28
	REAL-(64,16)	3.6	76	247	1439	0.31
NE	ALT	0.7	144	15533	214572	12.88
	RE	2.4	61	2873	6170	1.72
	REAL-(16,1)	3.1	176	327	2099	0.45
	REAL-(64,16)	5.7	99	395	2095	0.52
CAL	ALT	0.9	186	27524	315506	26.33
	RE	2.3	72	1881	5156	1.11
	REAL-(16,1)	3.2	224	331	1794	0.42
	REAL-(64,16)	6.2	122	378	1799	0.47
LKS	ALT	1.4	276	41832	622476	45.78
	RE	5.2	108	3959	8475	2.91
	REAL-(16,1)	6.7	335	470	2542	0.67
	REAL-(64,16)	11.4	183	431	2926	0.66
E	ALT	1.9	354	43539	674704	52.80
	RE	5.2	136	3486	7078	2.58
	REAL-(16,1)	7.1	427	401	2066	0.62
	REAL-(64,16)	13.9	231	382	1957	0.58
W	ALT	3.4	642	75682	669930	86.42
	RE	7.9	239	2849	6596	2.02
	REAL-(16,1)	11.3	771	456	3501	0.66
	REAL-(64,16)	24.1	413	415	2575	0.62
CTR	ALT	11.9	1540	154980	1859858	276.11
	RE	42.8	538	6827	12937	6.44
	REAL-(16,1)	54.7	1845	756	4523	1.56
	REAL-(64,16)	97.8	948	563	3054	1.44
USA	ALT	15.2	2417	276195	2910133	410.73
	RE	46.4	918	7311	13886	5.78
	REAL-(16,1)	61.5	2923	905	5510	1.41
	REAL-(64,16)	120.5	1575	670	3499	1.22

USA, with both travel times and travel distances as length functions), we generated 1000 random queries for each Dijkstra rank between 8 and 24. Figures 3 to 6 report the average running times and average number of nodes scanned for each graph and each length function.

Comparing the two plots in each figure to one another, we observe that the curves showing running times and the curves showing scan counts are very similar, but shifted by a constant that is related to the time per vertex scan of each method.

As random queries tend to involve pairs of vertices that are very far apart, our previous discussion applies to local queries with high Dijkstra rank. In particular, in this context ALT is the slowest of the four algorithms, and the REAL variants are the fastest (with very similar performance).

Now consider very local queries, with small Dijkstra rank. REAL-(16,1) scans the fewest vertices, but due to the higher overhead of accessing landmark information its running time can be slightly worse than that of RE. REAL-(64,16) mostly visits low-reach vertices and thus fails to take advantage of the landmark data. It scans about the same number of vertices as RE, but is slower due to the higher overhead. ALT clearly has the worst asymptotic performance as a function of the Dijkstra rank, but for small ranks it scans only slightly more vertices than RE. As the rank grows, ALT becomes worse than the other methods, and RE becomes consistently worse than REAL-(16,1). REAL-(64,16) improves, and catches up with REAL-(16,1) for large ranks.

In terms of query times and scan counts, REAL-(16,1) and REAL-(64,16) are the best options. REAL-(64,16) is somewhat worse for very local queries, but slightly better for higher Dijkstra ranks. Its main advantage, however, is its lower space requirement.

6.1.4. *Reach-aware landmarks.* We now study how landmark sparsity influences query times and space usage. For 16 and 64 landmarks, we vary the fraction of vertices for which landmark data is maintained. Tables 6 and 7 show that, as the sparsity increases, we get a substantial reduction in the memory overhead associated with landmarks. The number of vertex scans increases steadily (but slowly).

Running times increase substantially when the sparsity increases from 1 to 2, despite the fact that the number of scans barely changes. This is because sparsity 1 corresponds to the standard REAL algorithm: all distances to and from landmarks are available, and therefore the algorithm does not need to deal with proxies, which slow down the computation. With sparsity 2, proxies must be taken into account. Although the lower bounds have similar quality (as the number of scans shows), computing them is more expensive.

Curiously, when we further increase the sparsity the effect on the running time is minor, despite the increase in the number of vertex scans. With increased sparsity, a relatively greater portion of the vertices will be scanned at the beginning of the search, before the algorithm starts using A^* search. The algorithm will behave essentially like RE at the beginning, and we have seen that RE has significantly lower overhead per scan than REAL.

One can win in all measures of query performance. As already mentioned, compared to REAL-(16,1), which is not landmark-aware, REAL-(64,16) uses less space and often runs faster. Note that our current preprocessing algorithm runs landmark generation on the full graph; for the landmark-aware case, one could eliminate low-reach vertices to improve performance.

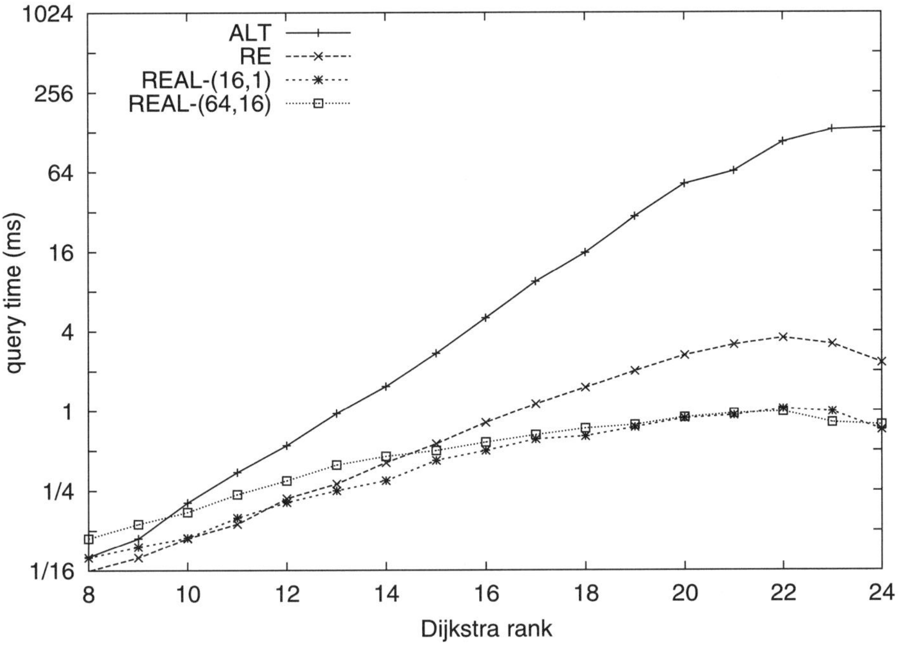

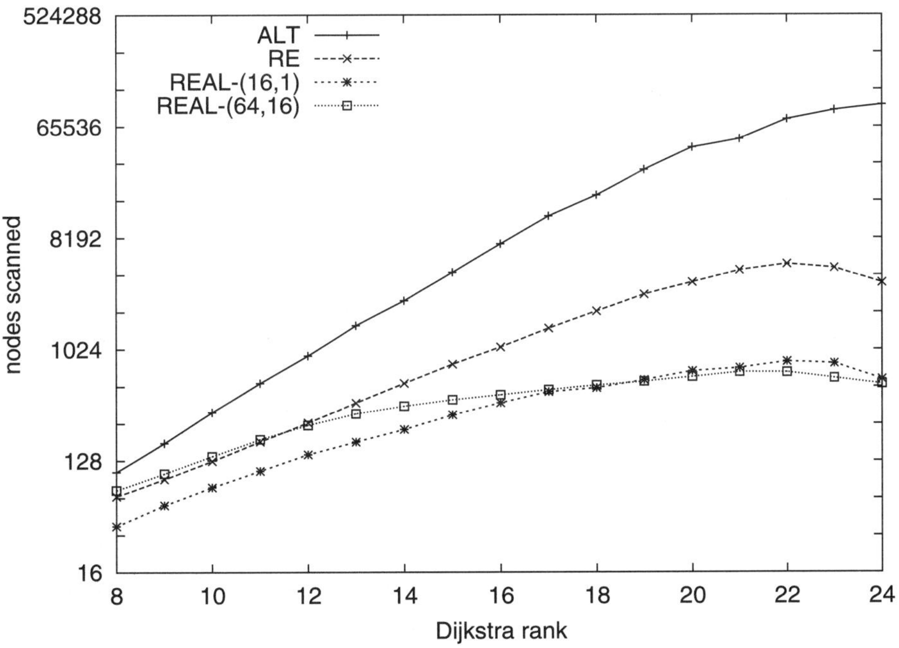

FIGURE 3. Local queries on Europe with travel times: running times and scan counts.

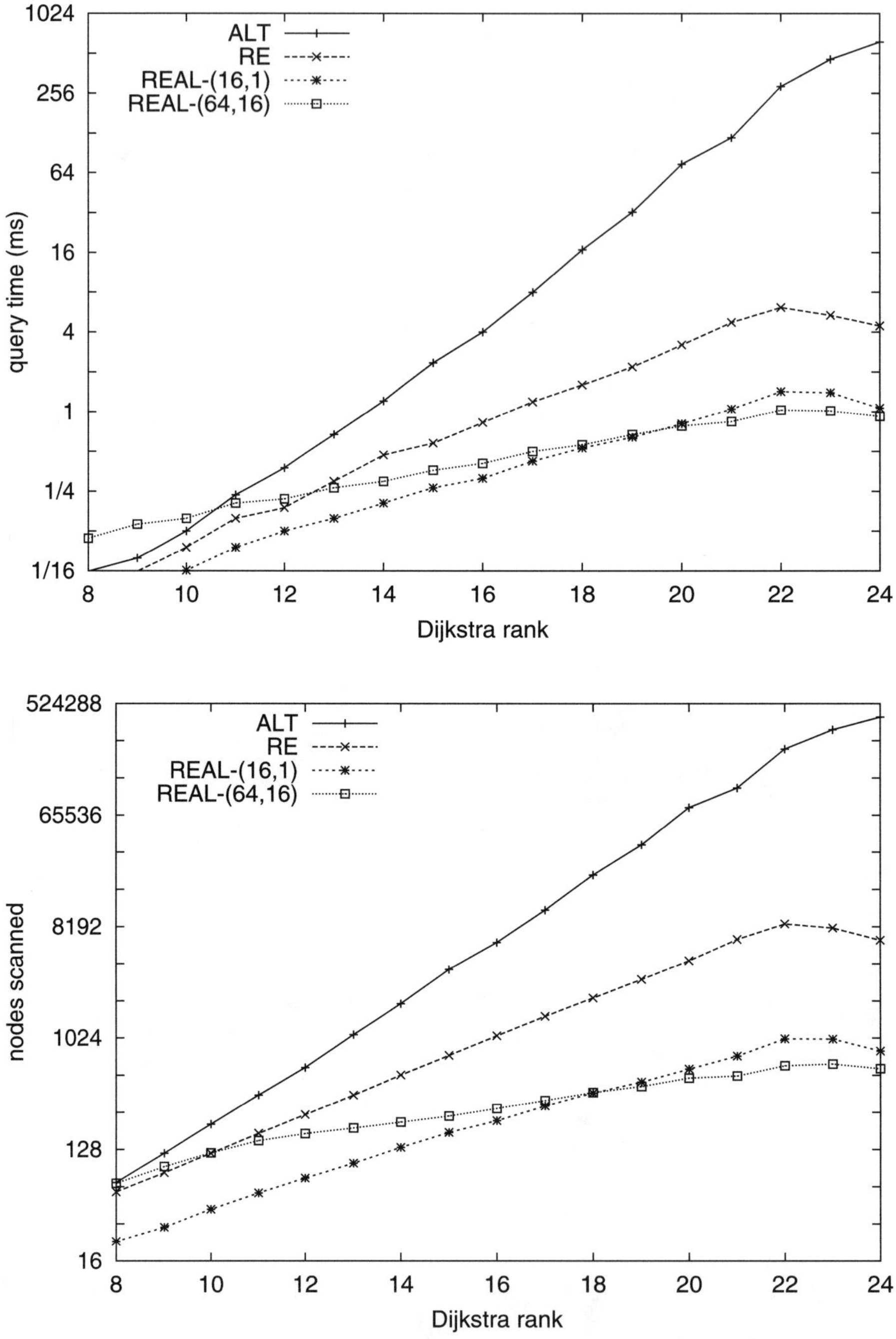

FIGURE 4. Local queries on Europe with travel distances: running times and scan counts.

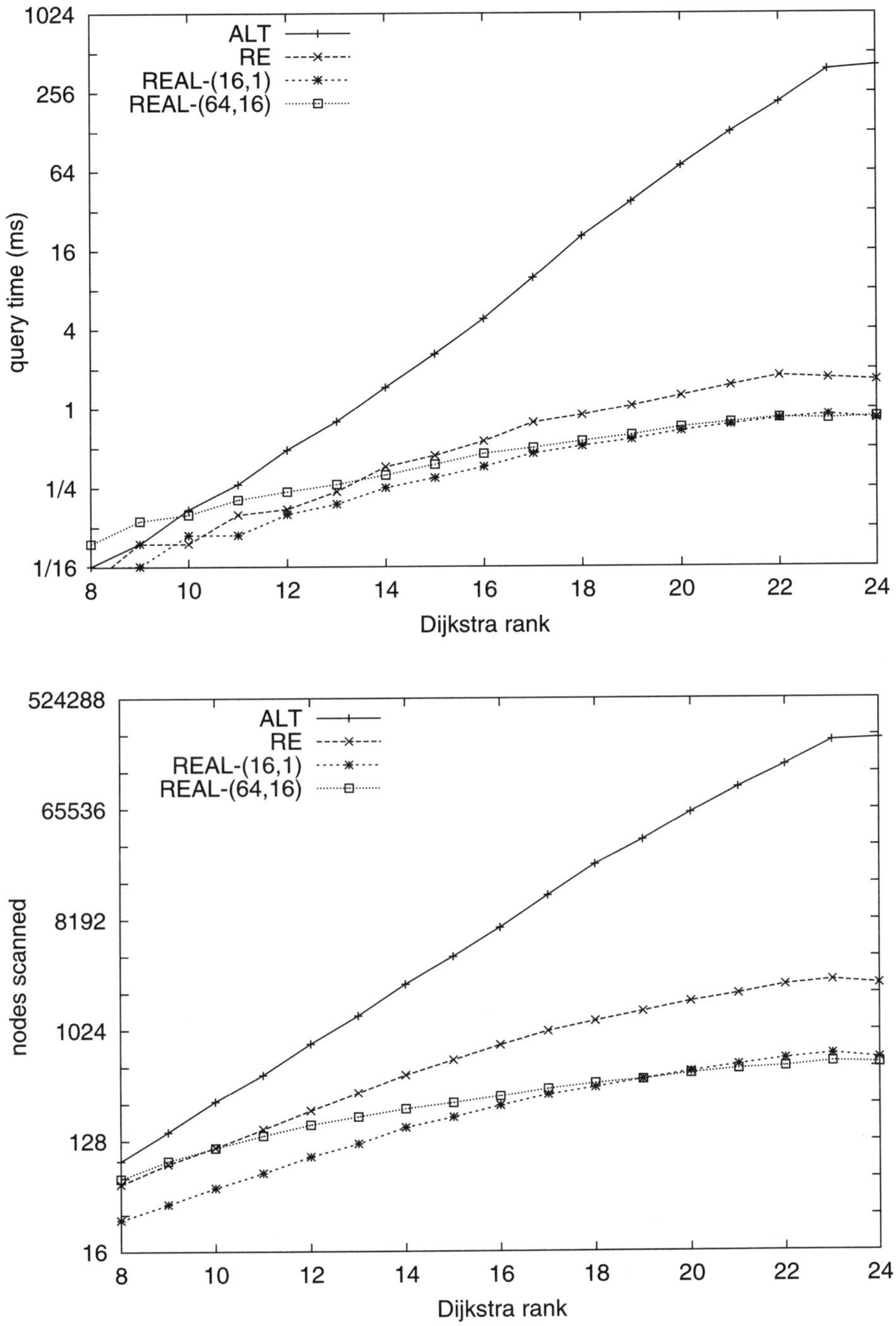

FIGURE 5. Local queries on USA with travel times: running times and scan counts.

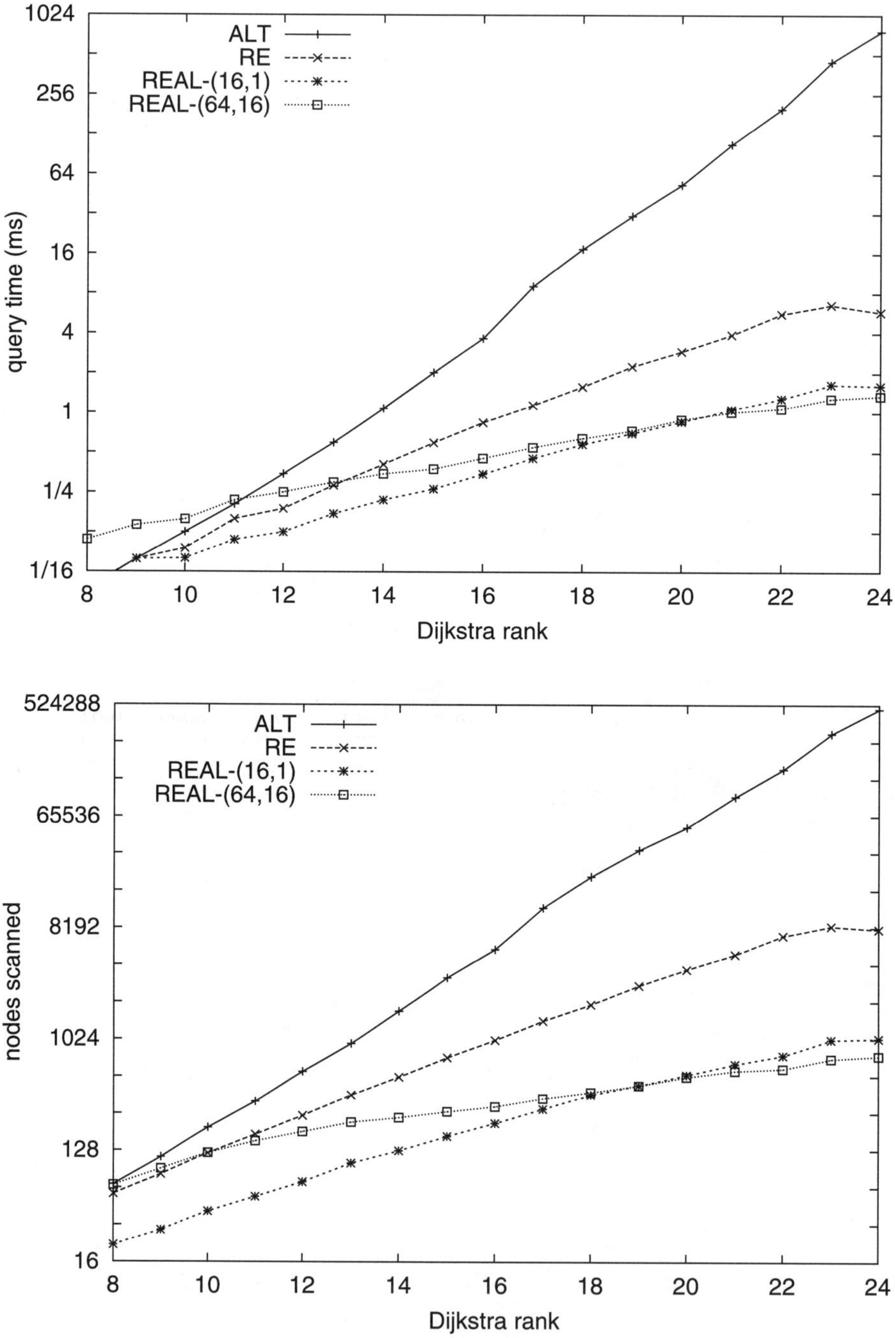

FIGURE 6. Local queries on USA with travel distances: running times and scan counts.

TABLE 6. Data for REAL on USA for various landmark/sparsity combinations.

METRIC	LAND.	SPAR.	PREP. TIME (min)	DISK SPACE (MB)	QUERY AVG SCANS	QUERY MAX SCANS	QUERY TIME (ms)
DISTANCES	16	1	61.5	2923	905	5510	1.45
		2	61.5	1960	915	5512	1.83
		4	61.5	1472	921	5696	1.84
		8	61.5	1216	933	5684	1.81
		16	61.5	1082	970	5578	1.80
		32	61.5	1014	1086	5443	1.81
		64	61.5	978	1292	5356	1.89
	64	4	120.5	3134	610	3526	1.20
		8	120.5	2109	624	3523	1.20
		16	120.5	1575	670	3499	1.16
		32	120.5	1300	810	3459	1.27
		64	120.5	1158	1049	3861	1.39
TIMES	16	1	45.5	3032	592	2668	0.81
		2	45.5	2001	600	2741	1.02
		4	45.5	1477	607	2660	0.97
		8	45.5	1206	618	2631	0.97
		16	45.5	1065	645	2696	1.00
		32	45.5	991	727	2843	1.03
		64	45.5	954	795	2968	1.05
	64	4	113.9	3229	497	2528	0.86
		8	113.9	2145	507	2536	0.81
		16	113.9	1579	538	2534	0.83
		32	113.9	1287	636	2495	0.88
		64	113.9	1137	713	2542	0.91

On a final note, observe that the number of landmark distances that must be stored depends on the ratio between the number of landmarks and the sparsity. In particular, when these ratios are the same, the number of distances stored will also be the same; this is the case of REAL-(16,8) and REAL-(64,32), for example. According to the tables, however, REAL-(64,32) actually requires slightly more space to represent the information. To understand why, recall (from Section 5.2.2) that landmark files are compressed. The compression ratio is better when vertices with similar identifiers are close to each other in the graph, which is generally true in the original graph. When vertices are partially reordered by reach, however, this is no longer true, especially for very high reaches (i.e., very low vertex identifiers): the compression ratio is much worse in the beginning of the file (high-reach vertices) than towards the end (low-reach vertices). REAL-(16,8) stores 16 files, each with $n/8$ pairs of distances (to and from the landmark); REAL-(64,32), on the other hand, must store 64 files, each with only $n/32$ pairs of distances. The average reach in the latter case is much higher, which explains the worse compression ratio.

6.1.5. *Shortcut generation.* We now examine how the performance of our algorithm is affected by the choice of shortcuts. Recall that we only allow a vertex to be bypassed if (among other criteria) the ratio of the number of new shortcuts created to the number of arcs eliminated it at most some constant c. In our algorithm, we set c to 0.5 in the first iteration, 1.0 in the second, and 1.5 for the remaining ones.

TABLE 7. Data for REAL on Europe for various landmark/sparsity combinations.

			PREP.	DISK	QUERY		
			TIME	SPACE	AVG	MAX	TIME
METRIC	LAND.	SPAR.	(min)	(MB)	SCANS	SCANS	(ms)
DISTANCES	16	1	41.1	1938	855	4867	1.20
		2	41.1	1326	865	4865	1.55
		4	41.1	1019	872	4633	1.52
		8	41.1	862	884	4659	1.44
		16	41.1	782	909	4779	1.47
		32	41.1	741	1007	4774	1.55
		64	41.1	720	1353	4636	1.67
	64	4	76.0	2026	515	2748	0.89
		8	76.0	1402	527	2711	0.89
		16	76.0	1084	562	2596	0.91
		32	76.0	920	692	2858	1.03
		64	76.0	837	1095	3214	1.22
TIMES	16	1	57.7	1869	714	3387	0.84
		2	57.7	1270	731	3386	1.12
		4	57.7	973	745	3421	1.16
		8	57.7	822	757	3440	1.12
		16	57.7	745	810	3467	1.11
		32	57.7	706	1038	3468	1.22
		64	57.7	686	1400	4754	1.47
	64	4	102.6	1944	534	3114	0.84
		8	102.6	1343	547	3107	0.84
		16	102.6	1037	610	2998	0.88
		32	102.6	880	876	3097	1.06
		64	102.6	800	1289	4754	1.31

We now consider what happens when the same (fixed) value of c is used in every iteration. We considered values of c from 0.5 to 2.5 and tested our four main graphs (USA and Europe with travel times and travel distances). For each case, Table 8 shows the preprocessing time, the number of shortcuts added (as a percentage of the original number of arcs), and the corresponding query results (aggregated over 1000 random pairs of vertices).

As anticipated, larger values of c usually result in more shortcuts being added, which speeds up both preprocessing (because the graph shrinks faster) and queries (because shortcuts reduce the reaches of eliminated arcs). The differences are most noticeable when c increases from 0.5 to 1.0; the results for $c = 2.0$ and $c = 2.5$, in contrast, are almost indistinguishable, and very close to those obtained with $c = 1.5$.

Recall that our original adaptive scheme uses $c = 0.5$ for the first iteration, 1.0 for the second, and 1.5 for the remaining ones. Comparing the results in Tables 2 and 3 with those in Table 8, we note that queries in the adaptive setting are very similar with those using $c = 1.5$ throughout the algorithm, and preprocessing is slightly slower. The main advantage of the adaptive scheme is that it adds fewer shortcuts. For USA, it adds 65.4% with travel times and 72.9% with travel distances; for Europe, the corresponding numbers are 62.2% and 74.9%. Note that

TABLE 8. Performance of RE for different shortcutting schemes. Only vertices with expansion at most c can be bypassed in each iteration.

GRAPH	METRIC	c	PR. TIME (min)	SHORTCUTS ADDED (%)	QUERY AVG SC.	QUERY MAX SC.	QUERY TIME (ms)
USA	TIMES	0.5	43.6	43.4	3893	7712	2.53
		1.0	29.5	64.7	2562	5189	1.84
		1.5	27.2	74.9	2427	4863	1.75
		2.0	26.8	77.9	2377	4837	1.77
		2.5	27.1	77.9	2375	4846	1.75
	DISTANCES	0.5	201.2	47.1	22201	42304	15.14
		1.0	71.1	72.9	10564	21812	7.73
		1.5	44.1	86.0	7265	13781	5.75
		2.0	39.4	90.2	6486	12252	5.33
		2.5	39.5	90.2	6492	12255	5.42
Europe	TIMES	0.5	155.7	35.0	10725	22437	6.00
		1.0	53.2	60.3	4924	9686	3.19
		1.5	44.8	68.6	4374	8484	3.05
		2.0	43.0	71.2	4253	8323	3.03
		2.5	43.1	71.3	4250	8323	3.03
	DISTANCES	0.5	256.5	41.5	31872	58167	18.77
		1.0	44.6	72.5	9922	18247	6.84
		1.5	30.8	82.8	7218	12983	5.26
		2.0	28.1	86.4	6521	11748	4.91
		2.5	27.6	86.5	6499	11682	4.94

these are close to the results obtained with $c = 1.0$ (fixed), but with better query performance and preprocessing time.

6.1.6. *Exact reaches and the refinement step.* Recall that the refinement step of the preprocessing algorithm recomputes δ highest reaches with an exact algorithm. In the previous experiments, we set $\delta = 2\lceil\sqrt{n}\rceil$. For the USA graph, in particular, we set δ to 9788. Table 9 shows the trade-off between preprocessing and query times as δ varies for the USA graph (with both travel times and travel distances as length functions). We tested δ values ranging from 0 to $10\lceil\sqrt{n}\rceil$; in each case, 100 000 random queries were run.

The table shows that increasing δ does improve queries, but significantly slows down the preprocessing routine. Using $\delta = 10\lceil\sqrt{n}\rceil$, queries are up to 20% faster than when no refinement step is used, but preprocessing is more than twice as slow. Recomputing all reaches would be ideal to speed up queries, but preprocessing would be prohibitively expensive. Even on Bay Area, a subgraph of USA with only 321 270 vertices, exact reach computation takes almost 2.5 hours with our new exact algorithm (the standard one takes more than 10 hours). But queries do become 40% faster with exact reaches. How much time, if any, to spend in the refinement phase of the preprocessing algorithm depends on the requirements of the application. We chose $\delta = 2\lceil\sqrt{n}\rceil$ for most of our experiments because its effect on the preprocessing time is negligible.

6.1.7. *Retrieving the shortest path.* The query times reported so far for RE and REAL consider only the task of finding the shortest path in the graph with shortcuts. Although this path has the same length as the corresponding path in the original

TABLE 9. Performance of RE on the USA graph as a function of δ (the number of vertices whose reaches are recomputed).

METRIC	δ	PR. TIME (min)	QUERY AVG SC.	MAX SC.	TIME (ms)
TIMES	0	25.1	2505	6176	1.84
	4894	26.3	2480	6171	1.83
	9788	27.7	2376	6048	1.77
	14682	29.7	2337	6000	1.75
	19576	32.5	2318	5978	1.72
	24470	36.0	2305	5952	1.70
	29364	39.8	2291	5941	1.71
	34258	44.0	2274	5911	1.68
	39152	49.4	2256	5880	1.70
	44046	55.2	2231	5847	1.66
	48940	60.3	2189	5767	1.63
DISTANCES	0	42.7	7394	15549	5.86
	4894	44.0	7391	15549	5.85
	9788	46.3	7260	15477	5.73
	14682	49.5	7213	15456	5.69
	19576	53.0	7166	15404	5.69
	24470	58.9	7087	15342	5.63
	29364	65.8	6945	15166	5.51
	34258	71.1	6638	14613	5.27
	39152	78.6	6501	14358	5.14
	44046	87.5	6422	14175	5.07
	48940	97.9	6376	14087	5.04

graph, it has much fewer arcs. On USA with travel times, for example, the shortest path between a random pair of vertices has more than 4500 vertices in the original graph, but fewer than 30 in the graph with shortcuts. With travel distances, these values increase to about 5000 and 43, respectively.

Our algorithm can retrieve the original path from the path with shortcuts in time proportional to the number of arcs on the original path. For this, it uses an *arc map*, which maps each shortcut arc to the two arcs it replaces. The arc map is built during the preprocessing step and has roughly the same size as the graph with shortcuts, but it is *not* included in the "disk space" column of our tables, since not all applications require it.

Since paths in the original graph have many more arcs than the paths found by RE or REAL, retrieving the original path can be relatively costly. To measure this, we reran the RE algorithm. After each query, we dumped the list of arc identifiers to an array, and at the same time computed the sum of the costs of these arcs. Even though we already know what the sum will be, this procedure is a good approximation of what an actual application might do. Retrieving the original list of arcs on the USA with travel times takes roughly one millisecond, which is comparable to the time REAL takes to actually determine the shortest path distance with travel times. With travel distances, REAL is comparatively slower, but retrieving the original path still takes about one millisecond.

On the Europe graph, the overhead of retrieving the shortest path is somewhat smaller, mainly because there are fewer arcs in the original paths: less than 1400 with travel times, and less than 3400 with travel distances. The paths with shortcuts have roughly 25 and 36 arcs, respectively. Retrieving the shortest paths takes 0.6 milliseconds with travel distances, and less than 0.2 milliseconds with travel times.

6.2. Grid Graphs. Computing driving directions was the main motivation for our work, and the experiments in the previous section show that we can solve this problem efficiently. In this section, we verify whether our algorithm is effective on other classes of graphs. It does run correctly on any input graph, since it does not use any property that is specific to road networks (such as being almost planar or having a known planar embedding). Its efficiency, however, may vary.

We tested our algorithm on grid graphs with arc lengths picked uniformly at random. Unlike road networks, these graphs do not have an obvious highway hierarchy. We used square 2-dimensional and cube-shaped 3-dimensional grids in our experiments. The 2-dimensional grids were generated using the `spgrid` generator, available at the 9th DIMACS Implementation Challenge download page. The 3-dimensional grids were generated by our own generator. Both families are directed (not symmetric), with a vertex connected to its neighbors in the grid by arcs of length chosen uniformly at random from the range $[1, n]$, where n is the number of vertices. We generated five grids of each size, and report the average results obtained; the only exception is the maximum number of vertices scanned, which is taken over all five instances.

TABLE 10. Data for 2-dimensional grids.

VERTICES	METHOD	PR. TIME (min)	DISK SPACE (MB)	QUERY AVG SC.	QUERY MAX SC.	QUERY TIME (ms)
65536	ALT	0.04	11.5	851	6563	1.19
	RE	1.36	3.4	2192	3666	1.59
	REAL-(16,1)	1.40	12.7	222	1013	0.34
	REAL-(64,16)	1.57	5.9	1987	3135	1.77
	BD	—	2.2	21358	51819	16.69
	D	—	2.2	33752	—	14.83
131044	ALT	0.10	23.8	1404	11535	2.56
	RE	3.08	6.8	3058	5072	2.67
	REAL-(16,1)	3.17	26.2	288	1632	0.52
	REAL-(64,16)	3.49	12.1	2359	3457	2.59
	BD	—	4.5	41682	103770	40.48
	D	—	4.5	66865	—	30.72
262144	ALT	0.21	48.3	2439	27936	4.36
	RE	6.82	13.8	4466	7210	4.09
	REAL-(16,1)	7.03	53.1	366	1969	0.67
	REAL-(64,16)	7.67	24.5	2666	3398	3.36
	BD	—	9.0	85587	205668	78.80
	D	—	9.0	134492	—	63.58
524176	ALT	0.26	96.6	6057	65664	6.28
	RE	7.77	27.7	6458	10049	4.75
	REAL-(16,1)	8.03	106.3	558	3189	0.89
	REAL-(64,16)	9.14	49.2	2823	3711	2.67
	BD	—	18.0	174150	416925	160.14
	D	—	18.0	275589	—	112.80

Table 10 shows computational results for 2-dimensional grids. The table includes results for ALT, RE, two versions of REAL, D (the reference implementation of Dijkstra's algorithm made available by the 9th DIMACS Implementation Challenge) and BD (our own implementation of the bidirectional version of Dijkstra's algorithm). Note that BD scans fewer vertices than D, but has higher overhead per scan, due to the fact that our implementation of BD is compatible with our more elaborate algorithms, whereas D is more restricted. We did not attempt to tune the algorithms for these graphs, except by setting the parameter c (defined in Section 4.3) to 1.0 for all iterations of the reach computation algorithm (the use of increasing values of c, starting at 0.5, is tuned for road networks). It is possible that additional tuning may help, especially if it takes into account the structure of these graphs.

Compared to BD, both RE and REAL are significantly faster. Queries and preprocessing times, however, are clearly not as good as those for road networks of similar size. But the fact that there is an improvement at all is noteworthy, given the absense of a clear hierarchy on grids with random edge weights.

For the graph sizes we consider, RE and ALT have comparable query times, but RE has better asymptotic performance. Preprocessing, however, is much more expensive for RE. REAL-(16,1) benefits from the performance improvements of both RE and ALT and therefore has much better query times (although its preprocessing times are also high, due to the reach computation).

Unlike on large road networks, REAL-(64,16) queries pay a significant performance penalty compared to REAL-(16,1). This is probably because the graphs are relatively small and well-structured, which makes increasing the number of landmarks not as advantageous as for large road networks. The fact that the relative performance gap between the two algorithms narrows as the graph size increases supports this conjecture. For the smallest grid in our test, REAL-(64,16) is more than five times slower than REAL-(16,1) and even slower than RE. For the largest grid, however, it is faster than RE and only about three times as slow as REAL-(16,1). REAL-(64,16) uses significantly less space than REAL-(16,1), however.

The experiments on 2-dimensional grids show that reaches help even when a graph does not have an obvious highway hierarchy, and that the applicability of REAL is not restricted to road networks. On the largest grid, it is over seven times faster than ALT, and two orders of magnitude faster than the bidirectional Dijkstra algorithm.

Next we discuss the results of running our algorithms on 3-dimensional grids, shown in Table 11. Since these graphs have higher average degree, we used $c = 2.0$ when generating shortcuts. The table shows that both reach-based and landmark-based algorithms are less effective than on 2-dimensional grids. ALT queries are only modestly slower, however, and ALT preprocessing time is not affected much. In contrast, RE preprocessing becomes asymptotically slower—the time roughly triples when the graph size doubles. With this rate of growth, it would take about two months to preprocess a grid comparable in size to the European road network.

Consider queries of RE and ALT. The average number of scans for RE is about seven times greater than for ALT. The running time is about five times greater, except for the largest problem, where it is greater by a factor of slightly over three. The average number of scans suggests that REAL-(16,1) has a small asymptotic

TABLE 11. Data for 3-dimensional grids.

VERTICES	METHOD	PR. TIME (min)	DISK SPACE (MB)	QUERY AVG SC.	QUERY MAX SC.	QUERY TIME (ms)
32768	ALT	0.02	4.5	472	3968	0.52
	RE	4.11	2.5	3141	7017	2.67
	REAL-(16,1)	4.13	5.5	349	1866	0.62
	REAL-(64,16)	4.20	3.4	3753	7729	3.42
	BD	—	1.6	5840	18588	4.10
	D	—	1.6	16747	—	6.90
64000	ALT	0.04	10.1	707	7560	0.91
	RE	12.23	4.9	4965	11711	4.89
	REAL-(16,1)	12.27	12.2	489	3380	0.95
	REAL-(64,16)	12.42	7.0	5872	13081	6.30
	BD	—	3.1	11338	36732	8.63
	D	—	3.1	31759	—	13.54
132651	ALT	0.09	23.5	1200	11282	1.84
	RE	38.18	10.3	8774	20467	10.05
	REAL-(16,1)	38.27	27.6	730	4584	1.80
	REAL-(64,16)	38.64	15.2	10159	21972	12.89
	BD	—	6.5	23738	79600	23.64
	D	—	6.5	66045	—	32.98
262144	ALT	0.22	49.6	2216	18157	5.31
	RE	113.15	20.3	14849	31819	18.84
	REAL-(16,1)	113.37	57.5	1159	6524	3.39
	REAL-(64,16)	114.17	30.7	16576	33169	24.78
	BD	—	12.8	48699	161036	52.31
	D	—	12.8	133552	—	89.65

advantage over ALT and RE. It is slightly slower than ALT on the smallest problem and a little faster on the largest one.

REAL-(64,16) is the slowest code and on average scans more vertices than RE. This might be explained by the very low quality of the lower bounds provided by sparse landmarks on graphs of small diameter.

These results show that reach-based methods are barely useful for 3-dimensional grids. Preprocessing does make queries faster for large graphs, but it is very expensive. Whether this is a basic limitation of the method, or a limitation of our preprocessing algorithm, is an interesting open question.

7. Concluding Remarks

The experimental analysis has shown that our algorithms are definitely practical for computing driving directions on large road networks. On the USA and European road networks, the average shortest path can be found while scanning less than 1000 vertices, which takes about one millisecond on a standard server. By applying techniques similar to those reported by Goldberg and Werneck [**23**], we can make our algorithm work from external memory, maintaining in RAM only information about vertices actually visited during the search. Since there are so few of those, a PocketPC implementation of RE (with the preprocessed data stored on a flash card) is practical enough for everyday use. Random queries take less than

10 seconds on a PocketPC with a 400 MHz ARM processor and 128 MB of RAM, and local queries (which should be much more common) are significantly faster.

An interesting direction of future research is to make the algorithms effective on a wider range of graphs. Although our methods work reasonably well on 2-dimensional grids, we have shown that they have only limited applicability to 3-dimensional grids, for instance.

Many other open problems remain. The performance of our best query algorithms depend crucially on the quality of the reaches available. Faster algorithms to compute exact reaches (or at least better upper bounds on reaches) would speed up our queries. More importantly, the problem of finding good shortcuts deserves a more detailed study. Our current algorithm uses a series of heuristics to determine when to add shortcuts, but there is no reason to believe that they find the best possible shortcuts. It is also desirable to get theoretical justification for the good practical performance of our algorithms.

Finally, it would be interesting to find other applications for the concepts and techniques we use. For example, reach information may also be useful for highway design: high-reach local roads are natural candidates for becoming highways with increased speed limits.

Recent work. A preliminary version of this paper [**20**] was presented at the 9th DIMACS Implementation Challenge [**8**]. Several other papers presented also dealt with the point-to-point shortest path problem. Lauther [**34**] and Köhler et al. [**29**] presented algorithms based on arc flags, but their running times (for both preprocessing and queries) are dominated by REAL and HH. Delling, Sanders, et al. [**7**] presented a variant of the partial landmarks algorithm in the context of highway hierarchies, but with only modest speedups (for technical reasons A* search cannot be combined naturally with HH). In a different paper, Delling, Holzer, et al. [**6**] showed how multi-level graphs could be used to perform random queries in less than 1 ms, but only after weeks of preprocessing time.

The fastest queries were achieved by algorithms based on *transit node routing*, a concept introduced by Bast et al. [**1**] and combined with highway hierarchies by Sanders and Schultes [**42**] (these papers were later merged [**2**]). The intuition is simple: when following the shortest path from a fixed source s to any point "far away," the path leaves the source's "local area" via one of a very limited number of *access nodes*. The set of all access nodes, considering all possible sources, are the transit nodes of the graph. On USA and Europe with travel times, there are only a few thousand of those. Preprocessing computes the access nodes and the distances between them, and most queries consist of a few table lookups. Both graphs can be processed in about three hours, and random queries take in $6\,\mu$s on average (local queries are slightly slower, at $20\,\mu$s). If preprocessing time is limited to about an hour, average query times are still only $11\,\mu$s (0.3 ms for local queries). The effectiveness of transit nodes depends strongly on the natural hierarchy of the underlying road network. If travel times are replaced by travel distances, preprocessing time increases to about eight hours, and average query times are close to 0.1 ms. Performance would probably be significantly worse on other graph classes, such as grids.

Even for travel distances, queries with transit-node routing are significantly faster than with REAL. Our method does appear to be more robust, however, to changes in the length function. Moreover, these approaches are not mutually exclusive. As Bast et al. observe in [**2**], a reach-based approach could be used instead of highway hierarchies to compute a suitable set of transit nodes and the corresponding distance tables. An actual implementation of the combined algorithm is an interesting topic for future research. It should be noted, however, that speeding up our algorithm would only make sense if a full description of the shortest path is not required; traversing the shortest path already costs about as much as finding its length.

Acknowledgements

We would like to thank Daniel Delling, Peter Sanders, and Dominik Schultes for many useful discussions. We also thank an anonymous referee for comments that helped improve the presentation of our paper.

References

[1] H. Bast, S. Funke, and D. Matijevic. TRANSIT: Ultrafast shortest-path queries with linear-time preprocessing. Presented at the 9th DIMACS Implementation Challenge, available at `http://www.dis.uniroma1.it/~challenge9/`. A revised version appears in this book, 2006.

[2] H. Bast, S. Funke, D. Matijevic, P. Sanders, and D. Schultes. In transit to constant time shortest-path queries in road networks. In *Proc. 9th International Workshop on Algorithm Engineering and Experiments*, pages 46–59. SIAM, 2006. Available at http://www.mpi-inf.mpg.de/ bast/tmp/transit.pdf.

[3] B. V. Cherkassky, A. V. Goldberg, and T. Radzik. Shortest Paths Algorithms: Theory and Experimental Evaluation. *Math. Prog.*, 73:129–174, 1996.

[4] L. J. Cowen and C. G. Wagner. Compact Roundtrip Routing in Directed Networks. In *Proc. Symp. on Principles of Distributed Computation*, pages 51–59, 2000.

[5] G. B. Dantzig. *Linear Programming and Extensions*. Princeton Univ. Press, Princeton, NJ, 1962.

[6] D. Delling, M. Holzer, K. Müller, F. Schulz, and D. Wagner. High-performance multi-level graphs. Presented at the 9th DIMACS Implementation Challenge, available at `http://www.dis.uniroma1.it/~challenge9/`. A revised version appears in this book, 2006.

[7] D. Delling, P. Sanders, D. Schultes, and D. Wagner. Highway hierarchies star. Presented at the 9th DIMACS Implementation Challenge, available at `http://www.dis.uniroma1.it/~challenge9/`. A revised version appears in this book, 2006.

[8] C. Demetrescu, A. V. Goldberg, and D. S. Johnson. 9th DIMACS Implementation Challenge: Shortest Paths. http://www.dis.uniroma1.it/∼challenge9/, 2006.

[9] C. Demetrescu and P. Italiano. A New Approach to Dynamic All Pairs Shortest Paths. *J. Assoc. Comput. Mach.*, 51:968–992, 2004.

[10] E. V. Denardo and B. L. Fox. Shortest-Route Methods: 1. Reaching, Pruning, and Buckets. *Oper. Res.*, 27:161–186, 1979.

[11] E. W. Dijkstra. A Note on Two Problems in Connexion with Graphs. *Numer. Math.*, 1:269–271, 1959.

[12] J. Doran. An Approach to Automatic Problem-Solving. *Machine Intelligence*, 1:105–127, 1967.

[13] D. Dreyfus. An Appraisal of Some Shortest Path Algorithms. Technical Report RM-5433, Rand Corporation, Santa Monica, CA, 1967.

[14] J. Fakcharoenphol and S. Rao. Planar graphs, negative weight edges, shortest paths, and near linear time. In *Proc. 42nd IEEE Annual Symposium on Foundations of Computer Science*, pages 232–241, 2001.

[15] M. L. Fredman and R. E. Tarjan. Fibonacci Heaps and Their Uses in Improved Network Optimization Algorithms. *J. Assoc. Comput. Mach.*, 34:596–615, 1987.

[16] G. Gallo and S. Pallottino. Shortest Paths Algorithms. *Annals of Oper. Res.*, 13:3–79, 1988.

[17] A. V. Goldberg. A Simple Shortest Path Algorithm with Linear Average Time. In *Proc. 9th Annual European Symposium Algorithms*, pages 230–241. Springer-Verlag, 2001.

[18] A. V. Goldberg. Shortest Path Algorithms: Engineering Aspects. In *Proc. 12th International Symposium on Algorithms and Computation, Lecture Notes in Computer Science*, pages 502–513. Springer-Verlag, 2001.

[19] A. V. Goldberg and C. Harrelson. Computing the Shortest Path: A* Search Meets Graph Theory. In *Proc. 16th ACM-SIAM Symposium on Discrete Algorithms*, pages 156–165, 2005.

[20] A. V. Goldberg, H. Kaplan, and R. F. Werneck. Better landmarks within reach. Presented at the 9th DIMACS Implementation Challenge, available at `http://www.dis.uniroma1.it/~challenge9/`. A revised version appears in this book, 2006.

[21] A. V. Goldberg, H. Kaplan, and R. F. Werneck. Reach for A*: Efficient Point-to-Point Shortest Path Algorithms. In *Proc. 8th International Workshop on Algorithm Engineering and Experiments*, pages 38–51. SIAM, 2006.

[22] A. V. Goldberg and C. Silverstein. Implementations of Dijkstra's Algorithm Based on Multi-Level Buckets. In P. M. Pardalos, D. W. Hearn, and W. W. Hages, editors, *Lecture Notes in Economics and Mathematical Systems 450 (Refereed Proceedings)*, pages 292–327. Springer Verlag, 1997.

[23] A. V. Goldberg and R. F. Werneck. Computing Point-to-Point Shortest Paths from External Memory. In *Proc. 7th International Workshop on Algorithm Engineering and Experiments*, pages 26–40. SIAM, 2005.

[24] R. Gutman. Reach-based Routing: A New Approach to Shortest Path Algorithms Optimized for Road Networks. In *Proc. 6th International Workshop on Algorithm Engineering and Experiments*, pages 100–111, 2004.

[25] P. E. Hart, N. J. Nilsson, and B. Raphael. A Formal Basis for the Heuristic Determination of Minimum Cost Paths. *IEEE Transactions on System Science and Cybernetics*, SSC-4(2):100–107, 1968.

[26] T. Ikeda, Min-Yao Hsu, H. Imai, S. Nishimura, H. Shimoura, T. Hashimoto, K. Tenmoku, and K. Mitoh. A Fast Algorithm for Finding Better Routes by AI Search Techniques. In *Proc. Vehicle Navigation and Information Systems Conference*, pages 2037–2044. IEEE, 1994.

[27] R. Jacob, M.V. Marathe, and K. Nagel. A Computational Study of Routing Algorithms for Realistic Transportation Networks. *Oper. Res.*, 10:476–499, 1962.

[28] P. Klein. Preprocessing an Undirected Planar Network to Enable Fast Approximate Distance Queries. In *Proc. 13th ACM-SIAM Symposium on Discrete Algorithms*, pages 820–827, 2002.

[29] E. Köhler, R. H. Möhring, and H. Schilling. Fast point-to-point shortest path computations with arc-flags. Presented at the 9th DIMACS Implementation Challenge, available at `http://www.dis.uniroma1.it/~challenge9/`. A revised version appears in this book, 2006.

[30] E. Köhler, R.H. Möhring, and H. Schilling. Acceleration of shortest path and constrained shortest path computation. In *Proc. 4th International Workshop on Efficient and Experimental Algorithms*, pages 126–138, 2005.

[31] Jr. L. R. Ford. Network Flow Theory. Technical Report P-932, The Rand Corporation, 1956.

[32] Jr. L. R. Ford and D. R. Fulkerson. *Flows in Networks*. Princeton Univ. Press, Princeton, NJ, 1962.

[33] U. Lauther. An Extremely Fast, Exact Algorithm for Finding Shortest Paths in Static Networks with Geographical Background. In *IfGIprints 22, Institut fuer Geoinformatik, Universitaet Muenster (ISBN 3-936616-22-1)*, pages 219–230, 2004.

[34] U. Lauther. An experimental evaluation of point-to-point shortest path calculation on road-networks with precalculated edge-flags. Presented at the 9th DIMACS Implementation Challenge, available at `http://www.dis.uniroma1.it/~challenge9/`. A revised version appears in this book, 2006.

[35] PTV Traffic Mobility Logistic. Western europe road network. http://www.ptv.de/, 2006.

[36] U. Meyer. Single-Source Shortest Paths on Arbitrary Directed Graphs in Linear Average Time. In *Proc. 12th ACM-SIAM Symposium on Discrete Algorithms*, pages 797–806, 2001.

[37] R.H. Möhring, H. Schilling, B. Schütz, D. Wagner, and T. Willhalm. Partitioning graphs to speed up dijkstra's algorithm. In *Proc. 4th International Workshop on Efficient and Experimental Algorithms*, pages 189–202, 2005.

[38] T. A. J. Nicholson. Finding the Shortest Route Between Two Points in a Network. *Computer J.*, 9:275–280, 1966.

[39] I. Pohl. Bi-directional Search. In *Machine Intelligence*, volume 6, pages 124–140. Edinburgh Univ. Press, Edinburgh, 1971.

[40] P. Sanders and D. Schultes. Highway Hierarchies Hasten Exact Shortest Path Queries. In *Proc. 13th Annual European Symposium Algorithms*, pages 568–579, 2005.
[41] P. Sanders and D. Schultes. Engineering Highway Hierarchies. In *Proc. 14th Annual European Symposium Algorithms*, pages 804–816, 2006.
[42] P. Sanders and D. Schultes. Robust, almost constant time shortest-path queries on road networks. Presented at the 9th DIMACS Implementation Challenge, available at `http://www.dis.uniroma1.it/~challenge9/`. A revised version appears in this book, 2006.
[43] D. Schultes. Fast and Exact Shortest Path Queries Using Highway Hierarchies. Master's thesis, Department of Computer Science, Universität des Saarlandes, Germany, 2005.
[44] F. Schulz, D. Wagner, and K. Weihe. Using Multi-Level Graphs for Timetable Information. In *Proc. 4th International Workshop on Algorithm Engineering and Experiments*, pages 43–59. LNCS, Springer, 2002.
[45] R. Sedgewick and J.S. Vitter. Shortest Paths in Euclidean Graphs. *Algorithmica*, 1:31–48, 1986.
[46] R. E. Tarjan. *Data Structures and Network Algorithms.* Society for Industrial and Applied Mathematics, Philadelphia, PA, 1983.
[47] M. Thorup. Undirected Single-Source Shortest Paths with Positive Integer Weights in Linear Time. *J. Assoc. Comput. Mach.*, 46:362–394, 1999.
[48] M. Thorup. Compact Oracles for Reachability and Approximate Distances in Planar Digraphs. In *Proc. 42nd IEEE Annual Symposium on Foundations of Computer Science*, pages 242–251, 2001.
[49] DC US Census Bureau, Washington. UA Census 2000 TIGER/Line files. `http://www.census.gov/geo/www/tiger/tigerua/ua.tgr2k.html`, 2002.
[50] D. Wagner and T. Willhalm. Geometric Speed-Up Techniques for Finding Shortest Paths in Large Sparse Graphs. In *Proc. 11th Annual European Symposium Algorithms*, pages 776–787, 2003.
[51] F. B. Zhan and C. E. Noon. Shortest Path Algorithms: An Evaluation using Real Road Networks. *Transp. Sci.*, 32:65–73, 1998.
[52] F. B. Zhan and C. E. Noon. A Comparison Between Label-Setting and Label-Correcting Algorithms for Computing One-to-One Shortest Paths. *Journal of Geographic Information and Decision Analysis*, 4:1–11, 2000.

MICROSOFT RESEARCH SILICON VALLEY, 1065 LA AVENIDA, MOUNTAIN VIEW, CA 94043, USA.
E-mail address: `goldberg@microsoft.com`

SCHOOL OF MATHEMATICAL SCIENCES, TEL AVIV UNIVERSITY, ISRAEL. PART OF THIS WORK WAS DONE WHILE THE AUTHOR WAS VISITING MICROSOFT RESEARCH SILICON VALLEY.
E-mail address: `haimk@post.tau.ac.il`

MICROSOFT RESEARCH SILICON VALLEY, 1065 LA AVENIDA, MOUNTAIN VIEW, CA 94043, USA.
E-mail address: `renatow@microsoft.com`

DIMACS Series in Discrete Mathematics
and Theoretical Computer Science
Volume **74**, 2009

Highway Hierarchies Star

Daniel Delling, Peter Sanders, Dominik Schultes, and Dorothea Wagner

ABSTRACT. We study two speedup techniques for route planning in road networks: highway hierarchies (HH) and goal directed search using landmarks (ALT). It turns out that there are several interesting synergies. Highway hierarchies yield a way to implement landmark selection more efficiently and to store landmark information more space efficiently than before. ALT gives queries in highway hierarchies an excellent sense of direction and allows some pruning of the search space. For computing shortest *distances* and *approximately* shortest travel times, this combination yields significant speedups (between a factor of 2.5 and 5) over HH alone, while for exact queries using the travel time metric only minor improvements are achieved. We also explain how to compute actual shortest paths very efficiently.

1. Introduction

Computing fastest routes in a road networks $G = (V, E)$ from a source s to a target t is one of the showpieces of real-world applications of algorithmics. In principle, we could use DIJKSTRA's algorithm [**4**]. But for large road networks this would be far too slow. Therefore, there is considerable interest in speedup techniques for route planning.

A classical technique that gives a speedup of around two for road networks is *bidirectional search* which simultaneously searches forward from s and backwards from t until the search frontiers meet. Most speedup techniques use bidirectional search as an (optional) ingredient.

Another classical approach is goal direction via $\mathbf{A}^*$ *search* [**10**]: lower bounds define a vertex potential that directs the search towards the target. This approach was recently shown to be very effective if lower bounds are computed using precomputed shortest path distances to a carefully selected set of about 20 **L***andmark* nodes [**6, 8**] using the **T**riangle inequality (*ALT*). Speedups up to a factor 30 over bidirectional DIJKSTRA can be observed.

A property of road networks worth exploiting is their inherent hierarchy. Commercial systems use information on road categories to speed up search. 'Sufficiently

---

2000 *Mathematics Subject Classification.* Primary 68R10; Secondary 90B20.

Key words and phrases. Shortest paths, graph, speed-up technique, hierarchy, goal-direction, preprocessing, routing, road network.

Partially supported by DFG grant SA 933/1-3. and by the Future and Emerging Technologies Unit of EC (IST priority – 6th FP), under contract no. FP6-021235-2 (project ARRIVAL).

far away' from source and target, only 'important' roads are used. This requires manual tuning of the data and a delicate tradeoff between computation speed and suboptimality of the computed routes. In a previous paper [**16**] we introduced the idea to *automatically* compute *highway hierarchies* that yield *optimal* routes *uncompromisingly quickly*. This was the first speedup technique that was able to preprocess the road network of a continent in realistic time and obtain large speedups (several thousands) over DIJKSTRA's algorithm. In [**17**] the basic method was considerably accelerated using many small measures and using *distance tables*: shortest path distances in the highest level of the hierarchy are precomputed. This way, it suffices to search locally around source and target node until the shortest path distance can be found by accessing the distance table.

A different hierarchy-based method—reach-based routing [**9**]—profits considerably from a combination with ALT [**5**]. The present state of affairs is that the combined method from [**5**] shows performance somewhat inferior to highway hierarchies with distance tables but without goal direction. Both methods turn out to be closely related. In particular, [**5**] uses methods originally developed for highway hierarchies to achieve fast preprocessing. Here, we explore the natural question how highway hierarchies can be combined with goal directed search in general and with ALT in particular.

1.1. Overview and Contributions. In the following sections we first review highway hierarchies in Section 2 (Algorithm HH) [**17**]. A new result presented there is a very fast algorithm for explicitly computing the shortest paths by precomputing unpacked versions of shortcut edges. Section 3 reviews Algorithm ALT [**6, 8**] and introduces refined algorithms for selecting landmarks. The main innovation there is restricting landmark selection to nodes on higher levels of the highway hierarchy.

The actual integration of highway hierarchies with ALT (Algorithm HH*) is introduced in Section 4. This is nontrivial in several respects. For example, we need incremental access to the distance tables for finding upper bounds and a different way to control the progress of forward and backward search. We also have to overcome the problem that search cannot be stopped when search frontiers meet. On the other hand, there are several simplifications compared to ALT. Abandoning the reliance on a stopping criterion allows us to use simpler, faster, and stronger lower bounds. Using distance tables obviates the need for dynamic landmark selection. Another interesting approach is to stop the search when a certain guaranteed solution quality has been obtained. There are several interesting further optimisations. In particular, we can be more space efficient than ALT by storing no landmark information on the lowest level of the hierarchy. We describe how the missing information can be reconstructed efficiently at query time. As a side effect, we introduce a way to limit the length of shortcuts. This measure turns out to be of independent interest since it also improves the basic HH algorithm. Note that Goldberg et al. [**7**] use similar techniques as we do in order to reduce the memory consumption of landmarks when combined with reach-based routing. They have already briefly mentioned this idea in [**5**].

Section 5 reports extensive experiments performed using road networks of Western Europe and the USA. Section 6 summarises the results and outlines possible future work.

1.2. More Related Work. There are several other approaches to goal directed search. Our first candidate for combination with highway hierarchies were *Precomputed Cluster Distances* [**13**]. PCDs allow the computation of upper and lower bounds based on precomputed distances between partitions of the road networks. These lower bounds cannot be used for A^* search since they can produce negative reduced edge weights so that DIJKSTRA's algorithm is no longer applicable. The search space can still be pruned by discontinuing search at node v if the lower bound from v to t indicates that the best upper bound seen so far cannot possibly be improved when passing through v. An advantage of PCDs over landmarks is that they need less space. We did not implement this however since PCDs are rather ineffective for search in the lower levels of the hierarchy and since our distance table optimisation from [**17**] is already very effective for pruning search at the higher levels of the hierarchy. In contrast, landmarks can be used together with A^* search and thus can direct the search towards the target already in the lower levels of the hierarchy.

An important family of speedup techniques [**20, 14, 12**] associates information with each edge e. This information specifies a superset of the nodes reached via e on some shortest path. *Geometric containers* [**20**] require node coordinates and store a simple geometrical object containing all the nodes reached via a shortest path. *Edge flags* partition the graph into regions. For each edge e and each region R one bit specifies whether there is a shortest path via e into region R [**14, 12**]. Both techniques alone already contain both direction information and hierarchy information so that very big speedups, comparable to highway hierarchies, can be achieved. However, so far these methods would have forbiddingly large preprocessing times for the largest available road networks. Therefore these approaches looked not so interesting for a first attempt to combine goal directed search with highway hierarchies.

2. Highway Hierarchies

The basic idea of the highway hierarchies approach is that outside some local areas around the source and the target node, only a subset of 'important' edges has to be considered in order to be able to find the shortest path. The concept of a *local area* is formalised by the definition of a neighbourhood node set[1] $\mathcal{N}(v)$ for each node v. Then, the definition of a *highway network* of a graph $G = (V, E)$ that has the property that all shortest paths are preserved is straightforward: an edge $(u, v) \in E$ belongs to the highway network iff there are nodes $s, t \in V$ such that the edge (u, v) appears in the canonical shortest path[2] $\langle s, \dots, u, v, \dots, t\rangle$ from s to t in G with the property that $v \notin \mathcal{N}(s)$ and $u \notin \mathcal{N}(t)$.

The size of a highway network (in terms of the number of nodes) can be considerably reduced by a contraction procedure: for each node v, we check a *bypassability criterion* that decides whether v should be *bypassed*—a operation that removes the node and creates shortcut edges (u, w) representing paths of the form $\langle u, v, w\rangle$. The

[1] In [**17**], we give more details on the definition of neighbourhoods. In particular, we distinguish between a forward and a backward neighbourhood. However, in this context, we would like to slightly simplify the notation and concentrate on the concepts that are important to understand the subsequent sections. The implementation, however, is based on [**17**] and not simplified.

[2] For each connected node pair (s, t), we select a unique *canonical shortest path* in such a way that each subpath of a canonical shortest path is canonical as well. For details, we refer to [**16**].

graph that is induced by the remaining nodes and enriched by the shortcut edges forms the *core* of the highway network. The bypassability criterion takes into account the degree of the node v and the number of shortcuts that would be created if v was bypassed: the net increase of the number of edges due to a bypass operation should be small. For details, we refer to [**17**].

A *highway hierarchy* of a graph G consists of several levels $G_0, G_1, G_2, \ldots, G_L$. Level 0 corresponds to the original graph G. Level 1 is obtained by computing the highway network of level 0, level 2 by computing the highway network of the core of level 1 and so on.

2.1. Highway Query. In [**16**], we show how the highway hierarchy of a given road network can be constructed efficiently. After that, we can use the *highway query algorithm* [**17**] to perform s-t queries. It is an adaptation of the bidirectional version of DIJKSTRA's algorithm. The search starts at s and t in level 0. When the neighbourhood of s or t is left, we switch to level 1 and continue the search. Similarly, we switch to the next level if the neighbourhood of the entrance point to the current level is left (Figure 1). When the core of some level has been entered, we never leave it again: in particular, we do not follow edges that lead to a bypassed node; instead, we use the shortcuts that have been created during the construction.

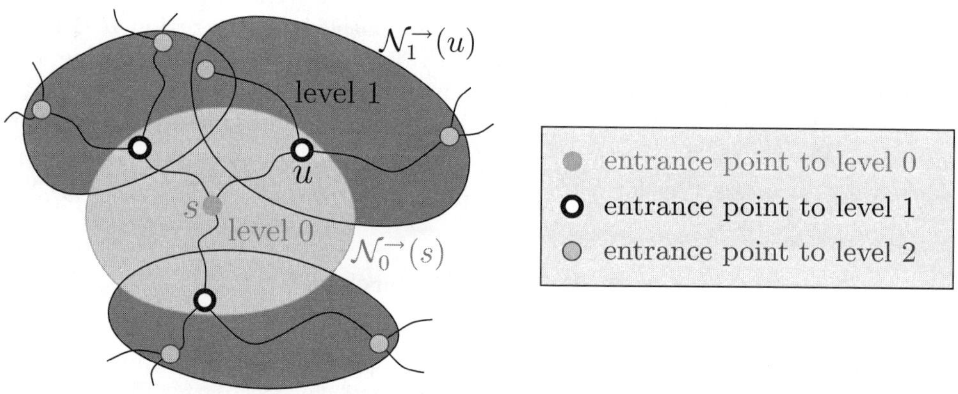

FIGURE 1. A schematic diagram of a highway query. Only the forward search started from the source node s is depicted.

At this point, we can observe two interesting properties of the highway query algorithm. First, it is *not* goal-directed. In fact, the forward search 'knows' nothing about the target and the backward search 'knows' nothing about the source, so that both search processes work completely independently and spread into all directions. Second, when both search scopes meet at some point, we cannot easily abort the search—in contrast to the bidirectional version of DIJKSTRA's algorithm, where we can abort immediately after a common node has been settled from both sides. The reason for this is illustrated in Figure 2. In the upper part of the figure, the bidirectional query from a node s to a node t along a path P is represented by a profile that shows the level transitions within the highway hierarchy. To get a sequential algorithm, at each iteration we have to decide whether a node from the forward or the backward queue is settled. We assume that a strategy is used that favours the smaller element. Thus, both search processes meet in the middle, at

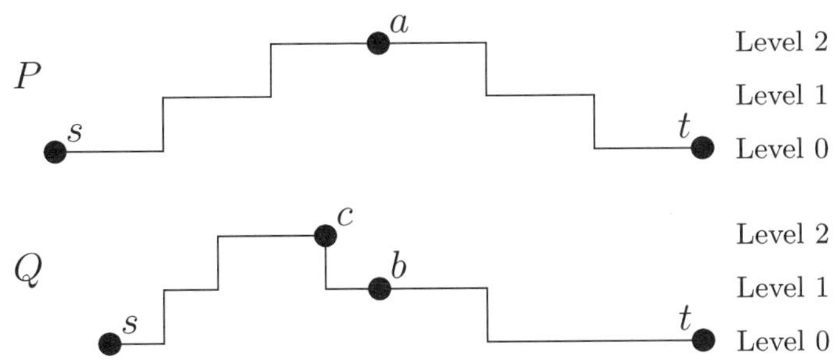

FIGURE 2. Schematic profile of a bidirectional highway query.

node a. When this happens, a path from s to t has been found. However, we have no guarantee that it is the shortest one. In fact, the lower part of the figure contains the profile of a shorter path Q from s to t, which is less symmetric than the profile of P. Note that the very flexible definition of the neighbourhoods allows such asymmetric profiles. When a on P is settled from both sides, b has been reached on Q by the backwards search, but *not* by the forward search since a search process never goes downwards in the hierarchy: therefore, at node c, the forward search is not continued on the path Q. We find the shorter path Q not until the backward search has reached c—which happens *after* P has been found. Hence, it would be wrong to abort the search when a has been settled.

In [**16**], we introduced some rather complicated abort criteria, which we dropped in [**17**] since they did reduce the search space, but the evaluation of the criteria was too expensive. Instead, we use a very simple criterion: the forward (backward) search is not continued if the key of the minimum element of the forward (backward) queue is larger then the current upper bound (i.e., the length of the tentative shortest path).

2.2. Using a Distance Table. The construction of fewer levels of the highway hierarchy and the usage of a complete *distance table* for the core of the topmost level can considerably accelerate the query: whenever the forward (backward) search enters the core of the topmost level at some node u, u is added to a node set $\overrightarrow{I}$ ($\overleftarrow{I}$) and the search is not continued from u. Since all distances between the nodes in the sets $\overrightarrow{I}$ and $\overleftarrow{I}$ have been precomputed and stored in a table, we can easily determine the shortest path length by considering all node pairs $(u,v), u \in \overrightarrow{I}, v \in \overleftarrow{I}$, and summing up $d(s,u) + d(u,v) + d(v,t)$. For details, we refer to [**17**].

Using the distance table can be seen as extreme case of goal-directed search: from the nodes in the set $\overrightarrow{I}$, we directly 'jump' to the nodes in the set $\overleftarrow{I}$, which are close to the target. Thus, we can say that the highway query with the distance table optimisation works in two phases: a strictly non-goal-directed phase till the sets $\overrightarrow{I}$ and $\overleftarrow{I}$ have been determined, followed by a 'goal-directed jump' using the distance table.

2.3. Complete Description of the Shortest Path. So far, we have dealt only with the computation of shortest path *distances*. In order to determine a complete description of the shortest path, we have to a) bridge the gap between the forward and backward topmost-core entrance points and b) expand the used shortcuts to obtain the corresponding subpaths in the original graph.

Problem a) can be solved using a simple algorithm: We start with the forward core entrance point u. As long as the backward entrance point v has not been reached, we consider all outgoing edges (u, w) in the topmost core and check whether $d(u, w) + d(w, v) = d(u, v)$; we pick an edge (u, w) that fulfils the equation, and we set $u := w$. The check can be performed using the distance table. It allows us to greedily determine the next hop that leads to the the backward entrance point.

Problem b) can be solved without using any extra data (Variant 1): for each shortcut (u, v), we perform a search from u to v in order to determine the path in the original graph; this search can be accelerated by using the knowledge that the first edge of the path enters a component C of bypassed nodes, the last edge leads to v, and all other edges are situated within the component C.

However, if a fast output routine is required, it is necessary to spend some additional space to accelerate the unpacking process. We use a rather sophisticated data structure to represent unpacking information for the shortcuts in a space-efficient way (Variant 2). In particular, we do not store a sequence of node IDs that describe a path that corresponds to a shortcut, but we store only *hop indices*: for each edge (u, v) on the path that should be represented, we store its rank within the ordered group of edges that leave u. Since in most cases the degree of a node is very small, these hop indices can be stored using only a few bits (in a fixed-length encoding). The unpacked shortcuts are stored in a recursive way, e.g., the description of a level-2 shortcut may contain several level-1 shortcuts. Accordingly, the unpacking procedure works recursively.

To obtain a further speed-up, we have a variant of the unpacking data structures (Variant 3) that caches the complete descriptions—without recursions—of all shortcuts that belong to the topmost level, i.e., for these important shortcuts that are frequently used, we do not have to use a recursive unpacking procedure, but we can just append the corresponding subpath to the resulting path.

3. A^* Search Using Landmarks

In this section we explain the known technique of A^* search [**10**] in combination with landmarks. We follow the implementation presented in [**8**]. In Section 3.2 we introduce a new landmark selection technique called *advancedAvoid*. Furthermore, we present how the selection of landmarks can be accelerated using highway hierarchies.

The search space of DIJKSTRA's algorithm can be visualised as a circle around the source. The idea of goal-directed or A^* search is to push the search towards the target. By adding a potential $\pi : V \to \mathbb{R}$ to the priority of each node, the order in which nodes are removed from the priority queue is altered. A 'good' potential lowers the priority of nodes that lie on a shortest path to the target. It is easy to see that A^* is equivalent to DIJKSTRA's algorithm on a graph with *reduced costs*, formally $w_\pi(u, v) = w(u, v) - \pi(u) + \pi(v)$. Since DIJKSTRA's algorithm works only on nonnegative edge costs, not all potentials are allowed. We call a potential π *feasible* if $w_\pi(u, v) \geq 0$ for all $(u, v) \in E$. The distance from each node v of G to the target t is the distance from v to t in the graph with reduced edge costs minus the potential of t plus the potential of v. So, if the potential $\pi(t)$ of the target t is zero, $\pi(v)$ provides a *lower bound* for the distance from v to the target t.

Bidirectional A^*. At first glance, combining A^* and bidirectional search seems easy. Simply use a feasible potential π_f for the forward and a feasible potential π_r for the backward search. However, such an approach does not work due to the fact that the searches might work on different reduced costs, so that the shortest path might not have been found when both searches meet. This can only be guaranteed if π_f and π_r are *consistent*, meaning $w_{\pi_f}(u, v)$ in G is equal to $w_{\pi_r}(v, u)$ in the reverse graph. We use the variant of an average potential function [**11**] defined as $p_f(v) = (\pi_f(v) - \pi_r(v))/2$ for the forward and $p_r(v) = (\pi_r(v) - \pi_f(v))/2 = -p_f(v)$ for the backward search. By adding $\pi_r(t)/2$ to the forward and $\pi_f(s)/2$ to the backward search, p_f and p_r provide lower bounds to the target and source, respectively. Note that these potentials are feasible and consistent but provide worse lower bounds than the original ones.

ALT.. There exist several techniques [**18, 21**] to obtain feasible potentials using the layout of a graph. The ALT algorithm uses a small number of nodes—so called *landmarks*—and the triangle inequality to compute feasible potentials. Given a set $S \subseteq V$ of landmarks and distances $d(L, v), d(v, L)$ for all nodes $v \in V$ and landmarks $L \in S$, the following triangle inequalities hold:

$$d(u, v) + d(v, L) \geq d(u, L) \quad \text{and} \quad d(L, u) + d(u, v) \geq d(L, v)$$

Therefore, $\underline{d}(u, v) := \max_{L \in S} \max\{d(u, L) - d(v, L), d(L, v) - d(L, u)\}$ provides a feasible lower bound for the distance $d(u, v)$. The quality of the lower bounds highly depends on the quality of the selected landmarks.

Our implementation uses the tuning techniques of *active landmarks*, *pruning* and the enhanced stopping criterion. We stop the search if the sum of minimum keys in the forward and the backward queue exceed $\mu + p_f(s)$, where μ represents the tentative shortest path length and is therefore an upper bound for the shortest path length from s to t. For each s-t query only two landmarks—one 'before' the source and one 'behind' the target—are initially used. At certain checkpoints we decide whether to add an additional landmark to the active set, with a maximal amount of six landmarks. Pruning means that before relaxing an arc (u, v) during the forward search we also check whether $d(s, u) + w(u, v) + \pi_f(v) < \mu$ holds. This technique may be applied to the backward search easily. Note that for pruning, the potential function need not be consistent.

3.1. Landmark Selection. A crucial point in the success of a high speedup when using ALT is the quality of landmarks. Since finding good landmarks is hard, several heuristics [**6, 8**] exist. We focus on the best known techniques; *avoid* and *maxCover*.

Avoid. This heuristic tries to identify regions of the graph that are not well covered by the current landmark set S. Therefore, a shortest-path tree T_r is grown from a random node r. The *weight* of each node v is the difference between $d(v, r)$ and the lower bound $\underline{d}(v, r)$ obtained by the given landmarks. The *size* of a node v is defined by the sum of its weight and the size of its children in T_r. If the subtree of T_r rooted at v contains a landmark, the size of v is set to zero. Starting from the node with maximum size, T_r is traversed following the child with highest size. The leaf obtained by this traversal is added to S. In this strategy, the first root is picked uniformly at random. The following roots are picked with a probability proportional to the square of the distance to its nearest landmark.

MaxCover [**8**]. The main disadvantage of avoid is the starting phase of the heuristic. The first root is picked at random and the following landmarks are highly dependent on the starting landmark. MaxCover improves on this by first choosing a candidate set of landmarks (using avoid) that is about four times larger than needed. The landmarks actually used are selected from the candidates using several attempts with a local search routine. Each attempt starts with a random initial selection.

3.2. New Selection Techniques. In the following we introduce a new heuristic called *advancedAvoid* to select landmarks. Furthermore, we use the highway hierarchies to speed up the selection of landmarks.

AdvancedAvoid. Another approach to remedy for the disadvantages of avoid is to exchange the first landmarks generated by the avoid heuristic. More precisely, we generate k avoid landmarks, then take the first k' landmarks from the set S and generate k' new landmarks using avoid again. The advantage of advancedAvoid compared to maxCover is the computation time. While maxCover takes about five times longer than avoid, the selection of 16 advancedAvoid ($k' = 6$) landmarks on the road network of Western Europe takes about 45% more time than pure avoid.

Core Landmarks. The computation of landmarks is expensive. Calculating maxCover landmarks on the European network takes about 75 minutes, while constructing the whole highway hierarchy can be done in about 15 minutes. A promising approach is to use the highway hierarchy to reduce the number of possible landmarks: The level-1 core of the European road network has six times fewer

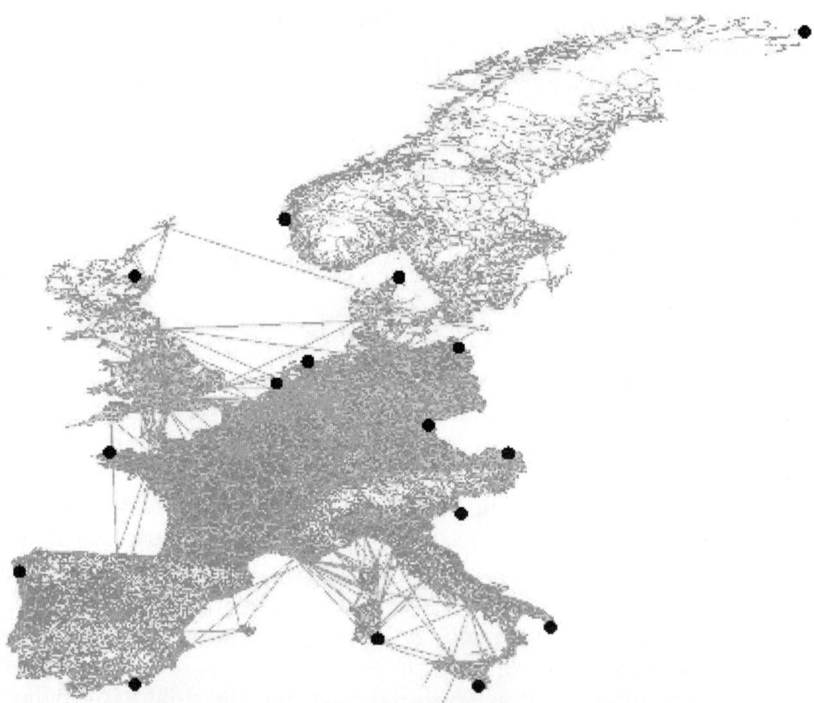

FIGURE 3. 16 advancedAvoid core 1 landmarks on the Western European road network

nodes than the original network and its construction takes only about three minutes. Using this level-1 core as possible positions for landmarks, the computation time for calculating landmarks (all heuristics) can be decreased. Note that using the nodes of higher level cores reduces the time for selecting landmarks even more. However, the core of a highway hierarchy shrinks towards the centre of the map and in [**6**], it has already been observed that good landmarks lie on the edge of a map (see Figure 3 for an example). Hence, using cores of a too high level would probably yield worse landmarks.

4. Combining Highway Hierarchies and A^* Search

Previously (see Section 2), we strictly separated the search phase to the topmost core from the access to the distance table: first, the sets of entrance points $\overrightarrow{I}$ and $\overleftarrow{I}$ into the core of the topmost level were determined, and afterwards the table look-ups were performed. Now we interweave both phases: whenever a forward topmost-core entrance point u is discovered, it is added to $\overrightarrow{I}$ and we immediately consider all pairs $(u, v), v \in \overleftarrow{I}$, in order to check whether the tentative shortest path length μ can be improved. (An analogous procedure applies to the discovery of a backward core entrance point.) This new approach is advantageous since we can use the tentative shortest path length μ as an upper bound on the actual shortest path length. In [**16, 17**], the highway query algorithm used a strategy that compares the minimum elements of both priority queues and prefers the smaller one in order to serialise forward and backward search. If we want to obtain good upper bounds very fast, this might not be the best choice. For example, if the source node belongs to a densely populated area and the target to a sparsely populated area, the distances from the source and target to the entrance points into the core of the topmost level will be very different. Therefore, we now choose a strategy that balances $|\overrightarrow{I}|$ and $|\overleftarrow{I}|$, preferring the direction that has encountered fewer entrance points. In case of equality (in particular, in the beginning when $|\overrightarrow{I}| = |\overleftarrow{I}| = 0$), we use a simple alternating strategy.

We enhance the highway query algorithm with goal-directed capabilities—obtaining an algorithm that we call *HH* search*—by replacing edge weights by *reduced costs* using potential functions π_f and π_r for forward and backward search. By this means, the search is directed towards the respective target, i.e., we are likely to find some s-t path very soon. However, just using the reduced costs only changes the *order* in which the nodes are settled, it does not reduce the search space. The ideal way to benefit from the early encounter of the forward and backward search would be to abort the search as soon as an s-t path has been found. And, as a matter of fact, in the case of the ALT algorithm [**6**]—even in combination with reach-based routing [**5**]—it can be shown that an immediate abort is possible without losing correctness if consistent potential functions are used (see Section 3). In contrast, this does not apply to the highway query algorithm since even in the non-goal-directed variant of the algorithm, we cannot abort when both search scopes have met (see Section 2).

Fortunately, there is another aspect of goal-directed search that can be exploited, namely *pruning*: finding any s-t path also means finding an upper bound μ on the length of the shortest s-t path. Comparing the lower bounds with the upper bound can be used to prune the search. In Section 3, the pruning of *edges* has already been mentioned. Alternatively, we can prune *nodes*: if the key of a settled node u is greater than the upper bound, we do not have to relax u's edges. Note that, using reduced costs, the key of u is the distance from the corresponding source to u plus the lower bound on the distance from u to the corresponding target.

Since we do not abort when both search scopes have met and because we have the distance table, a very simple implementation of the ALT algorithm is possible. First, we do not have to use consistent potential functions. Instead, we directly use the lower bound to the target as potential for the forward search and, analogously, the lower bound from the source as potential for the backward search. These potential functions make the searches approach their respective target faster than using consistent potential functions so that we get good upper bounds very early. In addition, the node pruning gets very effective: if one node is pruned, we can conclude that all nodes left in the same priority queue will be pruned as well since we use the same lower bound for pruning and for the potential that is part of the key in the priority queue. Hence, in this case, we can immediately stop the search in the corresponding direction.

Second, it is sufficient to select at the beginning of the query for each search direction only one landmark that yields the best lower bound. Since the search space is limited to a relatively small local area around source and target (due to the distance table optimisation), we do not have to pick more landmarks, in particular, we do not have to add additional landmarks in the course of the query, which would require flushing and rebuilding the priority queues. Thus, adding A^* search to the highway query algorithm (including the distance table optimisation) causes only little overhead per node.

However, there is a considerable drawback. While the goal-directed search (which gives good upper bounds) works very well, the pruning is not very successful when we want to compute *fastest* paths, i.e., when we use a travel time metric, because then the lower bounds are usually too weak. Figure 4 gives an example for this observation, which occurs quite frequently in practice. The first part of the shortest path from s to t is equal to the first part of the shortest path from s to the landmark u. Thus, the reduced costs of these edges are zero so that the forward search starts with traversing this common subpath. The backward search behaves in a similar way. Hence, we obtain a perfect upper bound very early; see Figure 4 (a). Still, the lower bound on $d(s,t)$ is quite bad: we have $d(s,u) - d(t,u) \leq d(s,t)$. Since staying on the motorway and going directly from s to u is much faster than leaving the motorway, driving through the countryside to t and continuing to u, the distance $d(s,t)$ is clearly underestimated.[3] The same applies to lower bounds on $d(v,t)$ for nodes v close to s. Hence, pruning the forward search does not work properly so that the search space still spreads into all directions before the process terminates; see Figure 4 (b). In contrast, the node s lies on the shortest path (in the reverse graph) from t to the landmark that is used by the backward search. (Since this landmark is very far away to the south, it has not been included in the figure.) Therefore, the lower bound is perfect so that the backward search stops immediately. However, this is a fortunate case that occurs rather rarely.

4.1. Approximate Queries. We pointed out above that in most cases we *find* a (near) shortest path very quickly, but it takes much longer until we *know*

[3]This negative effect is considerably weakened when a distance metric is used since the speed difference between the motorway and slower roads is not taken into account.

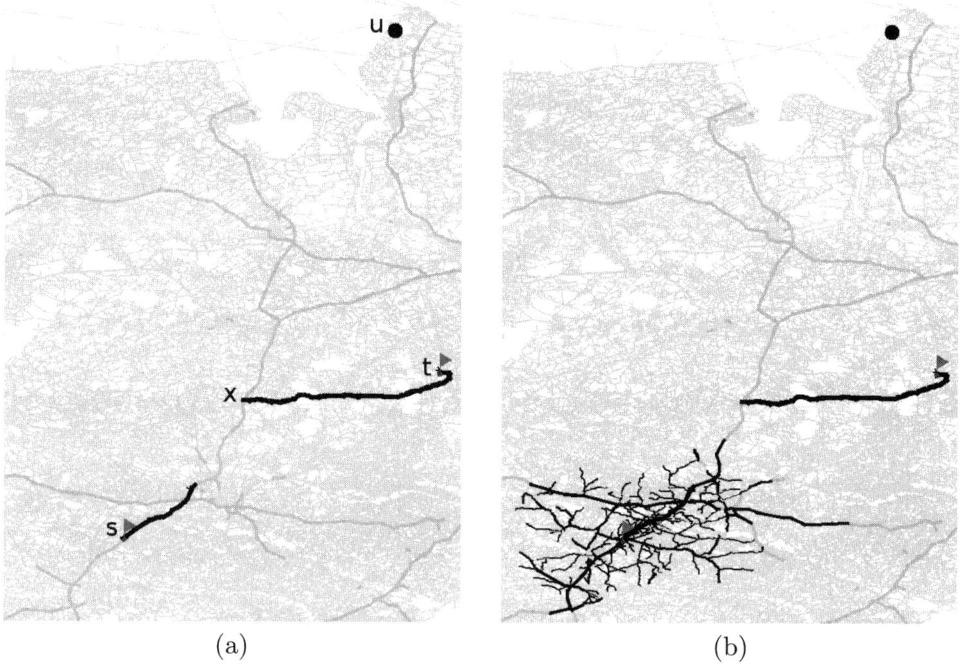

FIGURE 4. Two snapshots of the search space of an HH* search using a travel time metric. The landmark u of the forward search from s to t is explicitly marked. The landmark used by the backward search is somewhere below s and not included in the chosen clipping area. The search space is black, parts of the shortest path are represented by thick lines. In addition, motorways are represented by thick lines (dark grey). It is important to note that the shortest path from x to t is *not* a motorway, but a comparatively slow road.

that the shortest path has been found. We can adapt to this situation by defining an abort condition that leads to an approximate query algorithm: when a node u is removed from the forward priority queue and we have $(1+\varepsilon)\cdot(d(s,u)+\underline{d}(u,t)) > \mu$ (where $\varepsilon \geq 0$ is a given parameter), then the search is not continued in the forward direction. In this case, we may miss some s-t-paths whose length is $\geq d(s,u)+\underline{d}(u,t)$ since the key of any remaining element v in the priority queue is $\geq d(s,u)+\underline{d}(u,t)$ and it is a lower bound on the length of the shortest path from s via v to t. Thus, if the shortest path is among these paths, we have $d(s,t) \geq d(s,u)+\underline{d}(u,t) > \mu/(1+\varepsilon)$, i.e., we have the guarantee that the best path that we have already found (whose length corresponds to the upper bound μ) is at most $(1+\varepsilon)$ times as long as the shortest path. An analogous stopping rule applies to the backward search.

4.2. Optimisations.

Better Upper Bounds. We can use the distance table to get good upper bounds even earlier. So far, the distance table has only been applied to entrance points into the core V'_L of the topmost level. However, in many cases we encounter nodes that belong to V'_L earlier during the search process. Even the source and the target node could belong to the core of the topmost level. Still, we have to be careful since the distance table only contains the shortest path lengths within the topmost core and a path between two nodes in V'_L might be longer if it is restricted to the core of the topmost level than using all edges of the original graph. This is the reason why we have not used such a premature jump to the highest level before. But now, in order to just determine upper bounds, we could use these additional table look-ups. The effect is limited though because finding good upper bounds works very well anyway—the lower bounds are the crucial part. Therefore, the exact algorithm does without the additional look-ups. The approximate algorithm applies this technique to the nodes that remain in the priority queues after the search has been terminated since this might improve the result[4]. For example, we would get an improvement if the goal-directed search led us to the wrong motorway entrance ramp, but the right entrance ramp has at least been inserted into the priority queue.

Reducing Space Consumption. We can save preprocessing time and memory space if we compute and store only the distances between the landmarks and the nodes in the core of some fixed level k. Obviously, this has the drawback that we cannot begin with the goal-directed search immediately since we might start with nodes that do not belong to the level-k core so that the distances to and from the landmarks are not known. Therefore, we introduce an additional *initial query phase*, which works as a normal highway query and is stopped when all entrance points into the core of level k have been encountered. Then, we can determine the distances from s to all landmarks since the distances from s via the level-k core entrance points to the landmarks are known. Analogously, the distances from the landmarks to t can be computed. The same process is repeated for interchanged source and target nodes—i.e., we search forward from t and backward from s—in order to determine the distances from t to the landmarks and from the landmarks to s. Note that this second subphase can be skipped when the first subphase has encountered only bidirected edges.

The priority queues of the *main query phase* are filled with the entrance points that have been found during (the first subphase of) the initial query phase. We use the distances from the source or target node plus the lower bound to the target or source as keys for these initial elements. Since we never leave the level-k core during the main query phase, all required distances to and from the landmarks are known and the goal-directed search works as usual. The final result of the algorithm is the shortest path that has been found during the initial or the main query phase.

Limiting Component Sizes. Since the search processes from the source and target to the level-k core entrance points are often executed twice (once for each direction), it is important to bound this overhead. Therefore, we implemented a limit on the number of hops a shortcut may represent. By this means, the sizes of the components of bypassed nodes are reduced—in particular, the first contraction step tended to create quite large components of bypassed nodes so that it took a long time to leave such a component when the search was started from within it. Interestingly, this measure has also a very positive effect on the worst case analysis in [**17**]: it turned out that the worst case was caused by very large components of bypassed nodes in some sparsely populated areas, whose sizes now have been considerably reduced by the shortcut hops limit.

---

[4]In a preliminary experiment, the total error observed in a random sample was reduced from 0.096% to 0.053%.

Rearranging Nodes. Similar to [**7**], after the construction has been completed, we rearrange the nodes by core level, which improves locality for the search in higher levels and, thus, reduces the number of cache misses. By this means, speedups of up to 20% can be obtained.

5. Experiments

5.1. Environment, Instances, and Parameters. The experiments were done on one core of a single AMD Opteron Processor 270 clocked at 2.0 GHz with 4 GB main memory and 2 $\times$ 1 MB L2 cache, running SuSE Linux 10.0 (kernel 2.6.13). The program was compiled by the GNU C++ compiler 4.0.2 using optimisation level 3. We use 32 bit integers to store edge weights and path lengths. Results for the DIMACS Challenge benchmark can be found in Table 1.

TABLE 1. DIMACS Challenge [**1**] benchmarks for US (sub)graphs (query time [ms]).

	metric	
graph	time	dist
NY	29.6	28.5
BAY	34.7	33.3
COL	51.5	49.0
FLA	134.8	120.5
NW	161.1	146.1
NE	225.4	197.2
CAL	291.1	235.4
LKS	461.3	366.1
E	681.8	536.4
W	1 211.2	988.2
CTR	4 485.7	3 708.1
USA	5 355.6	4 509.1

We deal with the road network of Western Europe[5], which has been made available for scientific use by the company PTV AG. Only the largest strongly connected component is considered. The original graph contains for each edge a length and a road category, e.g., motorway, national road, regional road, urban street. We assign average speeds to the road categories, compute for each edge the average travel time, and use it as weight. In addition to this *travel time metric*, we perform experiments on variants of the European graph with a *distance metric* and the *unit metric*. We also perform experiments on the US road network (without Alaska and Hawaii), which has been obtained from the TIGER/Line Files [**19**]. Again, we consider only the largest strongly connected component. In contrast to the PTV data, the TIGER graph is undirected, planarised and distinguishes

[5] 14 countries: Austria, Belgium, Denmark, France, Germany, Italy, Luxembourg, the Netherlands, Norway, Portugal, Spain, Sweden, Switzerland, and the UK

TABLE 2. Properties of the used road networks.

	Europe	USA (Tiger)
#nodes	18 010 173	23 947 347
#directed edges	42 560 279	58 333 344
#road categories	13	4
average speeds [km/h]	10–130	40–100
neighbourhood size H (time)	60	70
neighbourhood size H (dist)	100, 200, 300, ...	
neighbourhood size H (unit)	80, 100, 120, ...	

only between four road categories. All graphs[6] have been taken from the DIMACS Challenge website [**1**]. Table 2 summarises the properties of the used networks.

At first, we report only the times needed to compute the shortest path distance between two nodes without outputting the actual route. These times are averages based on 10 000 randomly chosen (s,t)-pairs. In addition to providing average values, we use the methodology from [**16**] in order to plot query times (and error rates) against the 'distance' of the target from the source. In this context, the *Dijkstra rank* is used as a measure of distance: for a fixed source s, the Dijkstra rank of a node t is the rank w.r.t. the order which DIJKSTRA's algorithm settles the nodes in. Such plots are based on 1 000 random source nodes. In the last paragraph of Section 5.3, we also give the times needed to traverse the computed shortest paths.

Since it has turned out that a better performance is obtained when the preprocessing starts with a contraction phase, we practically skip the first construction step (by choosing neighbourhood sets that contain only the node itself) so that the first highway network virtually corresponds to the original graph. Then, the first real step is the contraction of level 1 to get its core. Note that in this case, distances within the core of level 1 are equal to the distances between level-1 core nodes in the original graph.

The shortcut hops limit (introduced in Section 4) is set to 10. The neighbourhood size H (i.e., the number of nearby nodes that belong to the neighbourhood of a node; introduced in [**16, 17**]) for the travel time metrics is set to 60 and 70 for the European and the US network, respectively. For the distance metric versions of both graphs, preliminary experiments indicate that using the linearly increasing sequence 100, 200, 300, ... as neighbourhood sizes to compute levels 2, 3, 4, ... of the hierarchy is a good choice. For the unit metric, we use $H = 80, 100, 120, \ldots$

5.2. Landmarks. We begin our experimental evaluation by analysing the quality of landmarks. Therefore, we evaluate the performance of *pure* ALT (without highway hierarchies) for different sets of landmarks. The evaluation of HH* is located in Section 5.3.

Preprocessing. First, we analyse the preprocessing of the ALT algorithm with different selection strategies on different cores of the highway hierarchy. We use 16 avoid, advancedAvoid and maxCover landmarks selected from the whole graph and

[6]Note that the experiments on the full TIGER graphs had been performed before the final versions of the DIMACS Challenge test instances, which use a finer edge costs resolution, were available. We did not repeat the experiments since we expect hardly any change in our measurement results.

TABLE 3. Overview of the preprocessing time for different selection strategies on the European and US network. All figures are given in minutes of computation time. For core-landmarks (indicated by cx where x depicts the level of the hierarchy used for selection), we report the time to construct the necessary highway information (*hh*), the time for selecting the landmarks (*sel*), and the time for computing the distances between all landmarks and all nodes (*dist*). Generating 16 maxCover landmarks on the whole graph requires more than 4 GB RAM. Therefore, these landmarks were generated on an AMD Opteron Processor 252 clocked at 2.6 GHz with 16 GB main memory.

		travel times				distances			
input	strategy	hh	sel	dist	total	hh	sel	dist	total
EUR	avoid	–	15.8	–	15.8	–	13.5	–	13.5
	adv.av.	–	23.2	–	23.2	–	19.2	–	19.2
	maxCov	–	88.3	–	88.3	–	75.3	–	75.3
	avoid-c1	2.7	2.5	6.3	11.5	2.7	2.1	4.2	9.0
	adv.av.-c1	2.7	3.6	6.3	12.6	2.7	3.0	4.2	9.9
	maxCov-c1	2.7	21.2	6.3	30.2	2.7	19.5	4.2	26.4
	avoid-c2	11.5	0.4	6.3	18.2	13.6	0.4	4.2	18.2
	adv.av.-c2	11.5	0.5	6.3	18.3	13.6	0.5	4.2	18.3
	maxCov-c2	11.5	3.3	6.3	21.1	13.6	2.4	4.2	20.2
	avoid-c3	13.7	0.1	6.3	20.1	20.1	0.1	4.2	24.4
	adv.av.-c3	13.7	0.1	6.3	20.1	20.1	0.1	4.2	24.4
	maxCov-c3	13.7	0.8	6.3	20.8	20.1	1.2	4.2	25.5
USA	avoid	–	20.5	–	20.5	–	18.3	–	18.3
	adv.av.	–	30.5	–	30.5	–	26.4	–	26.4
	maxCov	–	105.2	–	105.2	–	97.2	–	97.2
	avoid-c1	3.4	3.1	7.1	13.6	3.1	2.9	5.8	11.8
	adv.av.-c1	3.4	4.5	7.1	15.0	3.1	4.2	5.8	13.1
	maxCov-c1	3.4	28.4	7.1	38.9	3.1	28.2	5.8	37.1
	avoid-c2	14.9	0.5	7.1	22.5	17.4	0.6	5.8	23.8
	adv.av.-c2	14.9	0.7	7.1	22.7	17.4	0.9	5.8	24.1
	maxCov-c2	14.9	5.6	7.1	27.6	17.4	5.8	5.8	29.0
	avoid-c3	18.5	0.1	7.1	25.7	26.3	0.2	5.8	32.3
	adv.av.-c3	18.5	0.2	7.1	25.8	26.3	0.2	5.8	32.3
	maxCov-c3	18.5	1.2	7.1	26.8	26.3	1.5	5.8	33.6

from the core of levels 1–3. For advancedAvoid, we set $k' = 6$ (see Section 3.2). Table 3 gives an overview of the preprocessing of the ALT algorithm on the European and US network.

We observe that the time spent for selecting landmarks decreases significantly when switching to higher cores. Unfortunately, we have to compute the distances from and to all nodes in the original graph if we use core landmarks for the ALT algorithm (on the full graph these distances are computed during selection). In addition, we have to compute the highway information. Nevertheless, the computation of core 1 only takes about three minutes yielding a decrease of total preprocessing with regard to all selection techniques. With regard to preprocessing time, using avoid and advancedAvoid on the cores of level 2 or 3 does not seem reasonable while maxCover benefits from switching to higher cores.

Another advantage when switching to higher cores is memory consumption. While about 2.3 GB of RAM are needed for the distances from and to all nodes when selecting 16 avoid landmarks on the full graph, 384 MB are sufficient when using the core of level 1. Using the core-2 (core-3) even further reduces the memory consumption to 64 (17) MB. Note, that we use 32 bit integers for keeping the distances in the main memory.

Quality of Landmarks. Figure 5 gives an overview of the quality of landmarks. Therefor, we generate 10 different sets of 16 landmarks for each selection strategy, generated on the full graph, and on the cores up to a level of 3. In order to evaluate the quality of the generated landmarks, we logged the average search space for 1 000 random s-t ALT-queries on the road network of Western Europe and the US. The results are presented as box-and-whisker plot [**15**].

We see that for distances the quality of landmarks is almost independent of the chosen level of the hierarchy. Only when switching from level 2 to 3 we observe a mild increase of the search space when using advancedAvoid landmarks. However, for travel times on the European network an interesting phenomenon is that avoid gets better when switching from the whole graph to core 1 but gets worse and worse with higher levels on which landmarks are selected. On the US network, the search space reduces when switching to core 2 in combination with avoid landmarks. MaxCover is nearly independent of the chosen level on the European network while on the US network a slight loss of quality can be observed with higher levels.

There seem to be two counteracting effects here: On higher levels of the hierarchy, we lose information. For example, peripheral nodes that are candidates for good landmarks are dropped. On the other hand, concentrating on higher level edges in landmark selection heuristics could be beneficial since these are edges needed by many shortest paths.

In general, maxCover outperforms avoid and advancedAvoid regarding the average quality of the obtained landmarks. Nevertheless, in most cases the minimum average search space is nearly the same for all selection strategies within a core, while some sets of avoid and advancedAvoid landmarks lead to search spaces 25% higher than the worst maxCover landmarks. So, the maxCover routine seems to be more robust than avoid or advancedAvoid. Comparing avoid and advancedAvoid we observe just a mild improvement in quality. Thus, the additional computation time of advancedAvoid is not worth the effort.

Combining the results from Table 3 and Figure 5, another strategy seems promising: maxCover landmarks from the core of level 2 or 3 outperform avoid landmarks from the full graph and their computation, including the highway information, needs only additional 5 minutes compared to avoid landmarks from the full graph. For this reason, we use such landmarks for our further experiments.

Efficiency and Approximation. Table 4 indicates the efficiency of our implementation by reporting query times in comparison to the bidirectional variant of DIJKSTRA's algorithm. For comparison with approximate HH queries we also provide the results for an approximate ALT algorithm: Stop the query if the sum of the minimum keys in the forward and the backward queue exceed $\mu/(1+\varepsilon)+p_f(s)$ with $\varepsilon = 0.1$. This stopping criterion keeps the error rate below 10%.

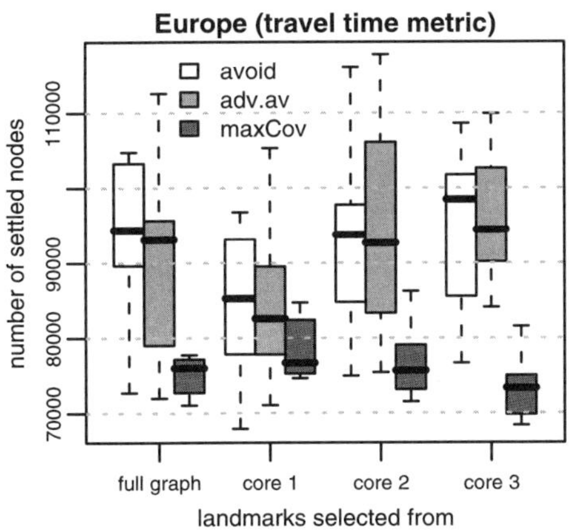

(a) Europe, travel times

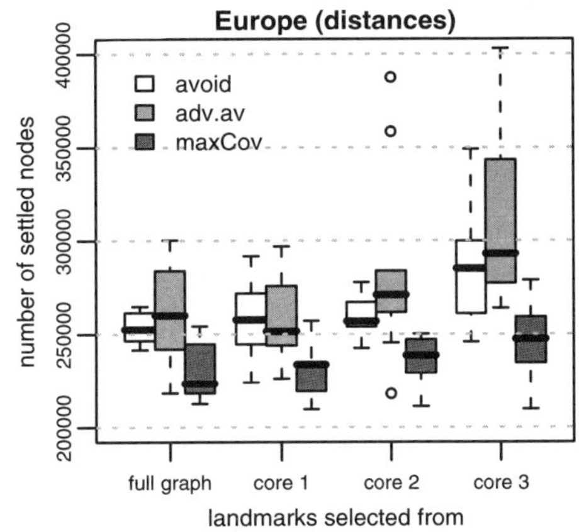

(b) Europe, distance

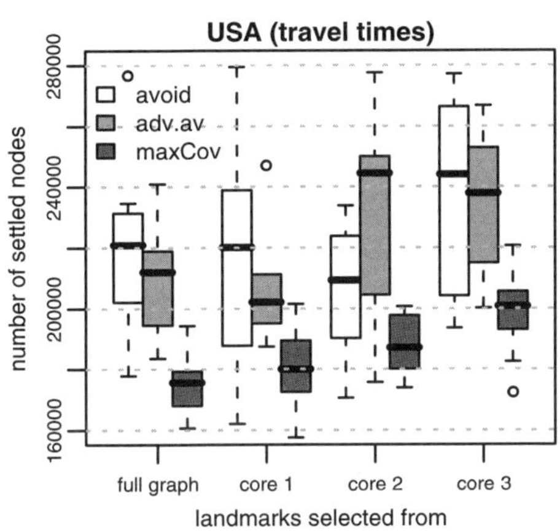

(c) USA, travel times

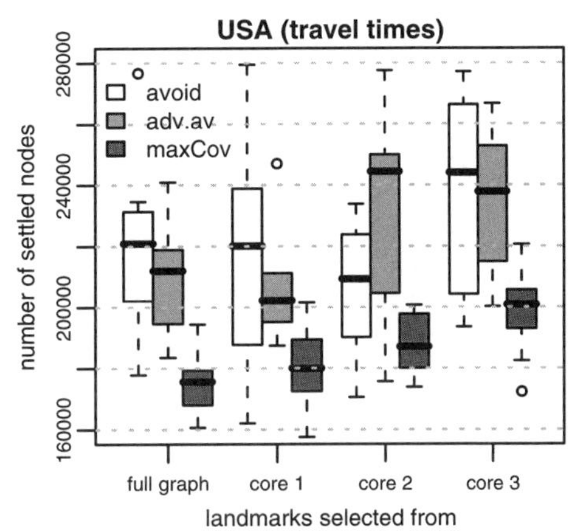

(d) USA, distance

FIGURE 5. Overview of the quality of landmarks. For each type of selection strategy, 10 different sets are generated. The quality of a landmark set is evaluated by the average number of nodes settled by the ALT algorithm for 1 000 random queries on the road networks of Western Europe and the US. The results are represented as box-and-whisker plot [**15**]: each box spreads from the lower to the upper quartile and contains the median, the whiskers extend to the minimum and maximum value omitting outliers, which are plotted individually.

Analysing the speedups compared to the bidirectional variant of DIJKSTRA's algorithm, we observe a search space reduction for Europe (travel times) by a factor of about 63.6. This reduction leads to a speedup factor of 49.0 concerning query times. For the USA (travel times), speedup concerning search space and query times is smaller than for Europe. We observe a factor of 38.5 for search space and 29.5 for query times. The reason for this discrepancy is the overhead for computing the potential and is also reported in [**6, 8, 5**].

TABLE 4. Comparison of the bidirectional variant of DIJKSTRA's algorithm, the ALT algorithm, and the approximate ALT algorithm concerning search space, query times and error rate. The landmarks are 16 maxCover core-3 landmarks. The figures are based on 1 000 random queries.

input	metric		bi.Dij.	ALT	approx.ALT
EUR	time	#settled nodes	$4.68 \cdot 10^6$	73 563	61 939
		query time [ms]	2 707	55.2	45.8
		inaccurate queries	–	–	12.1%
	dist	#settled nodes	$5.27 \cdot 10^6$	241 476	219 124
		query time [ms]	2 013	169.2	150.9
		inaccurate queries	–	–	33.7%
USA	time	#settled nodes	$7.42 \cdot 10^6$	192 938	182 426
		query time [ms]	3 808	129.2	116.9
		inaccurate queries	–	–	8.9%
	dist	#settled nodes	$8.11 \cdot 10^6$	281 335	263 375
		query time [ms]	3 437	177.1	163.5
		inaccurate queries	–	–	24.8%

For the distance metric on the European network we observe a reduction in search space of factor 21.8, leading to a speedup factor of 11.8. The corresponding figures for the US are 28.8 and 19.4. Thus, the situation is opposite to travel times. Here, speedups are better on the US network than on the European network. The higher speedups for travel times are due to the fact that for distances the advantage of taking fast highways instead of slow streets is smaller than for travel times. Since the difference between the slowest and fastest road category (see Table 2) is bigger for Europe, the ALT algorithm performs better on this network than on the US network when using travel times.

Comparing our results with the ones from [**5**] we have about 10% higher search spaces on the US network (travel times). This derives from the fact that on the US network with travel times the quality of maxCover landmarks slightly decreases when switching to higher cores (see Figure 5). Nevertheless, our average query times in this instance are 2.49 (129 ms to 322 ms) times faster, although we are using a slower computer. A reason for this is a different overhead factor, i.e., the time spent per settled node. While our implementation has an overhead of factor 1.3, the figures from [**5**] suggest an overhead of 2.

For the travel time metric, approximate queries perform only 20% better on Europe and 10% better on the US than exact ones. The percentage of inaccurate queries is 12% and 8%, respectively. For the distance metric, the speedup for approximate queries is even less and the percentage of inaccurate queries is much higher, namely 33.7% and 24.8% for the European and US network, respectively. These high numbers of wrong queries are due to the fact that for the distance metric there are more possibilities of short paths with similar lengths since the difference between taking fast highways and driving on slow streets fades. So, approximation for ALT adds only a small speedup not justifying the loss of correctness. For a detailed analysis of the approximation error see Table 10 and Figures 12–15 in Appendix A.

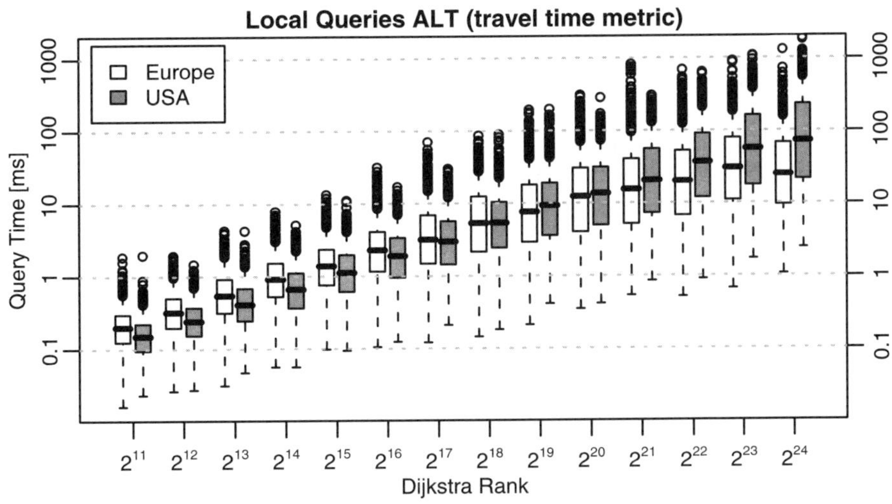

FIGURE 6. Comparison of the query times using the Dijkstra rank methodology on the road networks of Europe and the US. The landmarks are chosen from the level-3 core using maxCover. The results are represented as box-and-whisker plot [**15**]: each box spreads from the lower to the upper quartile and contains the median, the whiskers extend to the minimum and maximum value omitting outliers, which are plotted individually.

Local Queries. Figure 6 gives an overview of the query times in relation to the Dijkstra rank. The results for the distance metric are located in Appendix A (Figure 9).

The fluctuations in query time both between different Dijkstra ranks and with fixed Dijkstra rank are so big that we had to use a logarithmic scale. Even typical query times vary by an order of magnitude for large Dijkstra ranks. The slowest queries for most Dijkstra ranks are two orders of magnitude slower than the median query times.

An interesting observation is also that for small ranks ALT is faster on the network of the US whereas for ranks higher than 2^{21}, queries are faster on the European network. A plausible explanation seems to be the different geometry of the two continents. Queries within the (pen)insulae of Iberia, Britain, Italy, or Scandinavia lack landmarks in many directions. For example, a user in Scotland might have the queer experience, that queries in north-south direction are consistently faster than queries in east-west direction (see Figure 3). In contrast, long distance routes often have to go through bottlenecks which simplify search, as those bottlenecks are part of many long distance routes. In the US, such effects are rare.

5.3. Highway Hierarchies and A^* Search.

Default Settings. Unless otherwise stated, we use the following default settings. After the level-5 core has been determined, the construction of the hierarchy is stopped. A complete distance table is computed on the level-5 core. For the distance metric, we stop at the level-6 core instead. We use 16 maxCover landmarks that have been computed in the level-3 core. Landmark distances are stored only in the level-1 core. The approximate query algorithm uses a maximum error rate of 10%, i.e., $\varepsilon = 0.1$.

Using a Distance Table and/or Landmarks. As described in Section 2, using a distance table can be seen as adding a very strong sense of goal direction after the core of the topmost level has been reached. If the highway query algorithm (without distance table) is enhanced by the ALT algorithm, the goal direction comes into effect much earlier. Still, the most considerable pruning effect occurs in the middle of rather long paths: close to the source and the target, the lower bounds are too weak to prune the search. Thus, both optimisations, distance tables and ALT, have a quite similar effect on the search space: using either of both techniques, in the case of the European network with the *travel time metric*, the search space size is reduced from 1 662 to 916 (see Table 5). (Note that it is a coincidence that *exactly* the same number of settled nodes is achieved. Furthermore, we note that a slightly more effective reduction of the search space is obtained when all landmarks are used

TABLE 5. Comparison of all variants of the highway query algorithm using no optimisation ($\emptyset$), a distance table (DT), ALT, or both techniques. Values in parentheses refer to *approximate* queries. Note that the *disk space* includes the memory that is needed to store the original graph.

metric		$\emptyset$	DT	ALT	both
		Europe			
time	preproc. time [min]	17	19	20	22
	total disk space [MB]	886	1 273	1 326	1 714
	#settled nodes	1 662	916	916	686 (176)
	query time [ms]	1.16	0.65	0.80	0.55 (0.18)
dist	preproc. time [min]	47	47	50	49
	total disk space [MB]	894	1 506	1 337	1 948
	#settled nodes	10 284	5 067	3 347	2 138 (177)
	query time [ms]	8.21	4.89	3.16	1.95 (0.25)
unit	preproc. time [min]		24		27
	total disk space [MB]		925		1 368
	#settled nodes		1 714		1 249 (709)
	query time [ms]		1.18		0.99 (0.60)
		USA			
time	preproc. time [min]	23	26	27	28
	total disk space [MB]	1 129	1 574	1 743	2 188
	#settled nodes	1 966	1 098	1 027	787 (162)
	query time [ms]	1.18	0.73	0.80	0.60 (0.17)
dist	preproc. time [min]	55	57	59	59
	total disk space [MB]	1 140	1 721	1 754	2 335
	#settled nodes	9 706	5 477	2 784	2 021 (169)
	query time [ms]	7.10	4.95	2.52	1.74 (0.27)
unit	preproc. time [min]		29		32
	total disk space [MB]		1 981		2 542
	#settled nodes		1 665		1 072 (187)
	query time [ms]		1.29		0.89 (0.22)

to compute lower bounds instead of selecting only one landmark for each direction, namely to 903 instead of 916.) When we consider other aspects like preprocessing time, memory usage, and query time, we can conclude that the distance table is somewhat superior to the landmarks optimisation. Since both techniques have a similar point of application, a combination of the highway query algorithm with both optimisations gives only a comparatively small improvement compared to using only one optimisation. In contrast to the exact algorithm, the approximate variant reduces the search space size and the query time considerably—e.g., to 19% and 27% in case of Europe (relative to using only the distance table optimisation)—, while guaranteeing a maximum error of 10% and achieving a total error of 0.056% in our random sample of 1 000 000 (s,t)-pairs (refer to Table 7). Some results for US subgraphs can be found in Table 9 in Appendix A.

Using a *distance metric*, ALT gets more effective and beats the distance table optimisation since much better lower bounds are produced: the negative effect described in Figure 4 is weakened. Furthermore, in this case, a combination with both optimisations is worthwhile: the query time is reduced to 40% in case of Europe (relative to using only the distance table optimisation). While the highway query algorithm enhanced with a distance table has 7.5 times slower query times when applied to the European graph with the distance metric instead of using the travel time metric, the combination with both optimisations reduces this performance difference to a factor of 3.5—or even 1.4 when the approximate variant is used.

The performance for the *unit metric* ranks somewhere in between. Although computing shortest paths in road networks based on the unit metric seems kind of artificial, we observe a hierarchy in this scenario as well, which explains the surprisingly good preprocessing and query times: when we drive on urban streets, we encounter much more junctions than driving on a national road or even a motorway; thus, the number of road segments on a path is somewhat correlated to the road type.

Different Landmark Sets. In Table 6, we compare different sets of landmarks. Obviously, an increase of the number of landmarks improves the query performance. However, the rate of improvement is rather moderate so that using only 16 landmarks and thus, saving some memory and preprocessing time seems to be a good

TABLE 6. Comparison of the search spaces (in terms of number of settled nodes) of the highway query algorithm using different landmark sets. For each road network (with the travel time metric), the first column contains the search space size if the A^* search is *not* used. Values in parentheses refer to the search space sizes of approximate queries.

#landmarks	0	16		24		32	
	Europe						
core-1 avoid	916	687	(179)	665	(161)	651	(147)
core-3 maxCover		686	(176)	697	(177)	649	(140)
	USA						
core-1 avoid	1 098	808	(189)	762	(144)	736	(127)
core-3 maxCover		787	(162)	758	(134)	736	(121)

option. The quality of the selected landmarks is very similar for the two landmark selection methods that we have considered. Since the preprocessing times are similar as well, we prefer using the maxCover landmarks since they are slightly better.

Local Queries. In Figure 7, we compare the exact and the approximate HH* search in case of the European network with the travel time metric. (For the US network the results are similar. We refer to Figure 16 in Appendix A.) In the exact case, we observe a continuous increase of the query times: since the distance between source and target grows, it takes longer till both search scopes meet. For large Dijkstra ranks, the slope decreases. This can be explained by the distance table that bridges the gap between the forward and backward search for long-distance queries very efficiently, no matter whether we deal with a long or a very long path.

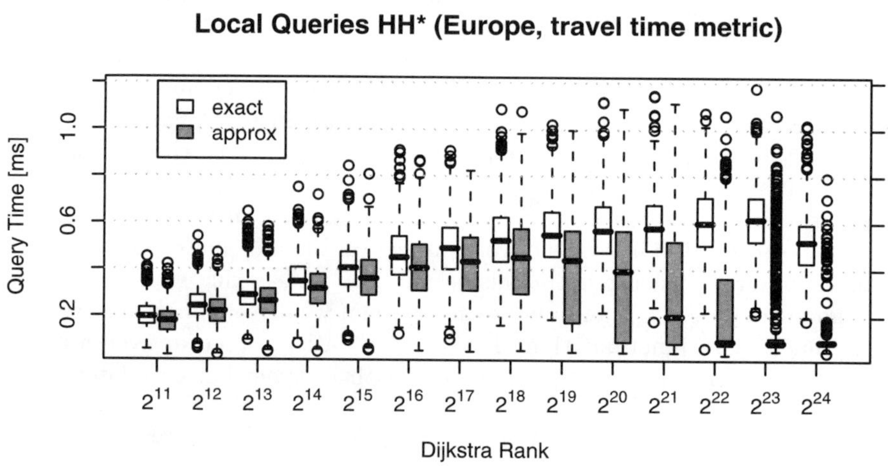

FIGURE 7. Comparison of the query times of the exact and the approximate HH* search using the Dijkstra rank methodology.

Up to a Dijkstra rank of 2^{18}, the approximate variant shows a very similar behaviour—even though at a somewhat lower level. Then, the query times *decrease*, reaching very small values for very long paths (Dijkstra ranks 2^{22}–2^{24}). This is due to the fact that the *relative* inaccuracy of the lower bounds, which is crucial for the stop condition of the approximate algorithm, is less distinct for very long paths: hence, most of the time, the lower bounds are sufficiently strong to stop very early. However, the large number and high amplitude of outliers indicates that sometimes goal direction does not work well even for approximate queries.

Approximation Error. Figure 8 shows the actual distribution of the approximation error for a random sample in the European network with the travel time metric, grouped by Dijkstra rank. (For the European network with the distance metric and the US network with both metrics, see Figures 17–19 in Appendix A.) For paths up to a moderate length (Dijkstra rank 2^{16}), at least 99% of all queries in the random sample returned an accurate result. Only very few queries approach the guaranteed maximum error rate of 10%. For longer paths, still more than 94%

of the queries give the correct result, and almost 99% of the queries find paths that are at most 2% longer than the shortest path. The fact that we get more errors for longer paths corresponds to the running times depicted in Figure 7: in the case of large Dijkstra ranks, we usually stop the search quite early, which increases the likelihood of an inaccuracy.

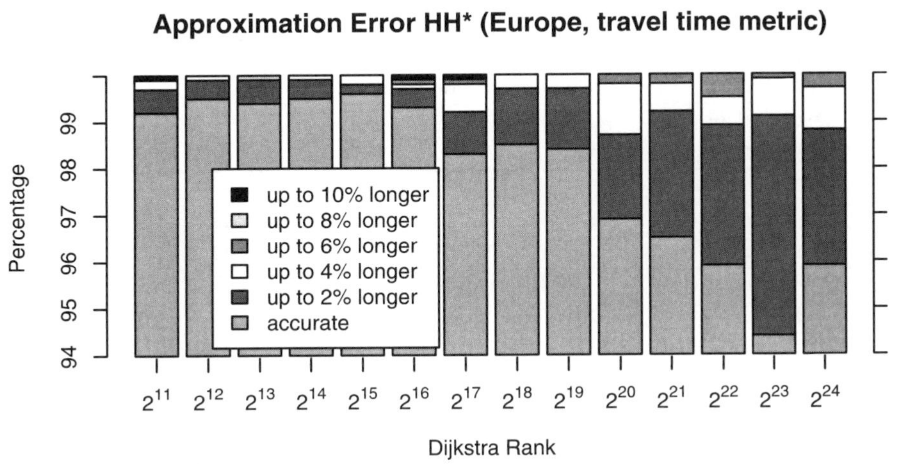

FIGURE 8. Actual distribution of the approximation error for a random sample, grouped by Dijkstra rank. Note that, in order to increase readability, the y-axis starts at 94%, i.e., at least 94% of all queries returned an accurate result.

While the approximate variant of the ALT algorithm gives only a small speedup (compare Figure 6 with Figure 10 in Appendix A) and produces a considerable amount of inaccurate results (in particular for short paths, see Figures 12 and 14), the approximate HH* algorithm is much faster than the exact version (in particular for long paths) and produces a comparatively small amount of inaccurate results. This difference is mainly due to the distance table, which allows a fast determination of upper bounds—and thus, in many cases early aborts—and provides accurate long-distance subpaths, i.e., the only thing that can go wrong is that the search processes in the local area around source and target do not find the right core entrance points.

In Table 7, we compared the effect of different maximum error rates ε. We obtained the expected result that a larger maximum error rate reduces the search space size considerably. Furthermore, we had a look at the actual error that occurs in our random sample: we divided the sum of all path lengths that were obtained by the approximate algorithm by the sum of the shortest path lengths. We find that the resulting total error is *very* small, e.g., only 0.056% in case of the European network with the travel time metric when we allow a maximum error rate of 10%. Similar to the results in Section 5.2, we observe that the total error and the percentage of inaccurate queries (see Figures 17 and 19) are much higher when using the distance metric instead of the travel time metric.

TABLE 7. Comparison of different maximum error rates ε. By the *total error*, we give the sum of the path lengths obtained by the approximate algorithm divided by the sum of the shortest path lengths. Note that these values are given in percent. This table is based on 1 000 000 random (s,t)-pairs (instead of the usual 10 000 pairs).

metric	ε [%]	0	1	2	5	10	20
		Europe					
time	#settled nodes	685	612	523	319	177	103
	total error [%]	0	0.0002	0.0015	0.018	0.056	0.112
dist	#settled nodes	2131	1302	843	333	184	143
	total error [%]	0	0.0112	0.0383	0.172	0.329	0.526
		USA					
time	#settled nodes	784	632	516	307	162	86
	total error [%]	0	0.0013	0.0073	0.034	0.082	0.144
dist	#settled nodes	2021	1101	672	277	169	134
	total error [%]	0	0.0108	0.0441	0.132	0.193	0.240

Complete Description of the Shortest Path. So far, we have reported only the times needed to compute the shortest path *distance* between two nodes. Now, we determine a complete description of the shortest path using the three algorithmic variants presented in Section 2.3. In Table 8 we give the additional preprocessing time and the additional disk space for the unpacking data structures. Furthermore, we report the additional time that is needed to determine a complete description

TABLE 8. Additional preprocessing time, additional disk space and query time that is needed to determine a complete description of the shortest path and to traverse it summing up the weights of all edges—assuming that the query to determine its lengths has already been performed. Moreover, the average number of hops—i.e., the average path length in terms of number of nodes—is given. These figures refer to experiments on the graphs with the travel time metric. Note that the experiments for Variant 1 have been performed without using a distance table for the topmost level.

	preproc. [s]	space [MB]	query [ms]	# hops (avg.)
	Europe			
Variant 1	0	0	16.70	1 370
Variant 2	71	112	0.45	1 370
Variant 3	75	180	0.17	1 370
	USA			
Variant 1	0	0	40.64	4 537
Variant 2	71	134	1.32	4 537
Variant 3	75	200	0.27	4 537

of the shortest path and to traverse[7] it summing up the weights of all edges as a sanity check—assuming that the distance query has already been performed. That means that the total average time to determine a shortest path is the time given in Table 8 plus the query time given in previous tables. We can conclude that even Variant 3 uses comparatively little preprocessing time and space. With Variant 3, the time for outputting the path remains considerably smaller than the query time itself and a factor 3–5 smaller than using Variant 2. The USA graph profits more than the European graph since it has paths with considerably larger hop counts, perhaps due to a larger number of degree-two nodes in the input. Note that due to cache effects, the time for outputting the path using preprocessed shortcuts is likely to be considerably smaller than the time for traversing the shortest path in the original graph.

6. Discussion

We have learned a few things about landmark A^* (ALT) that are interesting independently of highway hierarchies. We have explained why the lower bounds provided by ALT are often quite weak and why there are very high fluctuations in query performance. There are also considerable differences between Western Europe and the US. In Europe, we have *larger* execution times for local queries than in the US whereas for long range (average case) queries, times are *smaller*. Executing landmark selection on a graph where sparse subgraphs have been contracted is profitable in terms of preprocessing time even if we do not want highway hierarchies. Similarly, storing distances to landmarks only on this contracted graph considerably reduces the space overhead of ALT.

For highway hierarchies we have learned that they can also handle the case of travel distances. Compared to the case of travel times, space consumption is roughly the same whereas preprocessing time and query time increase by a factor of about 2–3.5 (when the combination with A^* search is applied). It is to be expected that any other cost metric that represents some compromise of travel time, distance, fuel consumption and tolls will have performance somewhere within this range. Highway hierarchies can be augmented to output shortest paths in a time below the time needed for computing the distances.

There is a complex interplay between highway hierarchies and the optimisations of distance tables and ALT. For exact queries using the travel time metric, distance tables are a better investment into preprocessing time and space than ALT. One incompatibility between highway hierarchies and ALT is that the search cannot be stopped when search frontiers meet. For approximate queries or for the distance metric, all three techniques work together very well yielding a speedup around four over highway hierarchies alone: Highway hierarchies save space and time for landmark preprocessing; distance tables obviate search in higher levels and allow simpler and faster ALT search with very effective goal direction. ALT provides good pruning opportunities for the distance metric and an excellent sense of goal direction for approximate queries yielding high quality routes most of the time while never computing very bad routes.

[7]Note that we do *not* traverse the path in the original graph, but we directly scan the assembled description of the path.

An interesting route of future research is to consider a combination of highway hierarchies with geometric containers or edge flags [**20, 14, 12**]. Highway hierarchies might harmonise better with these methods than with ALT because similar to highway hierarchies they are based on truncating search at certain edges. There is also hope that their high preprocessing costs might be reduced by exploiting the highway hierarchy.

Very recently, *transit node routing* (TNR) and related approaches [**3, 2**] have accelerated shortest path queries by another two orders of magnitude. Roughly, TNR precomputes shortest path distances to *access points* in a transit node set T (e.g., the nodes at the highest level of the highway hierarchy). During a query between "sufficiently distant" nodes, a distance table for T can be used to bridge the gap between the access points of source and target. However, TNR needs considerably more preprocessing time than the approach described in this paper. Furthermore, the currently best implementation of TNR uses highway hierarchies for preprocessing and local queries. It is likely that also landmarks might turn out to be useful in future versions of TNR. On the one hand, landmarks yield lower bounds that can be used for *locality filters* needed in TNR. On the other hand, the precomputed distances to access points could be used as landmark information for speeding up local search.

Acknowledgements. We would like to thank Timo Bingmann for work on visualisation tools. Two anonymous reviewers provided valuable suggestions.

References

1. 9th DIMACS Implementation Challenge, *Shortest Paths*, `http://www.dis.uniroma1.it/~challenge9/`, 2006.
2. H. Bast, S. Funke, D. Matijevic, P. Sanders, and D. Schultes, *In transit to constant time shortest-path queries in road networks*, Workshop on Algorithm Engineering and Experiments, 2007.
3. D. Delling, M. Holzer, K. Müeller, F. Schulz, and D. Wagner, *High-performance multi-level graphs*, 9th DIMACS Implementation Challenge [**1**], 2006, An updated version of the paper appears in this book.
4. E. W. Dijkstra, *A note on two problems in connexion with graphs.*, Numerische Mathematik **1** (1959), 269–271.
5. A. Goldberg, H. Kaplan, and R. Werneck, *Reach for A*: Efficient point-to-point shortest path algorithms*, Workshop on Algorithm Engineering & Experiments (Miami), 2006, pp. 129–143.
6. A. V. Goldberg and C. Harrelson, *Computing the shortest path: A* meets graph theory*, 16th ACM-SIAM Symposium on Discrete Algorithms, 2005, pp. 156–165.
7. A. V. Goldberg, H. Kaplan, and R. F. Werneck, *Better landmarks within reach*, 9th DIMACS Implementation Challenge [**1**], 2006, An updated version of the paper appears in this book.
8. A. V. Goldberg and R. F. Werneck, *Computing point-to-point shortest paths from external memory*, Workshop on Algorithm Engineering and Experimentation, 2005, pp. 26–40.
9. R. Gutman, *Reach-based routing: A new approach to shortest path algorithms optimized for road networks*, 6th Workshop on Algorithm Engineering and Experiments, 2004, pp. 100–111.
10. P. E. Hart, N. J. Nilsson, and B. Raphael, *A formal basis for the heuristic determination of minimum cost paths*, IEEE Transactions on System Science and Cybernetics **4** (1968), no. 2, 100–107.
11. T. Ikeda, M.Y. Hsu, H. Imai, S. Nishimura, H. Shimoura, T. Hashimoto, K. Tenmoku, and K. Mitoh, *A fast algorithm for finding better routes by AI search techniques*, Vehicle Navigation and Information Systems Conference. IEEE, 1994.
12. U. Lauther, *An extremely fast, exact algorithm for finding shortest paths in static networks with geographical background*, Geoinformation und Mobilität – von der Forschung zur praktischen Anwendung, vol. 22, IfGI prints, Institut für Geoinformatik, Münster, 2004, pp. 219–230.

13. J. Maue, P. Sanders, and D. Matijevic, *Goal directed shortest path queries using Precomputed Cluster Distances*, 5th Workshop on Experimental Algorithms (WEA), LNCS, no. 4007, Springer, 2006, pp. 316–328.
14. R. H. Möhring, H. Schilling, B. Schütz, D. Wagner, and T. Willhalm, *Partitioning graphs to speed up Dijkstra's algorithm*, 4th International Workshop on Efficient and Experimental Algorithms, 2005, pp. 189–202.
15. R Development Core Team, *R: A Language and Environment for Statistical Computing*, `http://www.r-project.org`, 2004.
16. P. Sanders and D. Schultes, *Highway hierarchies hasten exact shortest path queries*, 13th European Symposium on Algorithms, LNCS, vol. 3669, Springer, 2005, pp. 568–579.
17. P. Sanders and D. Schultes, *Engineering highway hierarchies*, 14th European Symposium on Algorithms, LNCS, vol. 4168, Springer, 2006, pp. 804–816.
18. R. Sedgewick and J. S. Vitter, *Shortest paths in Euclidean space*, Algorithmica **1** (1986), 31–48.
19. U.S. Census Bureau, Washington, DC, *UA Census 2000 TIGER/Line Files*, `http://www.census.gov/geo/www/tiger/tigerua/ua_tgr2k.html`, 2002.
20. D. Wagner and T. Willhalm, *Geometric speed-up techniques for finding shortest paths in large sparse graphs*, 11th European Symposium on Algorithms, LNCS, vol. 2832, Springer, 2003, pp. 776–787.
21. T. Willhalm, *Engineering shortest path and layout algorithms for large graphs*, Ph.D. thesis, Universität Karlsruhe (TH), Fakultät für Informatik, 2005.

Appendix A. Further Experiments

TABLE 9. Performance of HH* (using a distance table and landmarks) for US subgraphs with travel time metric. For small graphs, we deviate from the default settings: the landmark selection takes place in the core of the level given in column 2, the construction of the highway hierarchy is stopped at the core of the level given in column 3.

graph	landm. sel. core level	dist. table core level	preproc. time [min]	total disk space [MB]	#settled nodes	query time [ms]
NY	2	3	0:55	140	334	0.22
BAY	2	3	0:24	40	329	0.20
COL	2	3	0:29	49	327	0.19
FLA	3	3	1:08	115	354	0.22
NW	3	4	1:06	87	509	0.33
NE	3	4	2:14	169	526	0.36
CAL	3	4	2:23	176	519	0.35
LKS	3	4	4:25	398	543	0.39
E	3	5	4:07	255	650	0.46
W	3	5	7:22	453	695	0.50
CTR	3	5	23:12	1 132	762	0.73

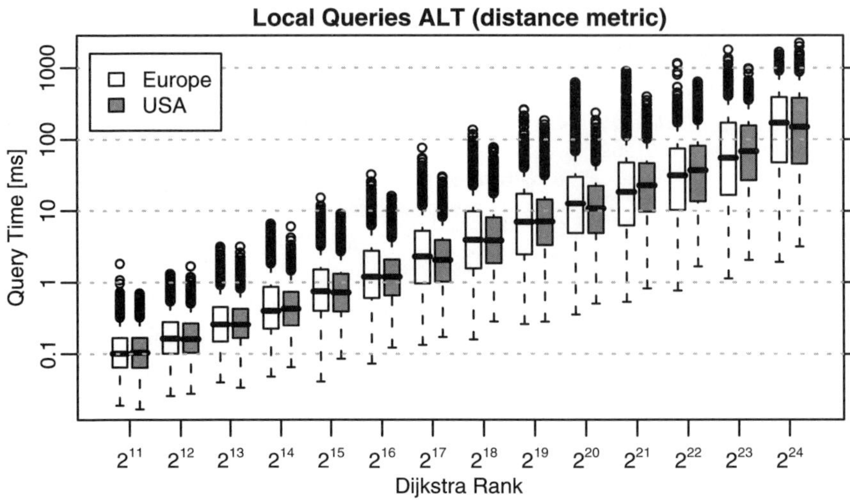

FIGURE 9. Comparison of the query times on the road network of Western Europe and the USA using the ALT algorithm. The landmarks are chosen from the core-3 using maxCover.

TABLE 10. Comparison of the exact and approximate ALT algorithm. The landmarks are taken from the full graph. The figures are based on 1 000 random queries on 10 different sets of 16 landmarks.

input	metric	landmarks	#settled nodes exact	#settled nodes approx.	inaccurate queries min	– max
EUR	time	avoid	93 520	81 582	9.8%	– 11.9%
		adv.av.	86 340	74 706	9.3%	– 12.6%
		maxCover	75 220	63 112	10.7%	– 11.7%
	dist	avoid	253 552	225 618	31.5%	– 38.4%
		adv.av.	256 511	227 779	30.9%	– 38.0%
		maxCover	230 110	203 564	31.3%	– 34.9%
USA	time	avoid	220 333	206 165	7.4%	– 10.1%
		adv.av.	210 703	194 920	7.6%	– 9.6%
		maxCover	175 359	161 230	7.6%	– 9.6%
	dist	avoid	308 823	289 701	24.8%	– 29.9%
		adv.av.	302 521	282 410	24.3%	– 29.3%
		maxCover	282 162	265 091	27.3%	– 22.3%

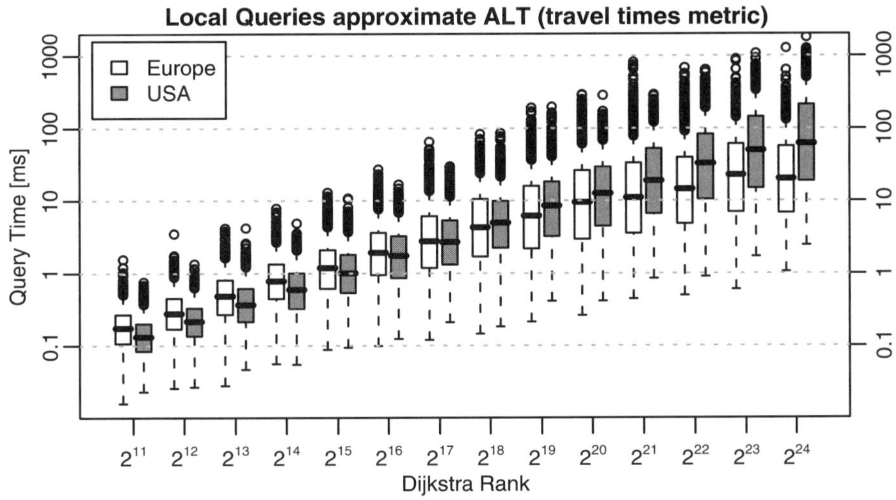

FIGURE 10. Comparison of the query times on the road network of Western Europe and the USA using the approximate ALT algorithm. The landmarks are chosen from the core-3 using max-Cover.

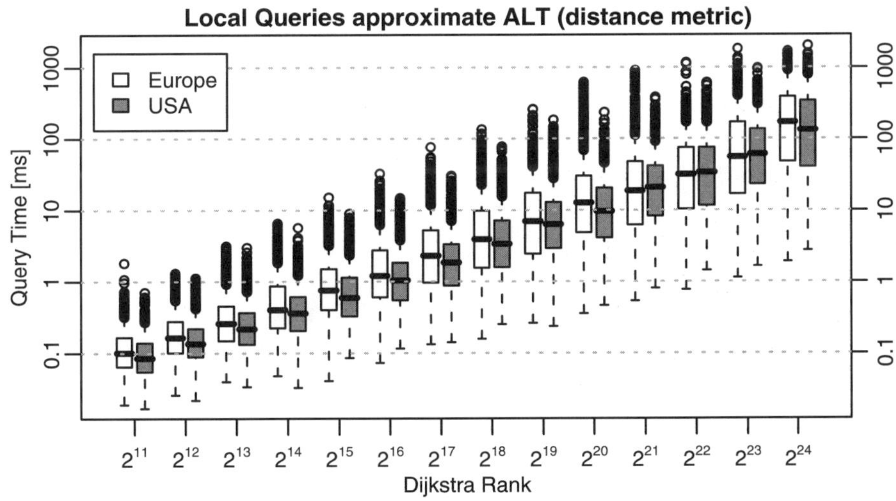

FIGURE 11. Comparison of the query times on the road network of Western Europe and the USA using the approximate ALT algorithm. The landmarks are chosen from the core-3 using max-Cover.

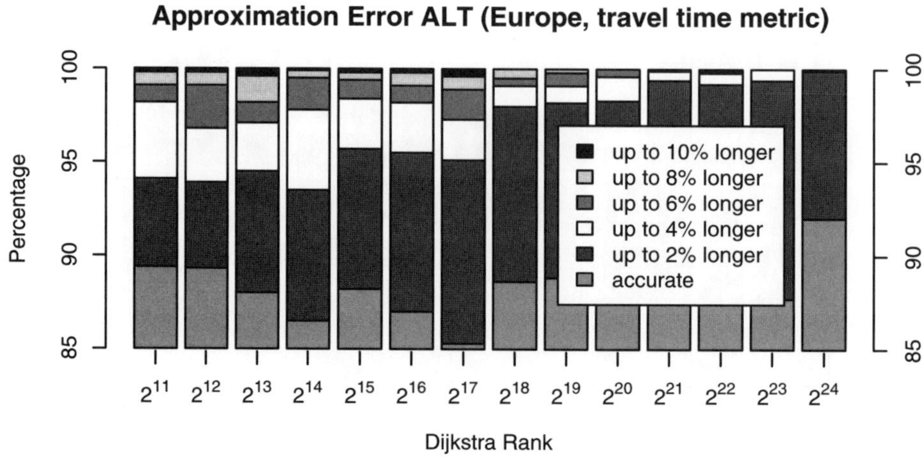

FIGURE 12. Actual distribution of the approximation error for a random sample, grouped by Dijkstra rank. Note that, in order to increase readability, the y-axis starts at 85%, i.e., at least 50% of all queries returned an accurate result.

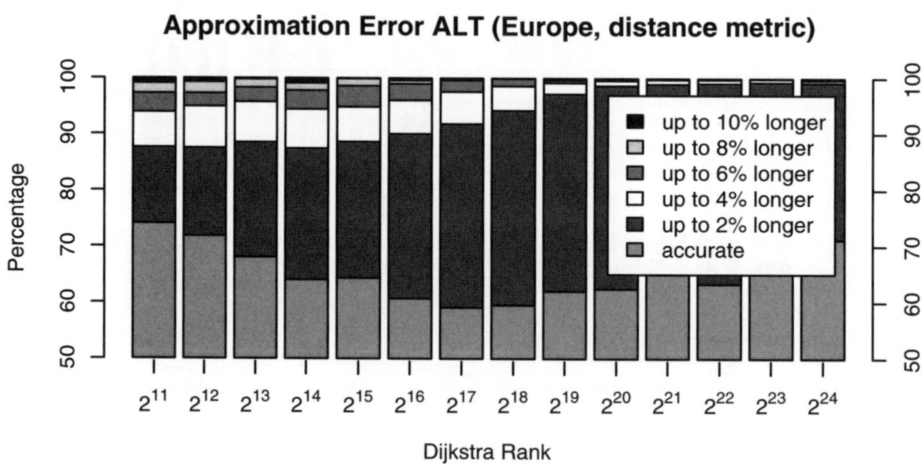

FIGURE 13. Actual distribution of the approximation error for a random sample, grouped by Dijkstra rank. Note that, in order to increase readability, the y-axis starts at 50%, i.e., at least 50% of all queries returned an accurate result.

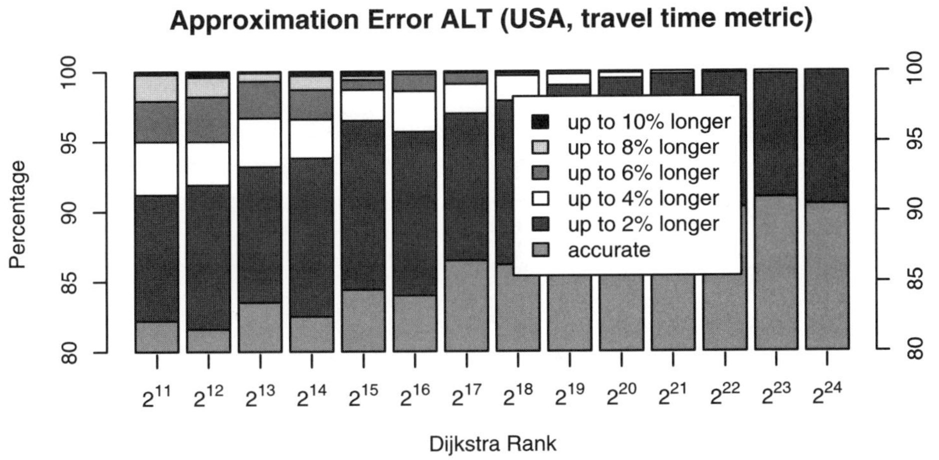

FIGURE 14. Actual distribution of the approximation error for a random sample, grouped by Dijkstra rank. Note that, in order to increase readability, the y-axis starts at 80%, i.e., at least 80% of all queries returned an accurate result.

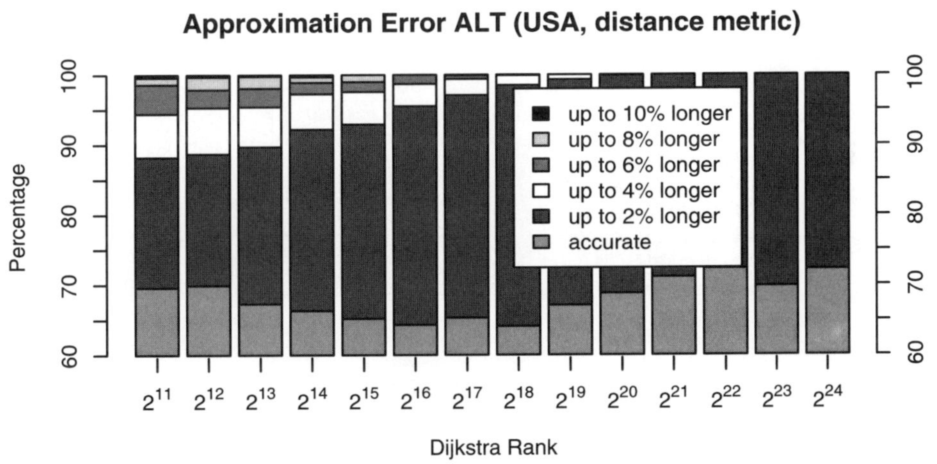

FIGURE 15. Actual distribution of the approximation error for a random sample, grouped by Dijkstra rank. Note that, in order to increase readability, the y-axis starts at 60%, i.e., at least 60% of all queries returned an accurate result.

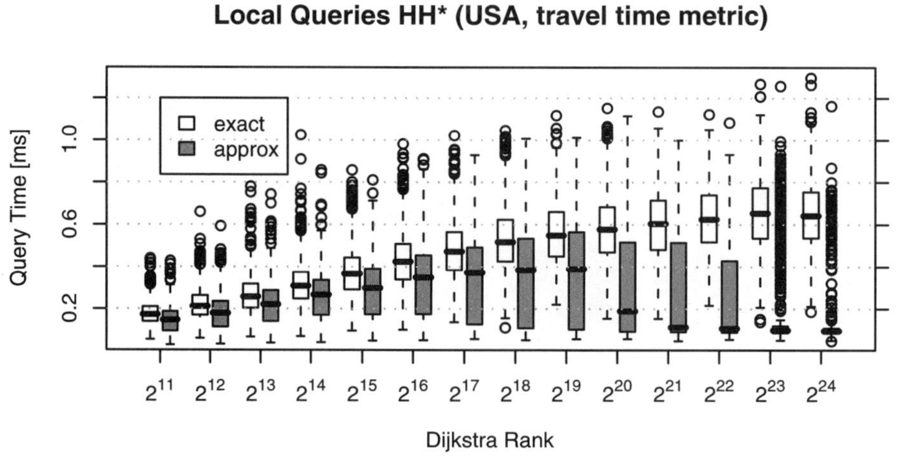

FIGURE 16. Comparison of the query times of the exact and the approximate HH* search.

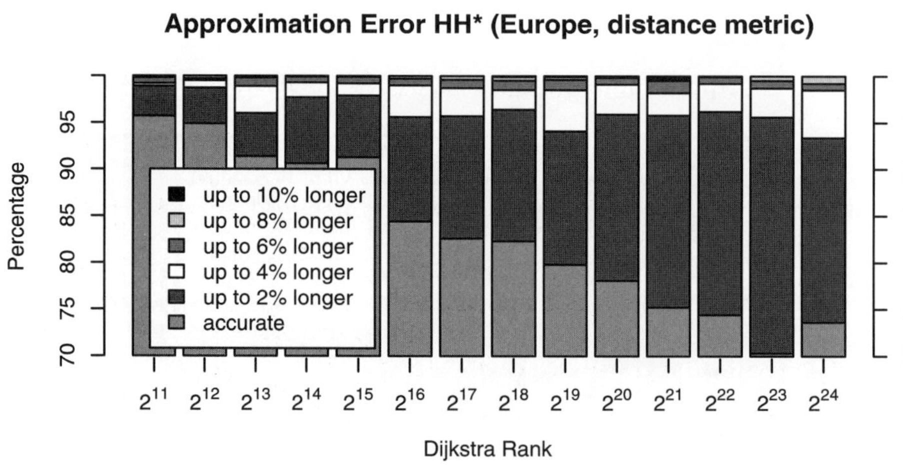

FIGURE 17. Actual distribution of the approximation error for a random sample, grouped by Dijkstra rank. Note that, in order to increase readability, the y-axis starts at 70%, i.e., at least 70% of all queries returned an accurate result.

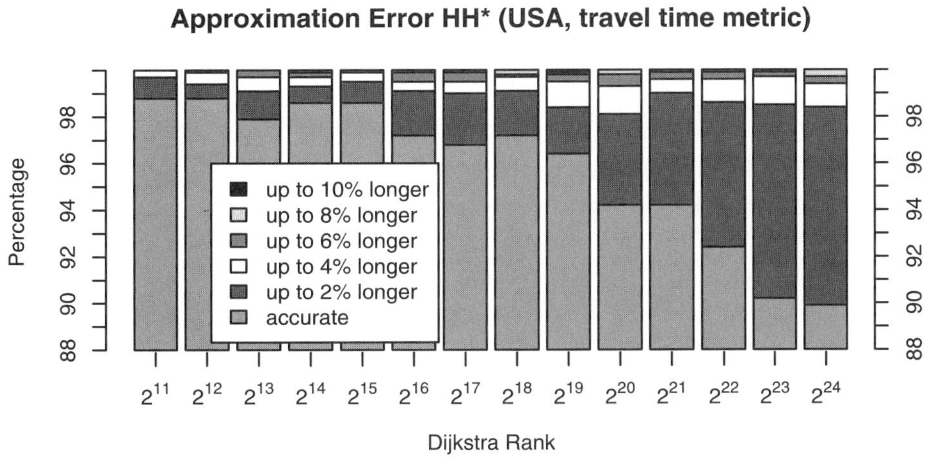

FIGURE 18. Actual distribution of the approximation error for a random sample, grouped by Dijkstra rank. Note that, in order to increase readability, the y-axis starts at 88%, i.e., at least 88% of all queries returned an accurate result.

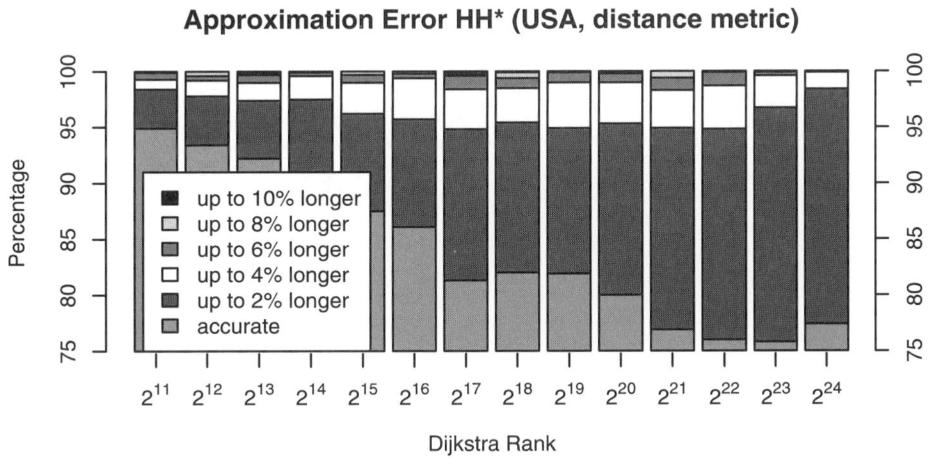

FIGURE 19. Actual distribution of the approximation error for a random sample, grouped by Dijkstra rank. Note that, in order to increase readability, the y-axis starts at 75%, i.e., at least 75% of all queries returned an accurate result.

DANIEL DELLING, UNIVERSITÄT KARLSRUHE (TH), FAKULTÄT FÜR INFORMATIK, POSTFACH 69 80, 76128 KARLSRUHE, GERMANY
E-mail address: delling@ira.uka.de

PETER SANDERS, UNIVERSITÄT KARLSRUHE (TH), FAKULTÄT FÜR INFORMATIK, POSTFACH 69 80, 76128 KARLSRUHE, GERMANY
E-mail address: sanders@ira.uka.de

DOMINIK SCHULTES, UNIVERSITÄT KARLSRUHE (TH), FAKULTÄT FÜR INFORMATIK, POSTFACH 69 80, 76128 KARLSRUHE, GERMANY
E-mail address: schultes@ira.uka.de

DOROTHEA WAGNER, UNIVERSITÄT KARLSRUHE (TH), FAKULTÄT FÜR INFORMATIK, POSTFACH 69 80, 76128 KARLSRUHE, GERMANY
E-mail address: wagner@ira.uka.de

DIMACS Series in Discrete Mathematics
and Theoretical Computer Science
Volume **74**, 2009

Ultrafast Shortest-Path Queries via Transit Nodes

Holger Bast, Stefan Funke, and Domagoj Matijevic

ABSTRACT. We introduce the concept of *transit nodes* as a means for preprocessing a road network such that point-to-point shortest-path queries can be answered extremely fast. We assume the road network to be given as a graph, with coordinates for each node and a travel time for each edge.

The transit nodes are a set of nodes, as small as possible, with the property that every *non-local* shortest path passes through at least one of these nodes. A path is called non-local if its source and target are at least a certain minimal euclidean distance apart. We precompute the lengths of the shortest paths between each pair of transit nodes, and between each node in the graph and its few, closest transit nodes. Then every non-local shortest path query becomes a simple matter of combining information from a few table lookups.

For the US road network, with about 24 million nodes and 29 million undirected edges, we achieve a worst-case query processing time of about 10 microseconds (not milliseconds) for 99% of all queries, namely the non-local ones. This improves over the best previously reported times by two orders of magnitude.

1. Introduction

The classical way to compute the shortest path between two given nodes in a graph with given edge lengths is Dijkstra's algorithm [**5**]. The asymptotic running time of Dijkstra's algorithm is $O(m + n \log m)$, where n is the number of nodes, and m is the number of edges [**6**]. For graphs with constant degree, like the road networks we consider in this paper, this is $O(n \log n)$. While it is still an open question, whether Dijkstra's algorithm is optimal for single-source single-target queries in general graphs, there is an obvious $\Omega(n + m)$ lower bound, because every node and every edge has to be looked at in the worst case. Sublinear query time hence requires some form of preprocessing of the graph. For general graphs, constant query time can only be achieved with superlinear space requirement; this is due to a recent result by Thorup and Zwick [**18**]. Like previous works, we therefore exploit special properties of road networks, in particular, that the nodes have low degree and that there is a certain hierarchy of more and more important roads, such that further away from source and target only the more important roads tend to be used on shortest paths.

This work is partially supported by the EU 6th Framework Programme under contract 001907 (DELIS).

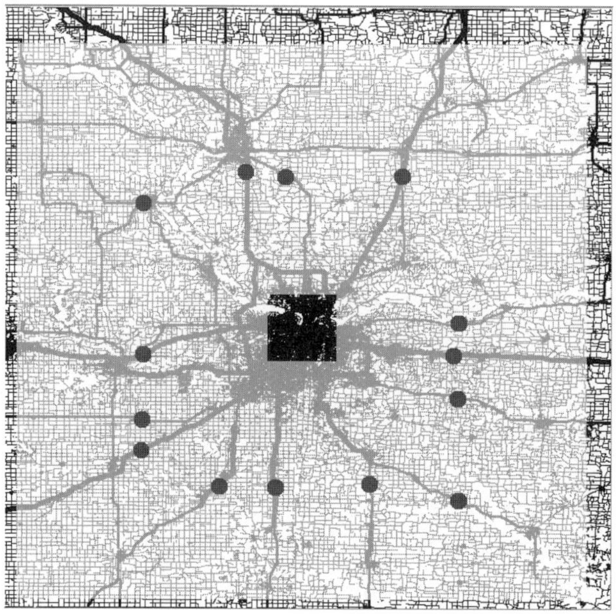

FIGURE 1. Transit nodes (red/bold dots) for a part of a city (center, dark) when travelling far (outside the light-gray area).

Our benchmark for most of this paper will be an undirected version of the US road network, which has about 24 million nodes and 29 million edges. On this network, a good implementation of Dijkstra's algorithm on a single state-of-the-art PC takes on the order of seconds, on average, for a random query. Note that for a random query, source and target are likely to be far away from each other, in which case Dijkstra's algorithm will settle a large portion of all nodes in the network before eventually reaching the target. For most of this paper, edge lengths will be travel times, so that shortest paths are actually paths with minimum travel time. We will continue to speak of *shortest*, however, because that is more familiar and to stress the wider applicability of our transit node idea. At the end of the paper we will also present results for unit edge lengths and when the length of an edge is the distance along the corresponding road segment, and results for the road network of Western Europe.

2. Our results

We present a new algorithm, named TRANSIT, which can answer non-local shortest path queries extremely fast, by combining information from a small number of lookups in a table. On the US road network, we achieve an average query processing time of around 10 microseconds (not milliseconds) for 99 % of all queries, when only the length (travel time) of the shortest path is required. The remaining 1 % of the queries are local in the sense that source and target are geometrically very close to each other. We also provide a simple algorithm for dealing with the few local queries efficiently. However, the focus of this work is on the non-local queries. In fact, we prefer to view our transit node approach as a *filter*: the vast majority of all queries can be processed extremely fast, leaving only a small fraction of local queries, which can be processed by any other method. Note that already Dijktra's algorithm can process the local queries by orders of magnitudes faster than arbitrary random queries.

Our processing times for the non-local queries beat the best previously reported figure of about 1 millisecond, due to Sanders and Schultes [**15**], *by two orders of magnitude*. When the full path, with all its edges, is to be output, we achieve an average query processing time of about 5 milliseconds on the US road network. We remark that all of the previous, more sophisticated algorithms use some form of path compression, which does not easily allow them to output the edges along the shortest path without using extra memory.

The basic idea of TRANSIT is as follows. For a given road network, compute a small set of *transit nodes* with the property that every shortest path that covers a certain not too small euclidean distance passes through at least one of these transit nodes. For every node in the given graph, then compute a set of *closest transit nodes*, with the property that every shortest path starting from that node and passing through a transit node at all (which it will if it goes sufficiently far), will pass through one of these closest transit nodes. These sets of closest transit nodes turn out to be very small: about 10 on average for our choice of transit nodes on the US road network. This allows us to precompute, for each node, the distances to each of its closest transit nodes. Also, the overall number of transit nodes turns out to be small enough so that we can easily precompute and store the distances between all pairs of transit nodes.

A non-local shortest path query can then easily be answered as follows. For a given source node *src* and target node *trg*, fetch the precomputed sets of closest transit nodes T_{src} and T_{trg}, respectively. For each pair of transit nodes $t_{src} \in T_{src}$ and $t_{trg} \in T_{trg}$ compute the length of the shortest path passing through these nodes, which is $d(src, t_{src}) + d(t_{src}, t_{trg}) + d(t_{trg}, trg)$. Note that all three distances in this sum have been precomputed. The minimum of these $|T_{src}| \cdot |T_{trg}|$ lengths is the length of the shortest path.

Given an algorithm for length-only shortest path queries, one can easily compute the edges along the shortest path using a few length-only shortest path queries per edge on the shortest path. To see this, assume we have already found a portion of the shortest path from the source to a node u. To find the next edge on the path, we simply launch a length-only shortest path query for each of the adjancent nodes of u. Given the length of the portion of the shortest path we already know, its total length, and the length of the edges adjacent to u, it is then easy to tell which of these edges is next on the shortest path. For details and possible improvements, see Section 4.5.

We want to stress that there are natural applications, where length-only shortest path queries are good enough, and not all the edges along the path are required. For example, most car navigation systems merely have a local view of the road network (if any). In that case it suffices to know the next few edges on the shortest path, and these can be computed by just a few length-only shortest-path queries, as described above.

We decribe TRANSIT in more detail in Section 4.

3. Related Work

We give a quick survey of work directly relevant to the problem of preprocessing road networks for subsequent fast shortest-path querying.

Gutman in **[9]** proposes a general concept of *edge levels* (he called it *reach*, though). Consider an edge e that appears "in the middle" of a shortest path, – shortest with respect to travel time – between two nodes that are a certain distance d apart – distance with respect to some arbitrary other metric, e.g., euclidean distance. Then the level of e is the higher, the larger d is. Gutman defines levels with respect to euclidean distance, but he notes that *any* metric can be used for the discrimination of the "in the middle" property. He presents simple algorithms which compute upper bounds for the edge levels and instruments those to obtain more efficient exact shortest path queries on moderate-size road networks. Due to the use of the euclidean metric as classifying metric, his approach allows for several variants of Dijkstra, in particular a natural goal-directed (unidirectional) version as well as efficient one-to-many shortest path queries. The – compared to later work like **[14]** or **[8]** – less competitive running times, both for the preprocessing phase as well as the queries are mainly due to the lack of an efficient compression scheme. The latter is very important for obtaining fast running times since in particular the networks induced by higher level edges contain very long chains of degree-two nodes following which is quite expensive. They can be easily skipped by suitable shortcut/path compression edges, though.

Later Sanders and Schultes have adopted a different classifying metric for their so-called *Highway-Hierarchies* **[14]**. In an ordinary Dijkstra computation from a source *src*, say that the rth node settled has Dijkstra rank r with respect to *src*. Sanders and Schultes say that the level of an edge (u, v) is high if it is on a shortest path between some *src* and *trg* such that v has high Dijkstra rank with respect to *src* and u has high Dijkstra rank with respect to *trg*. They achieve a drastic improvement both in preprocessing time as well as in query times, mainly because of the use of the Dijkstra rank as classifying metric as well as a highly efficient compression and pruning scheme in the higher levels of the network. The output of the algorithm is a path containing compressed edges, though, and uncompressing those edges does require some additional time and space. Their variant is also inherently bidirectional, so both goal-direction as well as one-to-many queries are not easily added, though later work has tried to address these issues.

Goldberg et al. in **[8]** combine edge levels with a compression scheme and they use lower bounds, based on precomputed distances to a few landmarks vertices, to allow for a more goal-directed search. They report running times comparable to those of **[14]**. Their space consumption is somewhat higher though, because every node in the network has to store distances to all landmarks. A non-goal-directed version of their algorithm exhibits considerably less storage requirements at the cost of only slightly higher query time.

More recently, Sanders and Schultes **[15]** have presented the so far best combination of preprocessing and query time. They show how to preprocess the US road network in 15 minutes, for subsequent query times of, on the average, 1 millisecond. While we could not yet come close to their extremely fast preprocessing time, our length-only scheme beats their query time by two orders of magnitude.

Möhring et al. **[12, 10]**, based on previous work by Lauther **[11]**, explored *arc flags* as means to achieve very fast query times. Intuitively, an arc flag is a sign that says whether the respective edge is on a shortest path to a particular region of the graph. In an extreme case, an edge could have a sign to every node on the shortest path to which it lies. A shortest path query could then be answered by simply following the signs to the target without any detour. However, to precompute these perfect signs requires an all-pairs shortest-path computation, which takes quadratic time and would be infeasible already for a small portion of the whole US road network, say the network of California. It is shown in [**12, 10**] and [**11**] how to cut down on this preprocessing somewhat, by putting up signs to sufficiently large regions of the graph. The largest network considered in these works has about one million nodes [**12**]. In the initial stages of our work, we experimented with the arc flag approach too, and were not able to achieve query processing times competitive with those of [**15**] with a reasonable amount of preprocessing time and extra space.

Most recently, following the first appearance of our paper, Sanders and Schultes have combined the transit node idea with highway hierarchies [**16**]. They report to have worked independently on similar ideas, but with a five times larger number of closest transit nodes (called *access nodes* in their work) per node. Note that the average query time of any scheme based on the transit node idea grows *quadratically* in the average number of closest transit nodes per node. The idea of precomputing all-to-all distances between a small subset of all nodes was already used in [**15**], to terminate local searches when they ascended far enough in the hierarchy. Prompted by our formulation of the transit node idea and the observation that an average of about 10 closest transit nodes per node suffice for a road network like that of the US, Sanders and Schultes were able to develop their ideas further to achieve very fast processing times comparable to those we report in this paper. They achieve these processing times for both non-local and local queries. (We would get a similar result by using the original highway hierarchies as a fallback for the local queries, but their implementation is more integrated as it uses highway hierarchies both for the local queries and for the computation of transit nodes.) Their preprocessing is an order of magnitude faster than what we report in this paper. The price is a more complex algorithm and implementation, and an increased space consumption. More details on the comparison between both approaches, our simple geometric one and the one based on highway hierarchies, are given in a joint follow-up paper [**2**].

In retrospect, the work of [**13**] (which later became [**3**]) can be taken as another alternative to computing transit nodes. In a nutshell, they use a hierarchy of separators to partition a given road network (making use of its almost-planarity). Their separator nodes could be taken as transit nodes, in which case local queries would be those with both endpoints in the same component. However, just like for the early attempts of Sanders and Schultes, this approach gives rise to an inherently much larger number of closest transit nodes (access nodes), which implies one to two orders of magnitude larger preprocessing time, space consumption and query processing times.

4. The TRANSIT algorithm

4.1. Intuition. The basic intuition behind our approach is very simple: imagine you live in a big city and intend to travel long-distance by car. What you will observe is that irrespectively of where your final destination is (as long it is reasonably far away) and where exactly you live in the city, there will be few roads via which you will actually leave the urban area when travelling on a shortest path

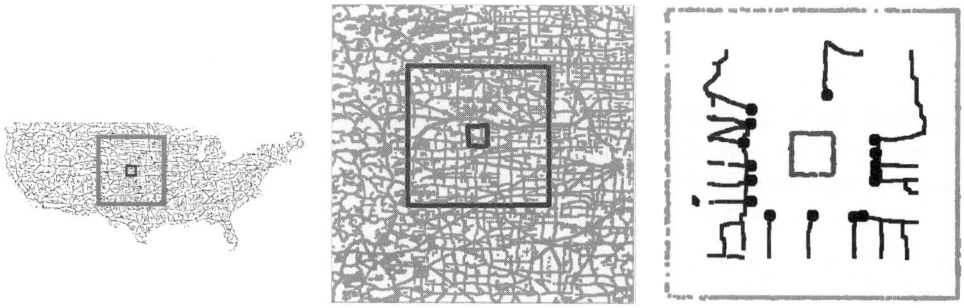

FIGURE 2. Transit neighborhood of a cell in a 64×64 subdivision of the US.

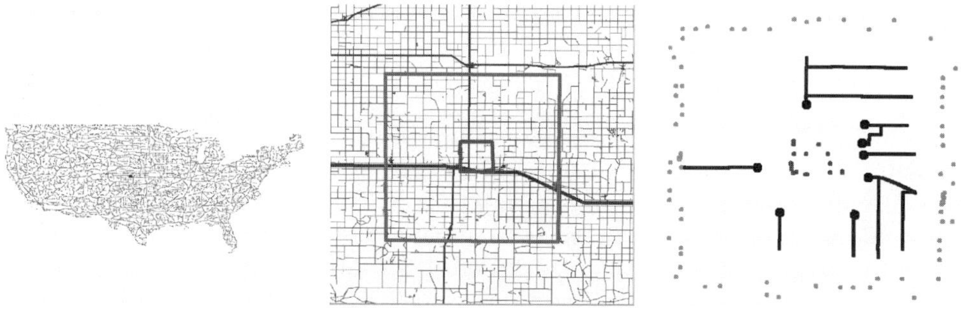

FIGURE 3. Transit neighborhood of a cell in a 1024×1024 subdivision of the US.

to your destination. In Figure 1 we have depicted these roads for the center part of a city. No matter where you start your journey inside the central region (in dark) – if your final destination lies outside the light-grey area and you travel on a shortest path, you will pass through one of the 14 marked roads (red/bold dots). This property, that long-distance trips (where the length is to be seen relative to the "starting region") pass through few *transit nodes*, is in fact to some degree invariable to scale. The example in Figure 1 shows the transit nodes for a cell in a 256×256 subdivision of the road network of the US; there are 14 of them. Figures 2 and 3 show transit nodes (or more precisely transit neighborhoods by which we compute transit nodes) for cells of a 64×64 and 1024×1024 subdivision of the US respectively. They exhibit 17 and 8 transit nodes respectively.

In essence our approach is then to construct a (geometric, in our case) subdivision of the network into cells and determine their transit nodes, such that the total number of transit nodes is small enough to allow us to precompute and store all pairwise distances between transit nodes in $O(n)$ space, i.e., in about the same amount of space as used for the original graph itself. Furthermore each node stores distances to the transit nodes of its resident cell. At query time a simple lookup yields the exact distance between any source-target pair provided they are not too close to each other.

4.2. Computing the Set of Transit Nodes. Consider the smallest enclosing *square* of the set of nodes (coming with x and y coordinate each), and the

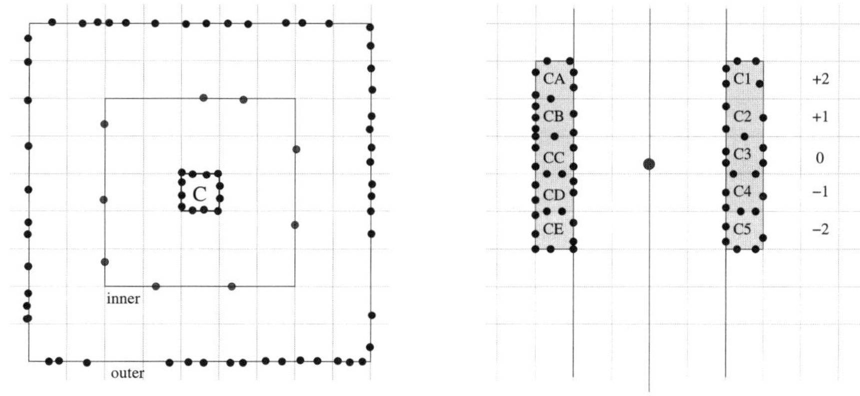

FIGURE 4. Definition and computation of transit nodes in the grid-based construction.

natural subdivision of this square into a *grid* of $g \times g$ equal-sized square cells, for some integer g. We define a set of transit nodes for each cell C as follows. Let S_{inner} and S_{outer} be the squares consisting of 5×5 and 9×9 cells, respectively, each with C at their center. Let E_C be the set of edges which have one endpoint inside C, and one outside, and define the set V_C of what we call *crossing nodes* by picking for each edge from E_C the node with the smaller id. Define V_{outer} and V_{inner} accordingly[1]. See the left side of Figure 4 for an illustration. The set of *closest transit nodes* for the cell C is now a set of nodes $T_C \subseteq V_{\text{inner}}$ with the property that for any pair of nodes p, q — one in V_C, one in V_{outer} — there exists a shortest path from p to q which passes through some node $v \in T_C$. Note that we also could have demanded that *all* shortest paths from p, q pass through some node in T_C, but this would have potentially increased the number of transit nodes with the only benefit of a slightly easier routine for reporting all shortest paths between a pair of nodes later on.

The overall set of transit nodes is just the union of these sets over all cells. It is easy to see that if two nodes are at least four grid cells apart in either horizontal or vertical direction, then the shortest path between the two nodes must pass through one of these transit nodes. By "four grid cells apart" we mean that between the grid cell containing the one node and the grid cell containing the other node there are at least four other grid cells. Also note that if a node is a transit node for some cell, it is likely to be a transit node for many other cells, each of them two cells away, too.

A naive way to compute these sets of transit nodes would be as follows. For each cell, compute all shortest paths between nodes in V_C and V_{outer}, and mark all nodes in V_{inner} that appear on at least one of these shortest paths. Figure 4 will again help to understand this. Such a naive computation is too time-consuming, though, for example for a 128×128 grid it required several days on the US network.

[1]That is, we consider the set of edges that have one endpoint inside $S_{\text{inner}}/S_{\text{outside}}$, the other outside. Note that those edges might not necessarily have endpoints in the cells directly adjacent to the crossing point with $S_{\text{inner}}/S_{\text{outside}}$.

As a first improvement, consider the following simple *sweep-line algorithm*, which runs Dijkstra computations within a radius of only *three* grid cells, instead of five, as in the naive approach. Consider one vertical line of the grid after the other, and for each such line do the following. Let v be one of the endpoints of an edge intersecting the line. We run a local Dijkstra computation for each such v as follows: let C_{left} be the set of cells two grid units left of v and which have vertical distance of at most 2 grid units to the cell containing v. Define C_{right} accordingly. See Figure 4, right; there we have $C_{\text{left}} = \{CA, CB, CC, CD, CE\}$ and $C_{\text{right}} = \{C1, C2, C3, C4, C5\}$. We start the local Dijkstra at v until all nodes on the boundary of the cells in C_{left} and C_{right} respectively are settled; we remember for all settled nodes the distance to v. This Dijkstra run settles nodes at a distance of roughly 3 grid cells. After having performed such a Dijkstra computation for all nodes v on the sweep line, we consider all pairs of boundary nodes (v_L, v_R), where v_L is on the boundary of a cell on the left and v_R is on the boundary of a cell on the right and the vertical distance between those cells is at most 4. We iterate over all potential transit nodes v on the sweep line and determine the set of transit nodes for which $d(v_L, v) + d(v, v_R)$ is minimal. With this set of transit nodes we associate the cells corresponding to v_L and v_R, respectively.

It is not hard to see that two such sweeps, one vertical and one horizontal, will compute exactly the set of transit nodes defined above (the union of all sets of closest transit nodes). The computation is space-efficient, because at any point in the sweep, we only need to keep track of distances within a small strip of the network. The consideration of all pairs (v_L, v_R) is negligible in terms of running time. As a further improvement, we first do the above computation for some *refinement* of the grid for which we actually want to compute transit nodes – let's say 128×128 is the grid we are finally aiming for. For some finer grid – say 256×256, we consider every second grid line (those also belonging to the 128×128 grid) and employ the computation described above to decide whether the respective boundary nodes are transit nodes in the *finer* grid. This computation is cheaper than in the coarser grid since the Dijkstra computations have to reach only half as far. Then, when computing the transit nodes for the coarser 128×128 grid, we can restrict ourselves to nodes from the sets of transit nodes computed for the finer grid and hence save Dijkstra computations. This easily generalizes to a sequence of refinements of 512×512, 1024×1024, ... grids where the finer grid essentially provides a "preselection" of the nodes that have to be considered for being a transit node in the coarser grid.

4.3. Computing the Distance Tables. For each node v, the distances to the closest transit nodes of its cell can be easily computed and memorized from the Dijkstra computations which had these transit nodes as source. In particular, each transit node thus knows the distance to all its (few) closest transit nodes. From this we can construct a graph with only the transit nodes as nodes, and an edge from each transit node to its closest transit nodes weighted by the respective distance. A standard all-pairs shortest-path computation on this auxiliary graph gives us the distances between each pair of transit nodes. Since the number of transit nodes is small (less than 8 000 for the US road network, using a 128×128 grid), this takes negligible time. The space consumption of these distance tables is discussed in Sections 4.7 and 4.8 below.

4.4. Shortest-path queries (length only). We next describe how to compute the *length* of the shortest path between a given source node *src* and a given target node *trg*, based on the preprocessing described in the previous two subsections. We here give a description for the scenario where we have precomputed only a single level of transit nodes. The extension to a hierachy of grids is straightforward, and will be explained in Section 4.7.

0. If *src* and *trg* are less than four grid cells (with respect to the grid used in the precomputation) apart, compute the distance from *src* to *trg* via an algorithm suitable for local shortest-path queries; a number of possibilities are described in Section 4.6. Otherwise, perform the following steps:
1. Fetch the lists T_{src} and T_{trg} of the closest transit nodes for the grid cells containing *src* and *trg*, respectively. Also fetch the lists of precomputed distances $d(src, t_{src}), t_{src} \in T_{src}$ and $d(trg, t_{trg}), t_{trg} \in T_{trg}$.
2. For each pair of $t_{src} \in T_{src}$ and $t_{trg} \in T_{trg}$ compute the sum of the lengths of the shortest path from *src* to t_{src}, from t_{src} to t_{trg}, and from t_{trg} to *trg*, which is $d(src, t_{src}) + d(t_{src}, t_{trg}) + d(t_{trg}, trg)$. Note that we may have $t_{src} = t_{trg}$, in which case $d(t_{src}, t_{trg}) = 0$.
3. Compute the length of the shortest path from *src* to *trg* as the minimum of the $|T_{src}| \cdot |T_{trg}|$ distances computed in step 2.

The algorithm is easily seen to be correct. Steps 1-3 will only be executed if source and target are more than four grid cells apart. Then, by the definition of the transit nodes in Section 4.2, the shortest path between source and target must pass through at least one transit node. But then, by the definition of closest transit nodes, the shortest path from *src* to *trg* will pass through one of the closest transit nodes of *src* as well as through one of the closest transit nodes of *trg*. The shortest path will therefore be among those tried in step 2, and we pick the shortest of these.

Since we have precomputed the distances from each node to its closest transit nodes and the distances between each pair of transit nodes, steps 1-3 take time $O(|T_{src}| \cdot |T_{trg}|)$. The average number of closest transit nodes of a node is a small constant — about 10 for the US road network.

4.5. Shortest-path queries (with edges). In this subsection, we describe how we can enhance the procedure given in the previous subsection to also output the edges along the shortest path from a given source node *src* to a given target node *trg*.

Assume that we have executed the procedure from the previous subsection, that is, we already know the length of the shortest path from *src* to *trg*. Assume that we have already found the part of the shortest path from *src* to some u (initially, $u = src$). Let $d(u, trg)$, which we can compute as $d(src, trg) - d(src, u)$, be the length of the part of the path which we have not found yet. Then the next node on the shortest path is that node v adjacent to u with the property that $d(u, trg) = c(u, v) + d(v, trg)$, where $c(u, v)$ is the length of the edge from u to v. This node can therefore be easily identified from the nodes adjacent to u, if only we can compute the distances $d(v, trg)$. But these are just instances of the problem we solved in the previous subsection: given two nodes, compute the length of the shortest path between them.

As described so far, the computation of $d(v, trg)$ would resort to the special algorithm for local shortest-path queries when v and *trg* are less than four grid cells apart. We can avoid this, if we compute the shortest path from *src* only until four grid cells away from *trg*, and, symmetrically, compute the shortest path from *trg* until four grid cells away from *src*. This will give us the full path if *src* and *trg* are at least eight grid cells apart, and parts of the path if they are more than four grid cells apart. For the remaining parts, or when *src* and *trg* are no more then four grid cells apart, we need to run the local algorithm.

This simple scheme can be improved in several ways. For example, we could store for each node, for each of its closest transit nodes, the index of the edge to that closest transit node. We would then obtain the next edge along the shortest path by a simple table lookup. The price would be a factor of two in the space consumption of the precomputed information.

Another idea would be to store for each transit node, the full path to each of its closest transit nodes. Using compression (edge ids along a shortest path typically do not differ much from one edge to the next, so some kind of gap encoding could be used), this could be achieved with relatively little extra space.

In our experiments, we restricted ourselves to length-only shortest-path queries.

4.6. Dealing with the Local Queries. If source and target are very close to each other (less than four grid cells apart in both horizontal and vertical direction for length-only shortest-path queries; less than eight grid cells apart in that way when computing the edges along the path), we cannot compute the shortest path via the transit nodes. This makes sense intuitively: there is hardly any hierarchy of roads in an area like, for example, downtown Manhattan, and a shortest path between two locations within the same such area will mostly consist of (small) roads of the same kind. In such a situation, no small set of transit nodes exist.

The good news is that most shortest-path algorithms are much faster when source and target are close to each other. In particular, Dijkstra's algorithm is about a thousand times faster for local queries, where source and target are at most four grid cells apart, for an 128×128 grid laid over the US road network, than for arbitrary random queries (most of which are long-distance). However, the non-local queries are roughly a million times faster and the fraction of local queries is about 1 %, so the average running time over all queries would be spoiled by the local Dijkstra queries.

Instead, we can use any of the recent sophisticated algorithms to process the local queries. Highway hierarchies, for example, achieve running times of a fraction of a millisecond for local queries, which would then only slightly affect the average processing time over all queries. The drawback is that we would need the full implementation of another method, and that this method requires additional space and precomputation time.

For our experiments in Section 5, we used a simple extension of Dijkstra's algorithm using geometric edge levels and shortcuts, as outlined in Section 3. This extension uses only six additional bytes per node. An edge $e = (p, q)$ has level l if lies on a shortest path from s to t, and both p and q are at least $f(l)$ far away from both s and t in euclidean distance along that path. Here $f(l)$ is a monotonically increasing function. For each node u, we insert at most two shortcuts as follows: consider the unique level, if any, where u lies on a chain of degree-2 nodes (degree with respect to edges of that level) *for the first time*; on that level insert a shortcut from u to the two endpoints of this chain. In each step of the Dijkstra computation for a local query, then consider only edges above a particular level (depending on the current euclidean distance from source and target), and make use of any available shortcuts suitable for that level. This algorithm requires an additional 5 bytes per node.

4.7. Multi-Level Grid. In our implementation as described so far, there is an obvious tradeoff between the size of the grid and the percentage of local queries which cannot be processed via precomputed distances to transit nodes. For a very coarse grid, say 64×64, the number of transit nodes, and hence the table storing the distances between all pairs of transit nodes, would be very small, but the percentage of local queries would be as large as 10 %. For a very fine grid, say 1024×1024, the percentage of local queries is only 0.1 %, but now the number of transit nodes is so large, that we can no longer store, let alone compute, the distances between all pairs of transit nodes. Table 1 gives the exact tradeoffs, also with regard to preprocessing time, for the US road network. The average query processing time for the non-local queries is around 10 microseconds, independent of the grid size.[2]

	$\lvert\mathcal{T}\rvert$	$\lvert\mathcal{T}\rvert \times \lvert\mathcal{T}\rvert$/node	avg. $\lvert A\rvert$	non-local	preproc.
64×64	2 042	0.1	11.4	91.7%	498 min
128×128	7 426	1.1	11.4	97.4%	525 min
256×256	24 899	12.8	10.6	99.2%	638 min
512×512	89 382	164.6	9.7	99.8%	859 min
$1\,024 \times 1\,024$	351 484	2 545.5	9.1	99.9%	964 min

TABLE 1. Number $\lvert\mathcal{T}\rvert$ of transit nodes, space consumption of the distance table, average number $\lvert A\rvert$ of closest transit nodes per cell, percentage of non-local queries (averaged over 100 000 random queries), and preprocessing time to determine the set of transit nodes for the US road network (excluding the computation of all-pair distances between transit nodes), TIGER version (see Section 5.2 for the differences to the DIMACS version).

To achieve a small fraction of local queries and a small number of transit nodes at the same time, we employ a *hierarchy of grids*. We briefly describe the two-level grid, which we used for our implementation. The generalization to an arbitrary number of levels would be straightforward.

The first level is an 128×128 grid, which we precompute as described so far. The second level is an 256×256 grid. For this finer grid, we compute the set of all transit nodes as described, but we compute and store distances only between those pairs which are local with respect to the 128×128 grid. This is a fraction of about 1/200th of all the distances, and can be computed and stored in negligible time and space via standard hashing. Note that in this simple approach, the space requirement for the individual levels simply add up. A more sophisticated approach to multi-level transit node routing is described in [**2**].

[2]According to our experiments, the bulk of the processing time for the non-local queries is spent in step 2 (trying out all combinations) of the procedure described in Section 4.4 and not in step 1 (fetching the relevant information for source and target node), that is, caching effects do not seem to play a dominant role here.

Query processing with such a hierarchy of grids is straightforward. In a first step, determine the coarsest grid with respect to which source and target are at least four grid cells apart in either horizontal or vertical direction. Then compute the shortest path using the transit nodes and distances computed for that grid, just as described in Sections 4.4 and 4.5. If source and target are at most four grid cells apart with respect to even the finest grid, we have to resort to the special algorithm for local queries.

4.8. Reducing the Space Further. As described so far, for each level in our grid hierarchy, we have to store the distances from each node in the graph to each of its closest transit nodes. For the US road network, the average number of closest transit nodes per node is about 10, independent of the grid size, and most distances can be stored in two bytes. For a two-level grid, this gives about 40 bytes per node.

To reduce this, we implemented the following additional heuristic. We observed that it is not necessary to store the distances to the closest transit nodes for *every* node in the network. Consider a simplification of the road network where chains of degree 2 nodes are contracted to a single edge. In the remaining graph we greedily compute a *vertex cover*, that is, we select a set of nodes such that for every edge at least one of its endpoints is a selected node. Using this strategy we determine about a third of all nodes in the network to store distances to their respective closest transit nodes. Then, for the source/target node v of a given query we first check whether the node is contained in the vertex cover, if so we can proceed as before. If the node is not contained in the vertex cover, a simple local search along chains of degree 2 nodes yields the desired distances to the closest transit nodes. The average number of distances stored at a node reduces from 11.4 to 3.2 for the 128×128 grid of the US, without significantly affecting the query times[3]. The total space consumption of our grid data structure then decreases to 16 bytes per node.

5. Implementation and Experiments

5.1. Experimental results. We tested all our schemes on the US road network, publically available via `http://www.census.gov/geo/www/tiger`. This is an undirected graph with 24,266,702 nodes and 29,049,043 edges, and an average degree of 2.4. Edge lengths are travel times. We implemented our algorithms in C++ (compiled with gcc 3.3.5 -O3) and ran all our experiments on a Dual Opteron Machine with two 2.4 GHz processors, 8 GB of main memory, running Linux 2.6.14 (64 bit); only one processor was used. Table 2 gives a summary of our experimental results. Experiments on the DIMACS benchmark collections and for other edge lengths than travel time are provided in Section 5.2

TRANSIT achieves an average query time of 12 microseconds for 99% of all queries. Together with our simple algorithm for the local queries, described in Section 4.6, we get an average of 63 microseconds over all queries. This overall average time could be easily improved by employing a more sophisticated algorithm, e.g. the one from [**15**], for the local queries, however at the price of a larger space

[3]Observe that we do not have to perform twice or four times the number of lookups in the distance table since the number of transit nodes for either s or t typically does not change at all (the transit nodes of nearby nodes are most of the time exactly the same). Following the degree-2 chains and obtaining the distances to the transit nodes costs no time compared to the few hundred table lookups.

non-local (99%)	local (1%)	all	preproc.	space/node
12 μs	5112 μs	63 μs	15 h	**21 bytes**

TABLE 2. Average query time (in microseconds), preprocessing time (in hours), and space consumption (in bytes per node in addition to the original graph representation) for our new algorithm TRANSIT, for the US road network, TIGER version (see Section 5.2 for the differences to the DIMACS version).

requirement and a more complex implementation. The space consumption of our algorithm is 21 bytes per node, which comes from 16 bytes per node for the distance tables of the two grids (Section 4.7) plus 5 bytes per node for the edge levels and shortcuts for the local queries (Section 4.6).

If we also output the edges along the shortest path, our average query processing becomes about 5 milliseconds (which happens to be the average processing time for the local queries, too). This is still competitive with the processing times reported in [**15**] and its closest competitors [**14**] [**7**] [**8**]. All of these schemes do *not* output edges along the shortest path, though outputting actual paths for these schemes would incur mostly a slight penalty in terms of space.

Many previous works provided a figure that showed the dependency of the processing time of a query on the *Dijkstra rank* of that query, which is the number of nodes Dijkstra's algorithm would have to settle for that query. The Dijkstra rank is a fairly natural measure of the difficulty of a query. For TRANSIT, query processing times are essentially constant for the non-local queries, because the number of table lookups required varies little and is completely independent from the distance between source and target. Table 3 therefore gives details on which percentage of the queries with a given Dijkstra rank are local. Note that for both the 128×128 grid and the 256×256 grid, all queries with a Dijkstra rank of $2^9 = 512$ or less are local, while all queries with Dijkstra rank above $2^{21} \approx 2,000,000$ are non-local.

grid size	$\leq 2^9$	2^{10}	2^{11}	2^{12}	2^{13}	2^{14}
128×128	100%	100%	100%	99%	99%	99%
256×256	100%	99%	99%	99%	97%	94%

grid size	2^{15}	2^{16}	2^{17}	2^{18}	2^{19}	2^{20}	$\geq 2^{21}$
128×128	98%	94%	85%	64%	29%	5%	0%
256×256	84%	65%	36%	12%	1%	0%	0%

TABLE 3. Estimated fraction of queries which are local with respect to the given grid, for various ranges of Dijkstra ranks. The estimate for the column labeled 2^r is the average over 1000 random queries with Dijkstra rank in the interval $[2^r, 2^{r+1})$.

5.2. Results for the DIMACS benchmark data. We also conducted experiments with additional benchmark data as provided by the DIMACS shortest path challenge website [**1**]. We used the same kind of machine as specified at the beginning of the previous section. For the sake of comparability with the results of other authors, Table 4 gives the results of the DIMACS core benchmark on such a machine.

Clearly, the efficacy of our grid-based approach does not depend on the metric used for computing the shortest paths; that is, for a given road network and resolution of the grid – say 128×128 the fraction of all queries that are considered "long range" does not change when varying the edge weights. What *does* change, though, is the number of transit nodes necessary to provide correct answers to these long range queries. In particular, when the cost measure is changed from travel time along an edge to distance along an edge or unit distance, the property of road networks to canalize traffic is weakened, hence the number of transit nodes necessary for a certain grid size increases. Likewise the average number of closest transit nodes per node increases and hence the query times; the increase is more pronounced for the distance weights than for the unit weights. In our benchmarks for the additional datasets we restricted to one level of transit nodes and only report the results for the non-local queries, which, for all the experiments in Table 5 and 6, were 97% of all queries.

Table 5 shows our results for different metrics and (sub)networks of the road network of the US. The astute reader will notice a difference in the number of transit nodes as well as in the preprocessing and average query time between the figures of Table 1 (TIGER data) and Table 5 (DIMACS data). This difference is due to the fact that the conversion from road types to speeds (and hence travel times) which we used for the TIGER data is different from the conversion used for the DIMACS data. In our conversion the difference in speed between slow and fast roads is more pronounced, and hence the canalizing property of the network with our travel times is stronger (fast roads are even more attractive). For the CTR network with the distance metric, the number of transit nodes for the 128×128 grid was too large, so we provided the results for a 64×64 grid instead.

Table 6 shows our results for the road network of Western Europe ($n = 18,010,173$, $m = 42,560,279$)[4]. A particularity of this network is a number of very slow ferry connections. Without special treatment of the corresponding edges (we tried a few heuristics but then decided to leave the data as is), the preprocessing time goes up significantly. This is so, because whenever one of the local Dijkstra computations in our transit node precomputation (Section 4) has to settle a node that can only be reached via a very long (slow) path, then almost *all* nodes in the network will be settled in that computation. Like this, the ferry connections give rise to a significant number of very time-consuming global Dijkstra computations in our precomputation. Note that the straightforward heuristic of splitting up very long edges into many short edges does not solve this problem: there will still be nodes which are geometrically close but with a very long shortest path between them. In Table 6, note that the problem indeed does not occur for unit edge lengths (in which case a ferry connection costs just as much as any other edge), and that

---

[4]We have considered an undirected variant of this network where the edge weights of reverse edges are equalized by taking the maximum of both since our current implementation does not allow for directed edges.

graph	#nodes	#edges	metric time	metric distance
NY	264346	733846	59.47	62.09
BAY	321270	800172	66.08	72.17
COL	435666	1057066	96.44	100.48
FLA	1070376	2712798	238.27	257.97
NW	1207945	2840208	282.40	328.19
NE	1524453	3897636	407.42	457.07
CAL	1890815	4657742	469.54	544.74
LKS	2758119	6885658	731.10	836.44
E	3598623	8778114	1042.63	1241.10
W	6262104	15248146	1988.49	2401.79
CTR	14081816	34292496	8934.93	9906.62

TABLE 4. Query times (ms) for the DIMACS core experiment (Opteron 240, 2.4 GHz, Linux 2.6.14, gcc 3.3.5, 64bit)

it is worst for the travel time metric (relative to other edges, travel time along a ferry connection is worse than distance).

In general, the space-efficiency of our approach improves with growing network size, the reason for that being that there is only little correlation between the number of transit nodes necessary for a 128×128 grid and the size of the respective road network. In fact the number of transit nodes can be even *larger* for subnetworks if they exhibit a worse canalizing property or the respective subnetwork covers more area of the square grid area (as observed for some subnetworks of the US). For amortizing the cost of storing the all-pairs distance table over the transit nodes, a large network size is beneficial. In particular, if the complete road network of the *whole world* was available, the per-node space requirement to store a transit node data structure of the same granularity would be considerably lower than for the US road network and still the same fraction of queries could be processed via a few table lookups. In that case one could probably even afford to create and store transit nodes based on a 512×512 grid which would resolve 99.8% of all queries by fast table-lookups.

5.3. Graphical User Interface. We have gone to quite some pain to implement a relatively comfortable graphical user interface (GUI) for displaying our road networks plus a number of additional elements. The GUI is implemented in C++ using the gtkmm library, which gives instant response times for dragging and zooming also for large road networks like that of the US. The GUI runs in its own thread, so that user and redraw events can be interleaved with computation and other code.

The GUI supports seamless dragging and zooming with the mouse (wheel), as in tools like Google Maps. This is very convenient for navigating in a large network quickly, but that was also the part that cost us the most work. The graph has to be divided into relatively small chunks, and only those chunks must be drawn which are actually visible from the current perspective and position. Also, there have to be priorities between edges, because always drawing all edges tends to clutter up the display and is an efficiency problem, too. The GUI also supports the drawing

graph	metric	grid	#tr.nodes	closest	query time	preproc.
USA	time	128x128	10 084	14	17.8 μs	7 h
USA	dist	128x128	31 536	36	69.4 μs	9 h
USA	unit	128x128	17 699	22	30.3 μs	9 h
BAY	time	128x128	10 077	8	9.1 μs	20 min
BAY	dist	128x128	13 269	13	11.6 μs	20 min
BAY	unit	128x128	10 314	9	9.2 μs	20 min
CAL	time	128x128	15 087	9	8.9 μs	30 min
CAL	dist	128x128	21 230	16	16.0 μs	30 min
CAL	unit	128x128	15 747	11	10.6 μs	30 min
E	time	128x128	10 477	12	12.2 μs	1h
E	dist	128x128	23 842	26	46.0 μs	2h
E	unit	128x128	13 915	15	19.0 μs	1h
FLA	time	128x128	6 248	9	7.8 μs	10 min
FLA	dist	128x128	9 937	14	12.3 μs	10 min
FLA	unit	128x128	6 404	9	7.7 μs	10 min
LKS	time	128x128	7 447	12	12.2 μs	30 min
LKS	dist	128x128	20 222	30	46.1 μs	1h
LKS	unit	128x128	10 257	16	17.5 μs	1h
NE	time	128x128	11 542	11	11.1 μs	20 min
NE	dist	128x128	22 937	23	28.0 μs	40 min
NE	unit	128x128	13 675	13	13.1 μs	25 min
NW	time	128x128	19 429	10	10.2 μs	30 min
NW	dist	128x128	23 963	15	14.8 μs	35 min
NW	unit	128x128	19 096	11	11.3 μs	25 min
NY	time	128x128	19 133	12	10.1 μs	10 min
NY	dist	128x128	24 435	15	14.3 μs	15 min
NY	unit	128x128	18 598	12	10.3 μs	10 min
W	time	128x128	19 107	10	10.6 μs	2h
W	dist	128x128	36 214	19	22.8 μs	2h
W	unit	128x128	25 554	14	15.2 μs	1h
CTR	time	128x128	24 540	14	17.5 μs	6h
CTR	dist	64x64	24 359	39	88.2 μs	12 h
CTR	unit	128x128	40 282	20	32.0 μs	7.5h
COL	time	128x128	10 502	9	7.0 μs	5 min
COL	dist	128x128	13 199	14	11.5 μs	10 min
COL	unit	128x128	10 686	10	7.9 μs	5 min

TABLE 5. Results for (sub)networks of the US road network with three kinds of edge lengths: travel time, distance along the corresponding road segment, and unit length.

of custom objects, like cross hairs (to visualize important locations), arrows along roads (to visualize something like edge signs), etc.

6. Conclusions

Transit nodes are a simple, yet powerful idea: they reduce the shortest-path computation for all but a small fraction of local queries to a few table lookups. In this paper we have focused on presenting this idea and giving a simple geometric algorithm realizing it.

graph	metric	grid	#tr.nodes	closest	query time	preproc.
Europe	time	128x128	10 394	14	13 μs	58h
Europe	dist	128x128	20 126	38	56 μs	29h
Europe	unit	128x128	7 708	14	12 μs	17h

TABLE 6. Results for the road network of Western Europe (undirected, including ferry connections).

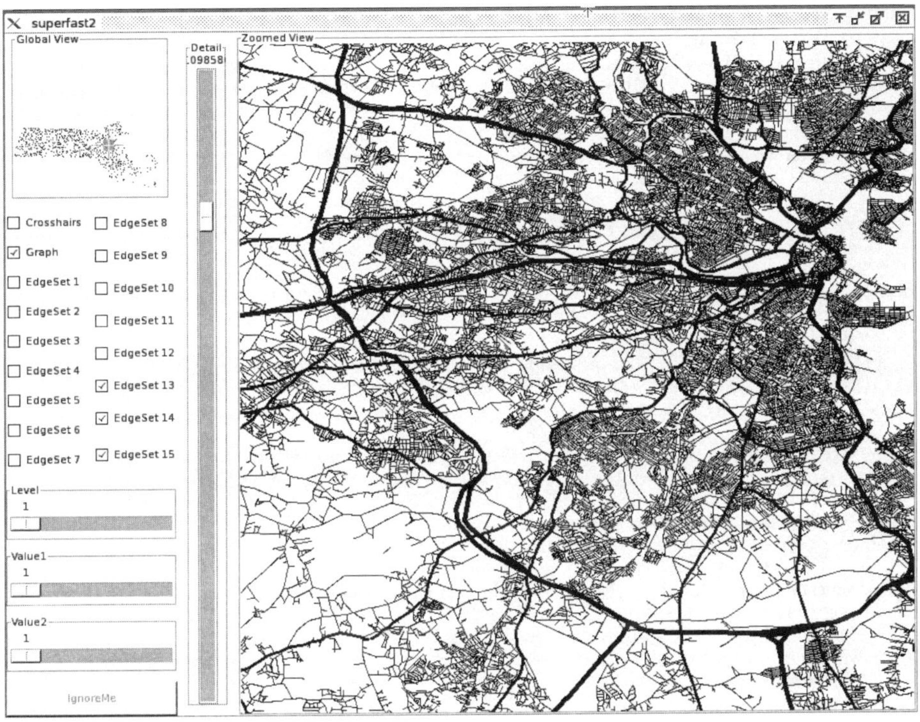

FIGURE 5. Screenshot of our interactive graphical user interface.

The algorithms in this paper work for undirected graphs. A generalization to directed graphs is not trivial but feasible. During the construction of the transit nodes one would have to distinguish between "incoming transit nodes", i.e., transit nodes that are visited by long paths *ending* in some node, and "outgoing transit nodes", i.e., transit nodes that are visited by long paths *starting* in some node. This can be taken care of by considering the reverse network during the construction step of the transit nodes. Of course, then the distance table is also not symmetric anymore and nodes would have to store "incoming" and "outgoing distances" to their closest transit nodes. The highway hierarchies from Sanders and Schultes, in particular their combination with the transit node idea [**16**], also work for directed graphs.

A more difficult open problem is how to design a data structure that yields similarly fast query times as our data structure but at the same time allows dynamic changes in the graph, like an increase of a few edge lengths due to a traffic jam.

Two solutions have recently been proposed in [**4**] and [**17**]; however, these do not achieve the ultrafast processing times reported in this paper.

Acknowledgements

We are grateful to the anonymous referees, especially one of them, for an extremely careful proof reading job and many constructive comments, which helped a lot in making the paper more precise and more readable.

References

[1] The 9th DIMACS Implementation Challenge: Shortest Paths; `http://www.dis.uniroma1.it/~challenge9/`.

[2] H. Bast, S. Funke, D. Matijevic, P. Sanders, and D. Schultes. In transit to constant time shortest-path queries in road networks. In *9th Workshop on Algorithm Engineering and Experiments (ALENEX'07)*, 2007.

[3] D. Delling, M. Holzer, K. Müller, F. Schulz, and D. Wagner. High-performance multi-level graphs. In *DIMACS Implementation Challenge Shortest Paths*, 2006. An updated version of the paper appears in this book.

[4] D. Delling and D. Wagner. Landmark-based routing in dynamic graphs. In *6th Workshop on Experimental Algorithms (WEA'07)*, pages 52–65, 2007.

[5] E. Dijkstra. A note on two problems in connexion with graphs. *Numerische Mathematik*, 1:269–271, 1959.

[6] M. L. Fredman and R. E. Tarjan. Fibonacci heaps and their uses in improved network optimization algorithms. *Journal of the ACM*, 34(3):596–615, 1987.

[7] A. Goldberg and C. Harrelson. Computing the shortest path: A* search meets graph theory. In *16th Symposium on Discrete Algorithms (SODA'05)*, pages 156–165, 2005.

[8] A. Goldberg, H. Kaplan, and R. Werneck. Reach for A*: Efficient point-to-point shortest path algorithms. In *8th Workshop on Algorithm Engineering and Experiments (ALENEX'06)*, 2006.

[9] R. Gutman. Reach-based routing: A new approach to shortest path algorithms optimized for road networks. In *6th Workshop on Algorithm Engineering and Experiments (ALENEX'04)*, 2004.

[10] E. Köhler, R. H. Möhring, and H. Schilling. Acceleration of shortest path and constrained shortest path computation. In *4th Workshop on Experimental and Efficient Algorithm (WEA'05)*, pages 126–138, 2005.

[11] U. Lauther. An extremely fast, exact algorithm for finding shortest paths in static networks with geographical background. In *Münster GI-Tage*, 2004.

[12] R. H. Möhring, H. Schilling, B. Schütz, D. Wagner, and T. Willhalm. Partitioning graphs to speed up dijkstra's algorithm. In *4th Workshop on Experimental and Efficient Algorithm (WEA'05)*, pages 189–202, 2005.

[13] K. Müller. Design and implementation of an efficient hierarchical speed-up technique for computation of exact shortest paths in graphs. Master's thesis, University of Karlsruhe, 2006.

[14] P. Sanders and D. Schultes. Highway hierarchies hasten exact shortest path queries. In *13th European Symposium on Algorithms (ESA'05)*, pages 568–579, 2005.

[15] P. Sanders and D. Schultes. Engineering highway hierarchies. In *14th European Symposium on Algorithms (ESA'06)*, pages 804–816, 2006.

[16] P. Sanders and D. Schultes. Robust, almost constant time shortest-path queries on road networks. In *DIMACS Implementation Challenge Shortest Paths*, 2006. An updated version of the paper appears in this book.

[17] D. Schultes and P. Sanders. Dynamic highway-node routing. In *6th Workshop on Experimental Algorithms (WEA'07)*, pages 66–79, 2007.

[18] M. Thorup and U. Zwick. Approximate distance oracles. *Journal of the ACM*, 51(1):1–24, 2005.

Max-Planck-Institute for Informatics, Saarbrücken, Germany
E-mail address: `bast@mpi-inf.mpg.de`

Max-Planck-Institute for Informatics, Saarbrücken, Germany
E-mail address: `funke@mpi-inf.mpg.de`

Max-Planck-Institute for Informatics, Saarbrücken, Germany
E-mail address: `dmatijev@mpi-inf.mpg.de`

DIMACS Series in Discrete Mathematics
and Theoretical Computer Science
Volume **74**, 2009

Robust, Almost Constant Time Shortest-Path Queries in Road Networks

Peter Sanders and Dominik Schultes

ABSTRACT. When you drive to somewhere 'far away', you will leave your current location via one of only a few 'important' traffic junctions. Recently, other research groups and we have largely independently developed this informal observation into *transit-node routing*, a technique for reducing quickest-path queries in road networks to a small number of table lookups. The contribution of our paper is twofold. First, we present a generic framework for transit-node routing that allows almost constant time routing for both global and local queries. Second, we develop a highly tuned implementation using *highway hierarchies* and *highway-node routing*. For the road maps of Western Europe and the United States, we obtain query times that are more than one million times faster than the best known algorithm for general networks. We also explain how to compute complete descriptions of shortest paths (and not just their lengths) very efficiently.

1. Introduction

Computing an optimal route in a road network between specified source and target nodes (i.e., places/intersections) is one of the showpieces of real-world applications of algorithmics. Besides the omnipresent application of car navigation systems and internet route planners, even faster route planning is needed for massive traffic simulations and optimisations in logistics systems. Beyond mere computational efficiency, the methods presented here also give quantitative insight into the structure of road networks and justify the way humans do route planning.

The classical algorithm for route planning—Dijkstra's algorithm [**8**]—iteratively visits all nodes that are closer to the source node than the target node before reaching the target. On road networks for a subcontinent like Western Europe or the USA, this takes about five seconds on a state-of-the-art workstation. Since this is too slow for many applications, commercial systems use heuristics that do not guarantee optimal routes. Therefore, there has been considerable interest in speedup techniques for computing *optimal* routes. In this paper, we present *transit-node*

2000 *Mathematics Subject Classification.* Primary 68R10; Secondary 90B20.

Key words and phrases. Shortest paths, routing, graph, road network, speed-up technique, hierarchy, preprocessing.

Partially supported by DFG grant SA 933/1-3.

routing, an approach that provides almost constant query times for road networks of any size.

1.1. Central Ideas. *Transit-node routing* is based on a simple observation intuitively used by humans: When you start from a source node s and drive to somewhere 'far away', you will leave your current location via one of only a few 'important' traffic junctions, called (forward) *access nodes* $\overrightarrow{A}(s)$. An analogous argument applies to the target t, i.e., the target is reached from one of only a few backward access nodes $\overleftarrow{A}(t)$. Moreover, the union of all forward and backward access nodes of all nodes, called *transit-node set* $\mathcal{T}$, is rather small. This implies that for each node the distances to/from its forward/backward access nodes and for each transit-node pair (u, v) the distance between u and v can be stored. For given source and target nodes s and t, the length of the shortest path that passes at least one transit node is given by

$$d_{\mathcal{T}}(s,t) = \min\{d(s,u) + d(u,v) + d(v,t) \mid u \in \overrightarrow{A}(s), v \in \overleftarrow{A}(t)\}.$$

Note that all involved distances $d(s, u)$, $d(u, v)$, and $d(v, t)$ can be directly looked up in the precomputed data structures. As a final ingredient, a *locality filter* $\mathcal{L} : V \times V \to \{\mathsf{true}, \mathsf{false}\}$ is needed that decides whether given nodes s and t are too close to travel via a transit node. $\mathcal{L}$ has to fulfil the property that $\neg\mathcal{L}(s,t)$ implies $d(s,t) = d_{\mathcal{T}}(s,t)$. Note that in general the converse need not hold since this might hinder an efficient realisation of the locality filter. Thus, *false positives*, i.e., "$\mathcal{L}(s,t) \wedge d(s,t) = d_{\mathcal{T}}(s,t)$", may occur.

The following algorithm can be used to compute $d(s,t)$:

1 **if** $\neg\mathcal{L}(s,t)$ **then** compute and return $d_{\mathcal{T}}(s,t)$;
2 **else** use any other routing algorithm.

Figure 1 gives a *schematic* representation of transit-node routing, while Figure 2 (first published in [**4**]) gives a *real-world* example.

Knowing the length of the shortest path, a complete description of it can be efficiently derived using iterative table lookups and precomputed representations of paths between transit nodes. Provided that the above observation holds and that the percentage of false positives is low, the above algorithm is very efficient since a large fraction of all queries can be handled in Line 1, $d_{\mathcal{T}}(s,t)$ can be computed using only a few table lookups, and source and target of the remaining queries in Line 2 are quite close. In fact, the remaining queries can be further accelerated by introducing additional levels of transit-node routing.

1.2. Related Work.

1.2.1. *Bidirectional Search.* A classical technique is *bidirectional search* which simultaneously searches forwards from s and backwards from t until the search frontiers meet. Many more advanced speedup techniques (including ours) use bidirectional search as an ingredient.

1.2.2. *Highway Hierarchies.* Commercial systems use information on road categories to speed up search. 'Sufficiently far away' from source and target, only 'important' roads are used. This requires manual tuning of the data and a delicate tradeoff between computation speed and suboptimality of the computed routes. In previous papers [**21, 22**] we introduced the idea to automatically compute highway hierarchies that yield optimal routes uncompromisingly quickly. The basic idea is to define a neighbourhood for each node to consist of its H closest neighbours. Now

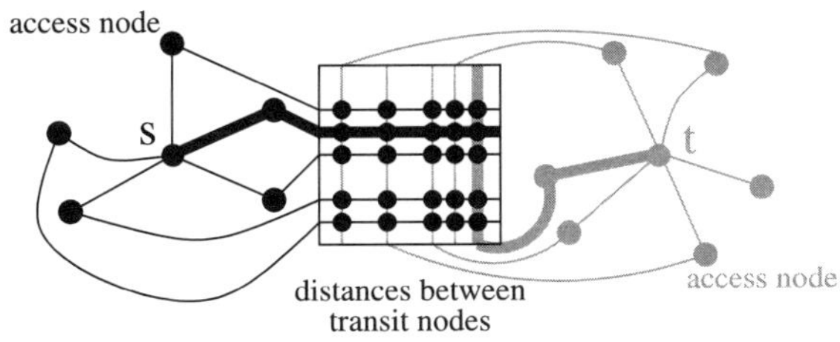

FIGURE 1. Schematic representation of transit-node routing.

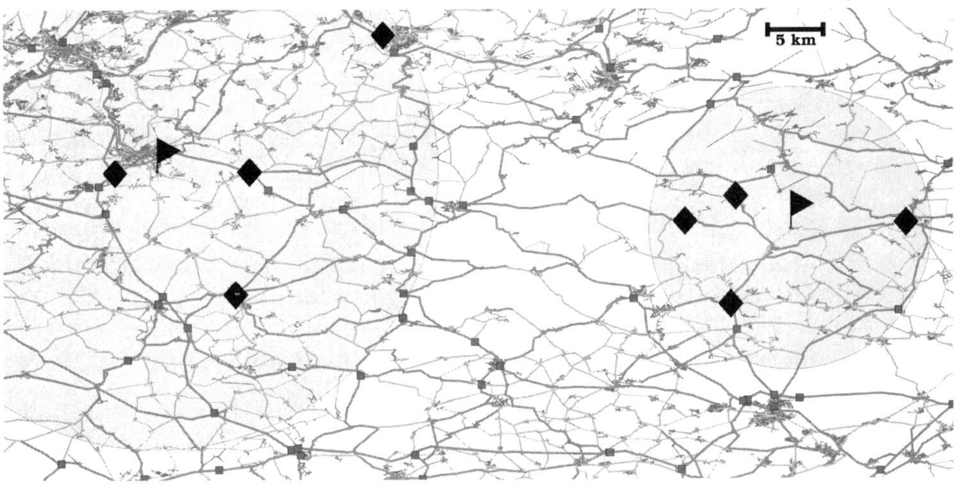

FIGURE 2. Finding the optimal travel time between two points (flags) somewhere between Saarbrücken and Karlsruhe amounts to retrieving the two times four *access nodes* (diamonds), performing 16 table lookups between all pairs of access nodes, and checking that the two disks defining the *locality filter* do not overlap. *Transit nodes* that do not belong to the access-node sets of the selected source and target nodes are drawn as small squares.

an edge (u, v) is a highway edge if there is some shortest path $\langle s, \ldots, u, v, \ldots t\rangle$ such that neither u is in the neighbourhood of t nor v is in the neighbourhood of s. This defines the first level of the highway hierarchy. After contracting the network to remove low degree nodes (which yields the so-called *core* of the highway network), the same procedure (identifying the highway network at the next level followed by contraction) is applied recursively. We obtain a hierarchy. The query algorithm is bidirectional Dijkstra with restrictions on relaxing certain edges. Roughly, far away from source or target, only high-level edges need to be considered. Highway hierarchies are successful (several thousand times faster than Dijkstra) because of the property of real-world road networks that for constant neighbourhood size H, the levels of the hierarchy shrink geometrically. One can view this as a *self-similarity*—each level of the hierarchy looks similar to the original network, just a constant factor smaller. Under certain (somewhat optimistic) assumptions, this self-similarity yields logarithmic query time in contrast to the superlinear query time of Dijkstra's algorithm.

1.2.3. *Reach-Based Routing.* Comparable effects can be achieved with the closely related technique of *reach-based routing.* Let $R(v) := \max_{s,t \in V} R_{st}(v)$ denote the *reach* of node v, where $R_{st}(v) := \min(d(s,v), d(v,t))$. Gutman [**13**] observed that a shortest-path search can be pruned at nodes with a reach too small to get to source or target from there. The basic approach was considerably strengthened by Goldberg et al. [**10**].

1.2.4. *Using Distance Tables.* In [**22**] transit-node routing is *almost* anticipated. Precomputed all-to-all distances on some sufficiently high level—say K—of the highway hierarchy are used to terminate the local searches when they ascended far enough in the hierarchy. The main differences to transit-node routing is that access nodes are computed online and that only distances within level K of the highway hierarchy (rather than distances in the underlying graph) are precomputed. The latter leads to considerably larger sets of access nodes (≈ 55 instead of 10) that made precomputing them appear much less attractive as it actually is. It was also not addressed, how to decide *when* the distance given by the distance table is the actual shortest-path distance.

1.2.5. *Separators.* Perhaps the most well known property of road networks is that they are almost planar, i.e, techniques developed for planar graphs will often also work for road networks. Queries accurate within a factor $(1+\epsilon)$ can be answered in near constant time using $O((n \log n)/\epsilon)$ space and preprocessing time [**27**]. Using $O(n \log^2 n)$ space and preprocessing time, query time $O(\sqrt{n} \log^2 n)$ can be achieved [**9, 15**] for directed planar graphs without negative cycles. A previous practical approach is the *separator-based multi-level method* [**26**]. The idea is to partition the graph into small components by removing a (hopefully small) set of separator nodes. These separator nodes together with edges representing precomputed paths between them constitute the next level of the graph.

1.2.6. *Highway-Node Routing.* [**25, 24**] is a generalisation of the separator-based multi-level method. It computes for a given sequence of node sets $V =: V_0 \supseteq V_1 \supseteq \ldots \supseteq V_L$ (which do *not* have to form separators) a hierarchy of *overlay graphs*: the level-ℓ overlay graph consists of the node set V_ℓ and an edge set E_ℓ that ensures the property that all distances between nodes in V_ℓ are equal to the corresponding distances in the underlying graph $G_{\ell-1}$. A bidirectional query algorithm takes advantage of the multi-level overlay graph by never moving downwards in the hierarchy—by that means, the search space size is greatly reduced.

1.2.7. *Separator-Based Transit-Node Routing.* Using more space and preprocessing time, separators can be used for transit-node routing. The separator nodes become transit nodes and the access nodes are the border nodes of the component of v. Local queries are those within a single component. Another layer of transit nodes can be added by recursively finding separators of each component. Independently from our work, Müller et al. have essentially developed this approach, using different terminology[1]. Note that their first results [**19**] were published before any other implementation of transit-node routing. However, it took some time till

[1]We chose to interpret their work using the transit-node terminology in order to point out similarities to our work.

reliable measurement data were available[2] [**6**]. An interesting difference to generic transit-node routing is that the required information for routing between any pair of components is arranged together. This takes additional space but has the advantage that the information can be accessed more cache efficiently (it also allows subsequent space optimisations).

Although separators of road networks have much better properties than the worst case bounds for planar graphs would suggest, separator-based transit-node routing needs about 4–8 times as many access nodes as our scheme (depending on the used metric) leading to much higher preprocessing times. The main reason for the difference in number of access nodes is that the separator approach does not take the 'sufficiently far away' criterion into account that is so important for reducing the number of access nodes in our approach, in particular in case of the travel time metric.

1.2.8. *Grid-Based Transit-Node Routing.* Bast, Funke and Matijevic proposed the transit-node routing approach based on a geometric grid [**2**]: The network is subdivided into uniform cells. Border nodes of these cells that are needed for 'long-distance' travel are used as access nodes. The union of all access nodes forms the transit-node set. As a locality filter it is sufficient to check whether source and target lie a certain number of cells apart.

They were the first to explicitly formulate the central observations and concepts of transit-node routing[3]. Our work was completed a few weeks later and has been accomplished largely independently from theirs except for the fact that their observation that about ten access nodes per node were sufficient motivated us to rethink our access node definition leading to a considerable reduction from around 55 to about ten, which made an implementation for large graphs much more practicable, accelerated our development process significantly and yielded very good query times. While most algorithms described in [**2**] cater to the specific grid-based approach, we prefer a more generic notion of transit-node routing and regard our implementation only as one possible (and very successful) instantiation of transit-node routing.

In a joint paper [**3**], both implementations are contrasted. One noticeable difference is that we deal with all types of queries in a highly efficient way, while the grid-based variant only answers non-local queries very quickly (which, admittedly, constitute a very large fraction of all queries if source and target are picked uniformly at random). The grid-based variant is designed for comparatively modest memory requirements, while our implementation has significantly smaller preprocessing and average query times. Note that our implementation would need considerably less memory if we concentrated only on undirected graphs and non-local queries as it is done in the grid-based implementation.

1.2.9. *Computing Distance Tables.* For given source and target node sets, a table containing the distances between all source-target node pairs can be computed

[2]In their implementation, the preprocessed data is stored on a hard disk. Using a more compact representation, the data would fit into main memory. Therefore, when measuring query times, it is justifiable to assume that the required data was in main memory. This situation makes performing experiments more difficult.

[3]In particular, they introduced the term 'transit node'. In a joint paper [**3**], we adopted some formulations and terms from [**2**] to describe the generic approach. For the sake of simplicity, we decided to keep these phrases in this paper.

very efficiently using a many-to-many shortest path algorithm [**16, 24**]. The development of this algorithm was another step on the way from the highway hierarchies enhanced by a distance table to transit-node routing since it allowed to compute distances in the original graph between all level-K nodes of the highway hierarchy.

1.2.10. *Geometry.* A tempting property of road networks is that nodes have a geographic position. Even if this information is not available, equally useful coordinates can be synthesised [**30**]. Interestingly, so far, successful geometric speedup techniques have always been beaten by related non-geometric techniques (e.g. [**14**] by [**11, 12**] or [**29**] by [**17, 18**]). We initially thought that our approach outperforming the grid-based approach to transit-node routing would turn out to be another instance of this phenomenon. However, currently it looks like we need a geometric locality filter for good performance. Arriving at this observation was our final step to a fully functional version of transit-node routing.

1.2.11. *Highway-Hierarchy-Based Transit-Node Routing.* Our first implementation of transit-node routing [**23, 4, 3**] was based on highway hierarchies. It is very similar to the implementation presented in Section 4 of this paper. The main difference is the fact that we now use highway-node routing instead of highway hierarchies, which leads to considerably smaller preprocessing times.

1.2.12. *Goal Direction.* Another interesting property of road networks is that they allow effective goal-directed search using **A*** *search* [**14**]: lower bounds define a vertex potential that directs search towards the target. This approach was recently shown to be very effective if lower bounds are computed using precomputed shortest-path distances to a carefully selected set of about 20 **L***andmark* nodes [**11, 12**] using the **T**riangle inequality (*ALT*). In combination with reach-based routing, this is one of the fastest known speedup techniques [**10**]. An interesting observation is that in transit-node routing, the access nodes could be used as landmarks (with aid of the distance tables). The resulting lower bound could be used for distinguishing local and global queries or for guiding local search.

1.3. Our Contributions. We have developed transit-node routing largely independently of the related work described in Sections 1.2.7 and 1.2.8—except for the influences already mentioned in the respective sections. One unique contribution of our work is that we introduce transit-node routing as a *generic framework*, which covers all existing implementations.

Furthermore, our framework extends transit-node routing to a *hierarchical approach* that consists of several levels: each level can have its own access nodes, an (only partly filled) distance table, and a locality filter. This way, all types of queries can be answered very efficiently. Finally, our concrete implementation outperforms the other implementations in most respects.

1.4. Outline. In Section 2, we present the key concepts of transit-node routing in a generic way. In Section 3, we instantiate the algorithm from the previous section by giving concrete access mappings, while a concrete choice of the transit-node sets is still *not* specified. Note that other instantiations of the generic algorithm that deviate from this section are possible (see Section 6.1). In Section 4, we give a concrete specialisation of the abstract instantiation of the previous section, determining transit-node sets using highway hierarchies (Section 1.2.2), performing the preprocessing using highway-node routing (Section 1.2.6) and the many-to-many algorithm based on highway-node routing (Section 1.2.9), and applying geometric circles to define the locality filters. Note that many other reasonable concrete instantiations are conceivable, which is the reason why we decided to specialise the generic algorithm from Section 2 in two steps instead of merging Sections 3 and 4. Experiments reported in Section 5 give average query times of about 4 μs and query times around 20 μs for slowest category of queries. Our main focus is on computing *quickest*-path[4] lengths. However, we also give some results on outputting a complete description of the quickest path and on computing travel *distances*.

2. A Generic Algorithm

For a given graph $G = (V, E)$, we consider $L + 1$ sets

$$V =: \mathcal{T}_0 \supseteq \mathcal{T}_1 \supseteq \ldots \supseteq \mathcal{T}_L$$

of *transit nodes*.[5] Moreover, for any level $\ell, 0 \leq \ell \leq L$, we consider

- a forward and a backward *access mapping* $\overrightarrow{A}_\ell : V \to 2^{\mathcal{T}_\ell}$ and $\overleftarrow{A}_\ell : V \to 2^{\mathcal{T}_\ell}$, which map a node to its forward and backward *access nodes*, respectively,
- a *locality filter* $\mathcal{L}_\ell : V \times V \to \{\mathsf{true}, \mathsf{false}\}$, which decides whether the distance between two nodes can be determined using only levels $\geq \ell$ of transit-node routing,
- a distance table $D_\ell : \mathcal{T}_\ell \times \mathcal{T}_\ell \to \mathbb{R}_0^+ \cup \{\infty\}$, which contains the correct distances between all node pairs from $\mathcal{T}_\ell \times \mathcal{T}_\ell$ except for the distances that can be computed using higher levels of transit-node routing,
- the distance $d_\ell : V \times V \to \mathbb{R}_0^+ \cup \{\infty\}$ that is obtained using level ℓ of transit-node routing, i.e., considering all access nodes to level ℓ and the distances between all pairs of these access nodes, and
- the minimal distance $d_{\geq \ell} : V \times V \to \mathbb{R}_0^+ \cup \{\infty\}$ that can be obtained using all levels $\geq \ell$.

To avoid some case distinctions, we introduce the following definitions:

- $\mathcal{T}_{L+1} := \emptyset$,
- $\overrightarrow{A}_0(u) := \overleftarrow{A}_0(u) := \{u\}$,
- $d_{\geq L+1}(s,t) := \infty$,
- $\min \emptyset := \infty$.

Now, for any level $\ell, 0 \leq \ell \leq L$, we give a precise definition of the three distance functions D_ℓ, d_ℓ, and $d_{\geq \ell}$:

$$D_\ell(s,t) := \begin{cases} d(s,t) & \text{if } d(s,t) < d_{\geq \ell+1}(s,t) \\ \infty & \text{otherwise} \end{cases} \tag{1}$$

$$d_\ell(s,t) := \min\{d(s,u) + D_\ell(u,v) + d(v,t) \mid u \in \overrightarrow{A}_\ell(s), v \in \overleftarrow{A}_\ell(t)\} \tag{2}$$

$$d_{\geq \ell}(s,t) := \min_{k \geq \ell} d_k(s,t) \tag{3}$$

---

[4]Note that we often use the term '*shortest* path' as a synonym for 'quickest path'.

[5]Note that in earlier publications **[23, 3, 4]**, the order of the levels (which we called 'layers' at that time) was reversed: the topmost transit-node set was denoted by $\mathcal{T}_1$, now it is denoted by $\mathcal{T}_L$. We have changed the order to blend well with other hierarchical approaches.

Note that the following equation is equivalent to (3):

$$d_{\geq \ell}(s,t) = \min(d_\ell(s,t), \min_{k \geq \ell+1} d_k(s,t)) \tag{4}$$

Obviously, all these distances are upper bounds on the actual shortest-path length, as stated in the following proposition:

PROPOSITION 1. *$D_\ell(s,t) \geq d(s,t)$, $d_\ell(s,t) \geq d(s,t)$, $d_{\geq \ell}(s,t) \geq d(s,t)$.*

We assume that all distances to/from forward/backward access nodes and all distances $D_\ell(s,t)$ have been precomputed. We can show that we always obtain the correct shortest-path length when we use all levels of transit-node routing:

LEMMA 1. *$d_{\geq 0}(s,t) = d(s,t)$.*

PROOF. Due to (2), we have $d_0(s,t) = d(s,s) + D_0(s,t) + d(t,t)$ since $\overrightarrow{A}_0(s) = \{s\}$ and $\overleftarrow{A}_0(t) = \{t\}$. If $d(s,t) < d_{\geq 1}(s,t)$, we have $d_0(s,t) = D_0(s,t) = d(s,t)$ (due to (1)) and thus, $d_{\geq 0}(s,t) = \min(d_0(s,t), d_{\geq 1}(s,t)) = d(s,t)$ by (4) and Proposition 1. Otherwise ($d(s,t) = d_{\geq 1}(s,t)$), we have $d_0(s,t) = D_0(s,t) = \infty$ (due to (1)) and, again, $d_{\geq 0}(s,t) = \min(d_0(s,t), d_{\geq 1}(s,t)) = d(s,t)$. □

Of course, using all levels is comparatively expensive. Therefore, we want to avoid accessing levels that are not needed to get the correct result. For the decision making we want to employ the already introduced locality filters. We require that

$$\neg \mathcal{L}_\ell(s,t) \rightarrow \big(d(s,t) = d_{\geq \ell}(s,t)\big). \tag{5}$$

Then, we can use the *transit-node routing* algorithm as specified in Figure 3 to efficiently compute the length of a shortest path from a given source node s to a given target node t.

input: source node s and target node t
output: distance $d(s,t)$

```
1  d' := ∞;
2  for ℓ := L downto 0 do
3      d' := min(d', d_ℓ(s,t));
4      assert d' = d_≥ℓ(s,t);
5      if ¬L_ℓ(s,t) then break;
6  return d';
```

FIGURE 3. The transit-node routing algorithm.

THEOREM 1. *Transit-node routing is correct.*

PROOF. If the condition in Line 5 is fulfilled at some point, we return $d' = d_{\geq \ell}(s,t) = d(s,t)$ due to (5). Otherwise, we return $d' = d_{\geq 0}(s,t) = d(s,t)$ according to Lemma 1. □

Practical Remarks. In the distance tables D_ℓ, it is sufficient to store only the non-infinity entries explicitly. For this purpose, we can use a space-efficient static hash table. Furthermore, as an alternative to precomputing the entries in D_0, we can use any other shortest-path algorithm to compute the distances D_0 on-the-fly when they are required.

3. An Abstract Instantiation

Covering-Paths Set. We consider a graph $G = (V, E)$, a node subset $V' \subseteq V$, a node $s \in V$, and a set $C \subseteq \{\langle s, \dots, u\rangle \mid u \in V'\}$ of paths in G.

DEFINITION 1. *The set C is a* covering-paths set *of s w.r.t. V' if for any node $t \in V'$ that can be reached from s, there is a node $u \in V'$ on some shortest s-t-path P such that $P|_{s\to u} \in C$, i.e.,*

$$P = \langle \underbrace{s, \dots, \overbrace{u}^{\in V'}}_{\in C}, \dots, \overbrace{t}^{\in V'} \rangle.$$

Access Mapping. For a node s and a level ℓ, consider a set C of covering paths of s w.r.t. $\mathcal{T}_\ell$ in G. (To obtain a very efficient algorithm, we might want to choose a *minimal* covering-paths set.) Let

$$\overrightarrow{A}_\ell(s) := \{v \mid P = \langle s, \dots, v\rangle \in C\}.$$

The backward access mapping is defined analogously, considering the reverse graph $\overleftarrow{G}$ instead of G.

Locality Filter. An explicit representation of a level-ℓ locality filter (storing n^2 bits) would need too much space for large graphs. Therefore, we look for a more space-efficient alternative. We want to identify node pairs (s, t) such that the distance $d(s, t)$ cannot be computed using transit-node routing in level ℓ or higher. For each of these pairs, we pick one *witness*, a particular node p on a shortest s-t-path. We make sure that both s and t memorise this witness p. Then, when we want to evaluate $\mathcal{L}_\ell(s, t)$, we just have to check whether s and t share a common witness. Note that this approach can lead to *false positives*, i.e., two nodes might share a common witness although their distance actually can be computed using transit-node routing in level ℓ or higher.

Beside paying attention to the memory requirements, we are also interested in fast preprocessing times. Therefore, we introduce the concept of *handing* computed data *down* from higher to lower levels: Let us consider some path

$$\langle s_0, \dots, s_1, \dots, p, \dots, t_1, \dots, t_0\rangle$$

with $s_0, t_0 \in \mathcal{T}_0$ and $s_1, t_1 \in \mathcal{T}_1$. Moreover, let us assume that we already know that $d(s_1, t_1)$ cannot be computed using level 2 or higher. Thus, we have some witness p and both s_1 and t_1 memorise this witness. Now, this witness is handed down from s_1 to s_0 and from t_1 to t_0. An equivalent formulation is to say that s_0 inherits the witness p from s_1. Now, if we want to decide whether $d(s_0, t_0)$ can be determined using level 2 or higher, the answer is 'no' since s_0 and t_0 share the common witness p. Note that by this means, the number of false positives may increase.

In the following, we work out the formal details of these ideas. The *level* $\ell(u)$ of a node $u \in V$ is $\max\{\ell \mid u \in \mathcal{T}_\ell\}$. Let us assume that we have some fixed strategy that picks for any two connected nodes s and t one particular node $p(s, t)$ on one particular shortest s-t-path. We define forward and backward node sets $\overrightarrow{K}_\ell : V \to 2^V$ and $\overleftarrow{K}_\ell : V \to 2^V$ in the following way: for any node s and any level $\ell < \ell(s) + 1$, $\overrightarrow{K}_\ell(s) := \emptyset$, for level $\ell = \ell(s) + 1$,

$$\overrightarrow{K}_\ell(s) := \{p(s, t) \mid t \in V \wedge \ell(s) = \ell(t) \wedge d(s, t) < d_{\geq \ell}(s, t)\} \tag{6}$$

and for any level $\ell > \ell(s) + 1$,

$$\overrightarrow{K}_\ell(s) := \bigcup_{u \in \overrightarrow{A}_{\ell-1}(s)} \overrightarrow{K}_\ell(u), \tag{7}$$

and analogously, for any node t and any level $\ell < \ell(t) + 1$, $\overleftarrow{K}_\ell(t) := \emptyset$, for level $\ell = \ell(t) + 1$,

$$\overleftarrow{K}_\ell(t) := \{p(s,t) \mid s \in V \wedge \ell(s) = \ell(t) \wedge d(s,t) < d_{\geq \ell}(s,t)\} \tag{8}$$

and for any level $\ell > \ell(t) + 1$,

$$\overleftarrow{K}_\ell(t) := \bigcup_{u \in \overleftarrow{A}_{\ell-1}(t)} \overleftarrow{K}_\ell(u). \tag{9}$$

Note that Equations 6 and 8 reflect the 'witness' idea, while Equations 7 and 9 reflect the 'handing down' idea.

Finally, we define the locality filter

$$\mathcal{L}_\ell(s,t) := \bigvee_{k \leq \ell} \left(\overrightarrow{K}_k(s) \cap \overleftarrow{K}_k(t) \neq \emptyset \right). \tag{10}$$

LEMMA 2. *Consider two nodes s and t with $d(s,t) \neq \infty$. If and only if there is some node $u \in \mathcal{T}_\ell$ on some shortest s-t-path P, then $d_{\geq \ell}(s,t) = d(s,t)$.*

PROOF. $\Leftarrow$) We have $d_{\geq \ell}(s,t) = d(s,t)$. This implies, by (3) and (2), that there is a level $k \geq \ell$, a node $u \in \overrightarrow{A}_k(s)$, and a node $v \in \overleftarrow{A}_k(t)$ such that $d(s,u) + D_k(u,v) + d(v,t) = d(s,t)$. Due to Proposition 1, we have $D_k(u,v) \geq d(u,v)$. We can conclude that u and v are nodes on a shortest s-t-path. Furthermore, we know that $u \in \overrightarrow{A}_k(s) \subseteq \mathcal{T}_k \subseteq \mathcal{T}_\ell$.

$\Rightarrow$) We pick the maximum level $k \geq \ell$ with the property that there is some node from $\mathcal{T}_k$ on some shortest s-t-path P. Let u and v denote the first and the last node from $\mathcal{T}_k$ on P, respectively. The case $u = v$ is possible. According to the definitions of the covering paths and the access mappings, there is a node $u' \in \overrightarrow{A}_k(s) \subseteq \mathcal{T}_k$ on a shortest s-u-path $\overrightarrow{P}$ and a node $v' \in \overleftarrow{A}_k(t) \subseteq \mathcal{T}_k$ on a shortest v-t-path $\overleftarrow{P}$. Consider the path

$$P' := \langle \overbrace{s, \ldots, u', \ldots, u}^{\overrightarrow{P}}, \ldots, \overbrace{v, \ldots, v', \ldots, t}^{\overleftarrow{P}} \rangle, \quad \text{with } \underbrace{u, \ldots, v}_{P|_{u \to v}}$$

which is a shortest s-t-path as well. According to (2), we have $d_k(s,t) \leq d(s,u') + D_k(u',v') + d(v',t)$. Due to our choice of k, we know that there is no node $x \in \mathcal{T}_{k+1}$ on any shortest u'-v'-path Q—otherwise, the same node x would be on the shortest s-t-path $P'|_{s \to u'} \circ Q \circ P'|_{v' \to t}$. From the part of this lemma that has already been proven, it follows that $d(u',v') < d_{\geq k+1}(u',v')$. Thus, by (1), $D_k(u',v') = d(u',v')$ and, consequently, $d_k(s,t) \leq d(s,u') + D_k(u',v') + d(v',t) = d(s,t)$. From $d_{\geq \ell}(s,t) \leq d_k(s,t) \leq d(s,t)$, it follows that $d_{\geq \ell}(s,t) = d(s,t)$ due to Proposition 1. □

LEMMA 3. *The locality filter specified in Equation 10 fulfils Equation 5.*

PROOF. Trivial for $d(s,t) = \infty$ (due to Proposition 1). For $d(s,t) \neq \infty$, we want to show the contraposition of Equation 5 and therefore assume that $d(s,t) \neq d_{\geq \ell}(s,t)$. Let k be the maximum level such that $d_{\geq k-1}(s,t) = d(s,t)$. Such a k must exist due to Lemma 1. The choice of k implies $k-1 < \ell$, $d_{\geq k}(s,t) \neq d(s,t)$, and $d_{k-1}(s,t) = d(s,t)$. Hence, there is some shortest s-t-path with nodes $u' \in \overrightarrow{A}_{k-1}(s)$ and $v' \in \overleftarrow{A}_{k-1}(t)$ on it. If $s \in \mathcal{T}_{k-1}$, we set $u := s$; otherwise, $u := u'$. Analogously, if $t \in \mathcal{T}_{k-1}$, we set $v := t$; otherwise, $v := v'$. In any case, we have $u, v \in \mathcal{T}_{k-1}$.

Lemma 2 and $d_{\geq k}(s,t) \neq d(s,t)$ imply that there is no shortest s-t-path that contains a node from $\mathcal{T}_k$. In particular, $u, v \notin \mathcal{T}_k$ and $d(u,v) < d_{\geq k}(u,v)$—otherwise, there would be a shortest u-v-path containing a node $x \in \mathcal{T}_k$ and thus, also a shortest s-t-path containing x. Since $u, v \in \mathcal{T}_{k-1} \setminus \mathcal{T}_k$, we have $\ell(u) = \ell(v) = k-1$. We can conclude that $p(u,v) \in \overrightarrow{K}_k(u) \cap \overleftarrow{K}_k(v)$ due to Equations 6 and 8. If $s \notin \mathcal{T}_{k-1}$, we have $\ell(s) < k-1$, which implies $p(u,v) \in \overrightarrow{K}_k(s)$ due to Equation 7: s inherits $p(u,v)$ from $u = u'$. Otherwise, we have $s = u$ so that $p(u,v) \in \overrightarrow{K}_k(s)$ holds as well. An analogous argument applies to $\overleftarrow{K}_k(t)$. Thus, $p(u,v) \in \overrightarrow{K}_k(s) \cap \overleftarrow{K}_k(t)$. Since $k \leq \ell$, $\mathcal{L}_\ell(s,t) = \mathsf{true}$ according to (10). □

3.1. Computing Access Nodes. Here, we describe how to determine the forward access nodes to the topmost level L. Analogous methods can be applied to compute forward and backward access nodes to different levels. From each node $u \in V$, we perform a Dijkstra search in G in order to determine the covering-paths set w.r.t. $\mathcal{T}_L$. We take each endpoint of a covering path as access node of u. In other words, we perform a Dijkstra search that can be stopped as soon as each path in the current partial shortest-path tree from the root u to a node v in the priority queue contains at least one settled node from $\mathcal{T}_L$; then, we can take on each branch the node $v \in \mathcal{T}_L$ closest to u—these nodes are the endpoints of a covering-paths set and, thus, they form the access-node set of u. Applied naively, this approach is rather inefficient. However, we can use two tricks to make it efficient.

First, we can prune the search at nodes from $\mathcal{T}_L$. However, in general, this variant does not yield a *minimal* access-node set, which would be preferable. Fortunately, the resulting set can be easily *reduced* if the distances between all transit nodes are already known: if an access node y can be reached from u via another access node w on a shortest path, we can discard y. Figure 4 gives an example.

Second, we can only determine the access node sets $\overrightarrow{A}_L(v)$ for all nodes $v \in \mathcal{T}_{L-1}$ and the sets $\overrightarrow{A}_{L-1}(u)$ for all nodes $u \in V$. Then, for each node $u \in V$, we can compute

$$\overrightarrow{A}_L(u) := \bigcup_{v \in \overrightarrow{A}_{L-1}(u)} \overrightarrow{A}_L(v).$$

Again, we can use the reduction technique to remove unnecessary elements from the set union. The idea to hand access nodes down can be extended to work across more than one level:

$$\overrightarrow{A}_L(u_0) := \bigcup_{u_1 \in \overrightarrow{A}_1(u_0)} \bigcup_{u_2 \in \overrightarrow{A}_2(u_1)} \cdots \bigcup_{u_{L-1} \in \overrightarrow{A}_{L-1}(u_{L-2})} \overrightarrow{A}_L(u_{L-1}). \tag{11}$$

LEMMA 4. *Handing down access nodes is correct, i.e., the resulting access-node set complies with the specification at the beginning of Section 3.*

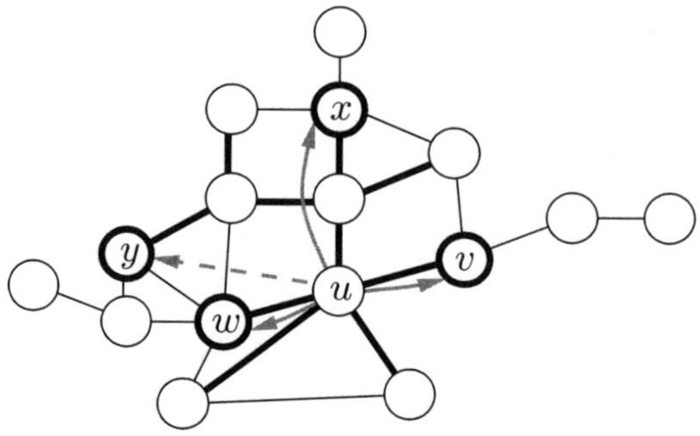

FIGURE 4. Example for the computation of access nodes including the first, but not the second 'trick'. Edge weights correspond to the lengths of the drawn line segments. The nodes v, w, x, and y belong to $\mathcal{T}_L$. The search is started from u. All thick edges belong to the search tree. All depicted nodes from $\mathcal{T}_L$ are endpoints of covering paths. However, y can be removed from this set since the path from u via w to y turns out to be shorter than the path that has been found. Thus, u has only three access nodes.

PROOF. We say that an access-node set $\overrightarrow{A}_\ell(u)$ is *proper* (i.e., it complies with the specification at the beginning of Section 3) iff there is a covering-paths set $C_\ell(u)$ of u w.r.t. $\mathcal{T}_\ell$ such that $\overrightarrow{A}_\ell(u) = \{v \mid P = \langle u, \dots, v\rangle \in C_\ell(u)\}$.

Assume that for some node u and some level $\ell > 0$, we have a proper access-node set $\overrightarrow{A}_{\ell-1}(u)$ (and thus, a corresponding covering-paths set $C_{\ell-1}(u)$) and that for each node $v \in \overrightarrow{A}_{\ell-1}(u)$, we have a proper access-node set $\overrightarrow{A}_\ell(v)$ (and thus, a corresponding covering-paths set $C_\ell(v)$). Let

$$\overrightarrow{A}_\ell(u) := \bigcup_{v \in \overrightarrow{A}_{\ell-1}(u)} \overrightarrow{A}_\ell(v)$$

and

$$C_\ell(u) := \{P = \langle u, \dots, v\rangle \mid P \in \mathcal{U}(G) \wedge v \in \overrightarrow{A}_\ell(u)\},$$

where $\mathcal{U}(G)$ denotes some fixed set of *canonical shortest paths* that contains for each connected pair $(s,t) \in V \times V$ exactly one unique shortest path from s to t such that $P = \langle s, \dots, s', \dots, t', \dots, t\rangle \in \mathcal{U}(G)$ implies that $P|_{s' \to t'} \in \mathcal{U}(G)$. We have to prove that $\overrightarrow{A}_\ell(u)$ is a proper access-node set. For that, it is sufficient to show that $C_\ell(u)$ is a covering-paths set of u w.r.t. $\mathcal{T}_\ell$.

Consider any node $t \in \mathcal{T}_\ell$ that can be reached from u. We have to show that there is a node $x \in \mathcal{T}_\ell$ on some shortest u-t-path P such that $P|_{u \to x} \in C_\ell(u)$.

Since $t \in \mathcal{T}_\ell \subseteq \mathcal{T}_{\ell-1}$, there is a node y on some shortest u-t-path P' such that $P'|_{u \to y} \in C_{\ell-1}(u)$ and thus, $y \in \overrightarrow{A}_{\ell-1}(u)$. Similarly, since $t \in \mathcal{T}_\ell$, there is a node x on some shortest y-t-path P'' such that $P''|_{y \to x} \in C_\ell(y)$ and thus, $x \in \overrightarrow{A}_\ell(y) \subseteq \mathcal{T}_\ell$. Let $P^c \in \mathcal{U}(G)$ denote the canonical shortest u-x-path. Set $P := P^c \circ P''|_{x \to t}$. Note that P is a shortest u-t-path. The definition of $\overrightarrow{A}_\ell(u)$ implies that $x \in \overrightarrow{A}_\ell(u)$. Hence, $P|_{u \to x} \in C_\ell(u)$.

By induction, this proof can be extended to multiple levels. □

3.2. Computing Distance Tables. To compute an all-pairs distance table, we can use the many-to-many algorithm mentioned in Section 1.2.9. Roughly, this algorithm first performs independent backward searches from all transit nodes and stores the gathered distance information in *buckets* associated with each node in the search space. Then, a forward search from each transit node scans all buckets it encounters and uses the resulting path length information to update a table of tentative distances.

For the topmost table D_L (where we always have $D_L(s,t) = d(s,t)$), this procedure can be applied directly. For all other tables D_ℓ, $\ell < L$, we have to respect that an explicit entry $D_\ell(s,t)$ is only required if $d(s,t) < d_{\geq \ell+1}(s,t)$—all other entries are ∞ and do not have to be explicitly stored. In order to be able to check this condition, the preprocessing of transit-node routing is done in a top-down fashion, i.e., we first compute the access nodes and the distance table for the topmost level before constructing level $L-1$, and so on. Thus, when we compute the table D_ℓ, we can already access $d_{\geq \ell+1}(s,t)$.

A naive application of the many-to-many algorithm is prohibitive for lower levels (probably even for level $L-1$). Fortunately, there is one simple trick based on Lemma 2: when performing backward and forward searches in order to compute table D_ℓ, $\ell < L$, we do not have to relax edges out of nodes $u \in \mathcal{T}_{\ell+1}$. By this measure, we only might miss shortest s-t-paths with a node from $\mathcal{T}_{\ell+1}$ on them. However, due to Lemma 2, we already know that in these cases $d_{\geq \ell+1}(s,t) = d(s,t)$ so that $D_\ell(s,t) = \infty$. Note that the computation of the distance table D_ℓ consists of the same local forward and backward searches as the computation of the access-node sets $\overrightarrow{A}_{\ell+1}$ and $\overleftarrow{A}_{\ell+1}$. Thus, it is sufficient to perform the respective search processes only once and extracting both the access nodes and the data required for the distance table computation.

3.3. Computing Locality Filters. As already mentioned, the preprocessing of transit-node routing is done in a top-down fashion. We compute the forward and backward node sets $\overrightarrow{K}_\ell$ and $\overleftarrow{K}_\ell$ first for all nodes in $\mathcal{T}_L$, then for the nodes in $\mathcal{T}_{L-1}$, and so on. For any $u \in \mathcal{T}_L$ and any level ℓ, we just have $\overrightarrow{K}_\ell(u) = \overleftarrow{K}_\ell(u) = \emptyset$. For a level $k < L$ and any node $u \in \mathcal{T}_k \setminus \mathcal{T}_{k+1}$, we 'inherit' the level-ℓ sets from the level-$(\ell-1)$ access nodes for $\ell > k+1$ according to Equations 7 and 9; for $\ell = k+1$, we apply Equations 6 and 8. In order to deal with the latter case, we have to determine all node pairs (s,t) such that $\ell(s) = \ell(t) = k$ and $d(s,t) < d_{\geq k+1}(s,t)$. This is exactly what we do when we compute the level-k distance table D_k. Hence, the computation of the sets $\overrightarrow{K}_{k+1}$ and $\overleftarrow{K}_{k+1}$ can be viewed as a byproduct of the computation of D_k.

After all sets $\overrightarrow{K}_\ell$ and $\overleftarrow{K}_\ell$ have been determined, the locality filters are defined according to Equation 10.

Faster Computation of Supersets. In spite of the trick mentioned in Section 3.2, the computation of a distance table can get expensive so that we might want to do without distance tables in the lower levels and use some shortest-path algorithm instead that computes the required distances on demand. In this case, the locality filters can no longer be obtained as a byproduct of the distance table computation so that we have to find a different way to compute them efficiently. Let us consider some level $k < L$ and two nodes s and t such that $\ell(s) = \ell(t) = k$. Consider a local forward search from s that determines covering paths of s w.r.t. $\mathcal{T}_{k+1}$ yielding a search tree $\overrightarrow{B}$ and, analogously, a local backward search from t yielding a search tree $\overleftarrow{B}$. We set

$$\overrightarrow{K}'_{k+1}(s) := \overrightarrow{B} \setminus \mathcal{T}_{k+1} \quad \text{and} \quad \overleftarrow{K}'_{k+1}(t) := \overleftarrow{B} \setminus \mathcal{T}_{k+1}.$$

LEMMA 5. *$\overrightarrow{K}'_{k+1}(s) \supseteq \overrightarrow{K}_{k+1}(s)$ and $\overleftarrow{K}'_{k+1}(t) \supseteq \overleftarrow{K}_{k+1}(t)$.*

PROOF. Consider a node u from $\overrightarrow{K}_{k+1}(s)$. According to Equation 6, there is a node t such that u is a node on some shortest s-t-path P and $d(s,t) < d_{\geq k+1}(s,t)$. Due to Lemma 2, we can conclude that there is no shortest s-t-path with a node from $\mathcal{T}_{k+1}$ on it; in particular, $u \notin \mathcal{T}_{k+1}$. Hence, the forward search is not pruned at any node on $P|_{s \to u}$ so that $u \in \overrightarrow{B} \setminus \mathcal{T}_{k+1}$, which implies $\overrightarrow{K}'_{k+1}(s) \supseteq \overrightarrow{K}_{k+1}(s)$. An analogous proof exists for $\overleftarrow{K}'_{k+1}(t) \supseteq \overleftarrow{K}_{k+1}(t)$. □

Obviously, locality filters that are based on these supersets are still correct in the sense that they fulfil Equation 5. However, the number of false positives increases. Note that the computation of the supersets $\overrightarrow{K}'_{k+1}(s)$ and $\overleftarrow{K}'_{k+1}(t)$ requires the same local searches as the computation of the access-node sets $\overrightarrow{A}_{k+1}(s)$ and $\overleftarrow{A}_{k+1}(t)$. Therefore, when dealing with supersets, the computation of the locality filters can be viewed as a byproduct of the computation of the access-node sets.

3.4. Trade-Offs. Instead of precomputing all access-node sets, distance tables, and locality filters, we can decide to compute only a part of the data required for transit-node routing and determine the remaining data on demand during the query. In case of the access nodes, we can postpone the local searches for the covering-paths set to query time. Moreover, it is sufficient to store for a node $u \in \mathcal{T}_\ell$ only the access nodes to level $\ell + 1$; then, during a query, access nodes to higher levels can be retrieved using Equation 11.

In case of the distance tables, we can—as already mentioned—omit the distance tables in the lowest levels and perform an explicit shortest-path search instead.

In case of the locality filters, we can postpone the application of Equations 7 and 9 until query time as well so that a node $u \in \mathcal{T}_\ell$ stores only $\overrightarrow{K}_{\ell+1}(u)$ and $\overleftarrow{K}_{\ell+1}(u)$.

Of course, postponing parts of the preprocessing reduces preprocessing time and memory consumption, but increases query time.

3.5. Outputting Complete Path Descriptions. Generally, in a graph with bounded degree (e.g., a road network) using a (near) constant time distance oracle, we can output a shortest path from s to t in (near) constant time per edge: Look for an edge (s, s') such that $d(s, s') + d(s', t) = d(s, t)$, output (s, s'). Continue by looking for a shortest path from s' to t. Repeat until t is reached.

In the special case of transit-node routing, we can speed up this process by two measures. Suppose the shortest path uses the access nodes $u \in \overrightarrow{A}_L(s)$ and $v \in \overleftarrow{A}_L(t)$. First, while reconstructing the path from s to u, we can determine the next hop by considering all adjacent nodes s' of s and checking whether $d(s, s') + d(s', u) = d(s, u)$. Usually[6], the distance $d(s', u)$ is directly available since u is also an access node of s'. Analogously, the path from v to t can be determined.

[6]In a few cases—when u is not an access node of s' (which can only happen if the shortest paths in the graph are not unique)—, we have to consider all access nodes u' of s' and check whether $d(s, s') + d(s', u') + d(u', u) = d(s, u)$. Note that $d(u', u)$ can be looked up in the topmost distance table.

Second, reconstructing the path from u to v can work on the overlay graph G_L of G with node set $\mathcal{T}_L$ rather than on the original graph G. Employing the same methods that are used to expand shortcuts in case of highway hierarchies [**7, 24**], we can precompute information that allows us to output the paths associated with each edge in G_L in time linear in the number of edges of G that it contains. Note that long distance paths will mostly consist of these precomputed paths so that the time per edge can be made very small. These techniques can be generalised to multiple levels.

4. A Concrete Instantiation

4.1. Specifying Transit Nodes. Nodes on high levels of a highway hierarchy (see Section 1.2.2) have the property that they are used on shortest paths far away from source and target. Hence, it is natural to use (the core of) some level K of the highway hierarchy for the transit-node set $\mathcal{T}_L$. Note that we have quite good (though indirect) control over the resulting size of $\mathcal{T}_L$ by choosing the appropriate neighbourhood sizes and the appropriate value for K. For further transit-node levels, we use (the cores of) lower levels of the highway hierarchy.

4.2. Computing Access Nodes. Access-node sets are computed exactly as described in Section 3.1 except for the fact that we use highway-node routing (see Section 1.2.6) to perform local searches in order to determine the covering-paths sets more efficiently.

This implies that before the actual preprocessing of transit-node routing is started, we have to construct a multi-level overlay graph using the transit-node sets as highway-node sets.

4.3. Computing Distance Tables. The topmost table is determined by a standard all-pairs shortest-path computation (using $|\mathcal{T}_L|$-times Dijkstra's algorithm) in the topmost overlay graph G_L. Note that for the topmost level, an application of the many-to-many algorithm using the same multi-level overlay graph would be virtually equivalent to executing just $|\mathcal{T}_L|$-times Dijkstra's algorithm.

All other distance tables, however, are computed as described in Section 3.2, i.e., using the many-to-many algorithm. At this, it is reasonable to employ an instantiation of the many-to-many algorithm that is based on the already constructed multi-level overlay graph.

4.4. Computing Locality Filters. An explicit and exact storage of the forward and backward node sets $\overrightarrow{K}_\ell$ and $\overleftarrow{K}_\ell$ would be very expensive w.r.t. memory consumption. Furthermore, we have to keep in mind that we need a very efficient operation that determines whether the intersection of two node sets is empty. For these reasons, we use *geometric circles* to represent supersets of the sets $\overrightarrow{K}_\ell$ and $\overleftarrow{K}_\ell$. We have already noted in Section 3.3 that using supersets of $\overrightarrow{K}_\ell$ and $\overleftarrow{K}_\ell$ still yields correct locality filters, only the number of false positives may increase.

We assume that a layout of the graph G is available, i.e., for each node in V we know its coordinates in the plane.[7] For each node u, we store forward and backward

[7]Even if this information is not available in the input, equally useful coordinates can be synthesised [**30**].

radii $\overrightarrow{\varrho}_\ell(u)$ and $\overleftarrow{\varrho}_\ell(u)$ such that

$$\overrightarrow{K}'_\ell := \{v \in V : ||v - u||_2 \le \overrightarrow{\varrho}_\ell(u)\} \supseteq \overrightarrow{K}_\ell$$

and, analogously,

$$\overleftarrow{K}'_\ell := \{v \in V : ||v - u||_2 \le \overleftarrow{\varrho}_\ell(u)\} \supseteq \overleftarrow{K}_\ell,$$

where $||v - u||_2$ denotes the Euclidean distance between u and v. An intersection test can be implemented very efficiently by comparing the distance between the two involved nodes with the sum of the radii of the relevant circles:[8]

$$\overrightarrow{K}'_\ell(s) \cap \overleftarrow{K}'_\ell(t) \neq \emptyset \quad \leftrightarrow \quad ||s - t||_2 \le \overrightarrow{\varrho}_\ell(s) + \overleftarrow{\varrho}_\ell(t). \tag{12}$$

Note that the application of Equations 7 and 9 to 'inherit' node sets is quite simple using geometric circles: we use

$$\overrightarrow{\varrho}_\ell(s) := \max\{||s - u||_2 + \overrightarrow{\varrho}_\ell(u) \mid u \in \overrightarrow{A}_{\ell-1}(s)\}$$

and an analogous assignment for $\overleftarrow{\varrho}_\ell(t)$. Figure 5 gives an example.

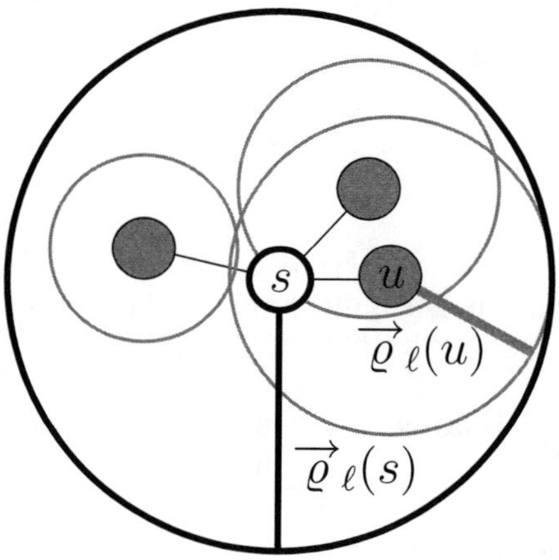

FIGURE 5. Example for the 'inheritance' of a geometric locality filter. The grey nodes constitute the set $\overrightarrow{A}_{\ell-1}(s)$.

Faster Evaluation. Combining Equations 10 and 12, we have

$$\mathcal{L}_\ell(s,t) := \bigvee_{k \le \ell} \Big(||s - t||_2 \le \overrightarrow{\varrho}_k(s) + \overleftarrow{\varrho}_k(t) \Big).$$

Thus, in order to evaluate $\mathcal{L}_\ell(s,t)$, we have to perform up to ℓ comparisons. We can easily do with only one comparison by precomputing

$$\overrightarrow{\varrho}'_\ell(s) := \max_{k \le \ell} \overrightarrow{\varrho}_k(s) \quad \text{and} \quad \overleftarrow{\varrho}'_\ell(t) := \max_{k \le \ell} \overleftarrow{\varrho}_k(t)$$

and using

$$\mathcal{L}_\ell(s,t) := \Big(||s - t||_2 \le \overrightarrow{\varrho}'_\ell(s) + \overleftarrow{\varrho}'_\ell(t) \Big).$$

Note that the number of false positives may increase.

[8]To avoid the expensive square root computation that is required to determine the Euclidean distance, we can alternatively square both sides of the inequality.

4.5. Hasty Inheritance. In order to accelerate the preprocessing, we have already made extensive use of the idea of handing down obtained data (access nodes, locality filters) to lower levels. Basically, for a node u in a level ℓ, we look for covering paths w.r.t. $\mathcal{T}_{\ell+1}$ and inherit the data stored at the endpoints of the covering paths.

We can think of a hastier approach: When we search for the covering paths of u and encounter a node v that has already been processed, i.e., that has already adopted the data from level $\ell+1$, we do not have to continue the search from v. Instead, we can directly inherit the data from v.

In our implementation, we use this technique when we hand data down from level 1 to level 0.

4.6. An Economical and a Generous Variant. In our experiments, we consider two different variants as illustrated in Figure 6.

Variant 'Economical' aims at a good compromise between space consumption, preprocessing time and query time. It uses three levels on top of the original graph (i.e., $L = 3$). We make extensively use of the options presented in Section 3.4. At each node $u \in \mathcal{T}_2$, we store the access nodes to level 3, and at each node $u \in V$, we store the access nodes to level 2. This means that the level-3 access nodes for nodes $u \notin \mathcal{T}_2$ have to be reconstructed during query using Equation 11. Level 1 is only used to accelerate the preprocessing (since it is faster to compute access nodes and locality filters only for a subset $\mathcal{T}_1 \subseteq V$, handing the data down to all nodes). We do not use level-1 access nodes or a level-1 distance table. Instead, we just set $\mathcal{L}_1(s,t) := \mathtt{true}$ for all node pairs (s,t) so that if the query reaches level 1, it is automatically forwarded to level 0.

We explicitly store the level-2 and level-3 distance tables. In level 0, instead of keeping a distance table, we perform a shortest-path query using highway-node routing.

The locality filters are dealt with analogously to the access nodes: at each node $u \in \mathcal{T}_2$, we store $\overrightarrow{\varrho}_3(u)$ and $\overleftarrow{\varrho}_3(u)$, and at each node $u \in V$, we store $\overrightarrow{\varrho}'_2(u)$ and $\overleftarrow{\varrho}'_2(u)$. The level-2 locality filter is determined with the fast but less precise method described at the end of Section 3.3.

Variant 'Generous' is tuned for very fast query times. As the economical variant, it uses three levels on top of the original graph (but in this case, level 1

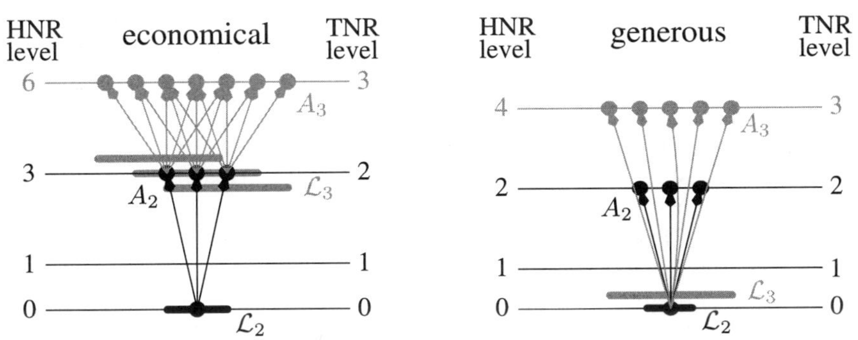

FIGURE 6. Two variants of transit-node routing (TNR) based on highway-node routing (HNR).

is not only used to accelerate the preprocessing). At each node u, we store the access nodes and the locality filters[9] required for the query in level 2 and 3. This allows direct access to these levels. For level 1, we store neither access nodes nor a locality filter. Instead, if required, we perform local searches to determine the access nodes and we use the trivial locality filter $\mathcal{L}_1(s,t) := \mathsf{true}$ for all node pairs (s,t). We explicitly store the level-1–3 distance tables, while we perform a shortest-path query in level 0 (if required). Note that having a level-1 distance table is a significant difference from the economical variant. Interestingly, the search for the level-1 access nodes already involves the search in level 0 so that no extra work is imposed by the level-0 search. This also explains why it is reasonable to just set $\mathcal{L}_1(s,t) := \mathsf{true}$.

4.7. Queries. Queries are performed in a top-down fashion. For a given query pair (s,t), first $\overrightarrow{A}_3(s)$ and $\overleftarrow{A}_3(t)$ are either looked up or computed depending on the used variant. Then table lookups in the top-level distance table yield a first guess for $d(s,t)$. Now, if $\neg\mathcal{L}_3(s,t)$, we are done. Otherwise, the same procedure is repeated for level 2. If even $\mathcal{L}_2(s,t)$ is true, we perform a bidirectional shortest-path search using highway-node routing that can stop if both the forward and backward search radius (i.e., the key of the minimum element in the respective priority queue) exceed the upper bound computed in levels 2 and 3. Furthermore, the search need not expand from any node $u \in \mathcal{T}_2$ since paths going over these nodes are covered by the search in levels 2 and 3. In the generous variant, the search is already stopped at the level-1 access nodes, and additional lookups in the level-1 distance table are performed.

4.8. Outputting Complete Path Descriptions. The general methods from Section 3.5 can be applied rather directly to our concrete instantiation in order to determine a complete description of the shortest path.

5. Experiments

5.1. Environment, Instances, and Parameters. The experiments were done on one core of a single AMD Opteron Processor 270 clocked at 2.0 GHz with 8 GB main memory and 2×1 MB L2 cache, running SuSE Linux 10.0 (kernel 2.6.13). The program was compiled by the GNU C++ compiler 4.0.2 using optimisation level 3. Results for the DIMACS Challenge benchmarks can be found in Table 1.

We deal with two road networks. The network of Western Europe[10] has been made available for scientific use by the company PTV AG. Only the largest strongly connected component is considered. The original graph contains for each edge a length and a road category, e.g., motorway, national road, regional road, urban street. We assign average speeds to the road categories, compute for each edge the average travel time, and use it as weight. In addition to this *travel time metric*, we perform experiments on variants of the European graph with a *distance metric* and the *unit metric*. The network of the USA (without Alaska and Hawaii) has been obtained from the TIGER/Line Files [**28**]. Again, we consider only the largest strongly connected component. In contrast to the PTV data, the TIGER graph

[9] i.e., the radii $\overrightarrow{\varrho}'_2(u)$, $\overleftarrow{\varrho}'_2(u)$, $\overrightarrow{\varrho}'_3(u)$, and $\overleftarrow{\varrho}'_3(u)$

[10] 14 countries: Austria, Belgium, Denmark, France, Germany, Italy, Luxembourg, the Netherlands, Norway, Portugal, Spain, Sweden, Switzerland, and the UK

TABLE 1. DIMACS Challenge [**1**] benchmarks for US (sub)graphs (query time [ms]).

graph	metric time	metric dist	graph	metric time	metric dist
NY	29.6	28.5	CAL	291.1	235.4
BAY	34.7	33.3	LKS	461.3	366.1
COL	51.5	49.0	E	681.8	536.4
FLA	134.8	120.5	W	1 211.2	988.2
NW	161.1	146.1	CTR	4 485.7	3 708.1
NE	225.4	197.2	USA	5 355.6	4 509.1

TABLE 2. Properties of the used road networks.

	Europe	USA (Tiger)
#nodes	18 010 173	23 947 347
#directed edges	42 560 279	58 333 344
#road categories	13	4
average speeds [km/h]	10–130	40–100

TABLE 3. Used neighbourhood sizes. For the travel time metric, we use a fixed neighbourhood size for the construction of all levels. For the other two metrics, we use linearly increasing sequences as neighbourhood sizes of the different levels.

metric	Europe	USA (Tiger)
time	30	40
dist	60, 120, 180, ...	60, 120, 180, ...
unit	40, 50, 60, ...	60, 120, 180, ...

is undirected, planarised and distinguishes only between four road categories. All graphs have been taken from the DIMACS Challenge website [**1**]. Table 2 summarises the properties of the used networks.

In Section 5.2 we report only the times needed to compute the shortest path distance between two nodes without outputting the actual route, while in Section 5.3, we also give the times needed to get a complete description of the shortest paths.

We apply the economical variant to the travel time, the distance, and the unit metric. In each case, in order to determine the highway-node sets (and consequently the transit-node sets) we construct a highway hierarchy according to the description in [**22, 24**]. The used neighbourhood sizes are represented in Table 3. Note that the construction of the highway hierarchy starts with an initial contraction step. We use the maverick factor $f = 2(i - 1)$ for the i-th iteration of the construction procedure, the contraction rate $c = 2$, and the shortcut hops limit 10.

In addition, we apply the generous variant to the travel time metric using the neighbourhood size $H = 90$.

5.2. Main Results.

5.2.1. *Preprocessing.* Table 4 gives the preprocessing times for all considered road networks, metrics, and variants. In addition, some key facts on the results

TABLE 4. Preprocessing statistics. The size $|D_2|$ of the level-2 distance table is given relative to the size of a complete $|\mathcal{T}_2| \times |\mathcal{T}_2|$ table. $|A_\ell|$ denotes $|\overrightarrow{A}_\ell \cup \overleftarrow{A}_\ell|$, i.e., the size of the union of forward and backward access nodes.

		level 3		level 2			overhead	time
metric	variant	$\lvert\mathcal{T}_3\rvert$	$\lvert A_3\rvert$	$\lvert\mathcal{T}_2\rvert$	$\lvert D_2\rvert$	$\lvert A_2\rvert$	[B/node]	[h]
				Europe				
time	eco	9 355	11.4	151 450	0.15%	5.3	99	0:25
	gen	9 458	11.3	293 209	0.14%	4.4	226	1:15
dist	eco	14 001	22.3	179 972	1.03%	8.8	301	2:42
unit	eco	10 923	12.7	212 014	0.28%	6.4	138	0:53
				USA (Tiger)				
time	eco	6 449	6.8	218 153	0.20%	5.2	121	0:38
	gen	10 261	6.1	449 945	0.08%	4.5	257	1:25
dist	eco	16 296	19.1	261 759	0.53%	7.5	280	3:37
unit	eco	10 901	12.5	239 029	1.00%	6.2	219	3:59

of the preprocessing, e.g., the sizes of the transit node sets, are presented. It is interesting to observe that for the travel time metric in level 2 the actual distance table size is at most 0.2% of the size a naive $|\mathcal{T}_2| \times |\mathcal{T}_2|$ table would have.

As expected, the distance metric yields more access nodes than the travel time metric (a factor 2–3) since not only junctions on very fast roads (which are rare) qualify as access nodes. The fact that we have to increase the neighbourhood size from level to level in order to achieve an effective shrinking of the highway networks leads to comparatively high preprocessing times for the distance metric.

5.2.2. *Random Queries Using the Travel Time Metric.* Table 5 summarises the average case performance of transit-node routing. For the travel time metric, the generous variant achieves average query times more than two orders of magnitude lower than those of highway-node routing or highway hierarchies (even when combined with goal-directed search [**7**]). Compared to Dijkstra's algorithm, we obtain a speedup of a factor 1.4 and 1.9 *million* in case of Europe and the USA, respectively. At the cost of about a factor three in query time, the economical variant saves around a factor of two in space and a factor of 2–3 in preprocessing time.

Finding a good locality filter is one of the biggest challenges of our instantiation of transit-node routing. The values in Table 5 indicate that our filter is suboptimal: for instance, only 0.0051% of the queries performed by the economical variant in the European network would require a local search to answer them correctly. However, the locality filter $\mathcal{L}_2$ forces us to perform local searches in 0.1364% of all cases. The high-quality level-2 filter employed by the generous variant is considerably more effective, still the percentage of false positives exceeds 90%.

5.2.3. *Random Queries Using the Distance or Unit Metric.* For the distance and unit metric, the situation is worse. A considerably larger fraction of the queries continues to level 2 and below. It is important to note that we have concentrated on the travel time metric—since we consider the travel time metric more important for practical applications—, and we spent comparatively little time to tune our approach for the distance and unit metric. Nevertheless, the current version shows feasibility and still achieves an improvement of a factor of at least 15 or 80 for the

TABLE 5. Query statistics w.r.t. 10 000 000 randomly chosen (s,t)-pairs. Each query is performed in a top-down fashion. For each level ℓ, we report the percentage of the queries that are not answered correctly in some level $\geq \ell$ and the percentage of the queries that are not stopped after level ℓ (i.e., $\mathcal{L}_\ell(s,t)$ is true). Note that only the generous variant can perform a query in level 1 (but, as the economical variant, it always continues to level 0).

		level 3 [%]		level 2 [%] (level 1 [%])		
metric	variant	wrong	cont'd	wrong	cont'd	query time
			Europe			
time	eco	0.57	3.36	0.0051	0.1364	11.0 μs
time	gen	0.25	1.55	0.0016 (0.00019)	0.0180 (0.0180)	4.3 μs
dist	eco	3.89	13.21	0.0121	0.4897	37.6 μs
unit	eco	1.06	5.23	0.0070	0.1731	13.1 μs
			USA (Tiger)			
time	eco	0.37	2.44	0.0045	0.1130	9.5 μs
time	gen	0.10	0.87	0.0010 (0.00009)	0.0124 (0.0124)	3.3 μs
dist	eco	1.04	5.35	0.0067	0.1587	86.1 μs
unit	eco	1.67	8.66	0.0099	0.2729	19.8 μs

distance or unit metric, respectively, compared to highway hierarchies combined with goal-directed search.

5.2.4. *Local Queries Using the Travel Time Metric.* Since the overwhelming majority of all cases is handled in the top level (more than 99% in case of the US network using the generous variant), the average case performance says little about the performance for more local queries which might be very important in some applications. Therefore, we use the methodology introduced in [**21**] to get more detailed information about the query time distributions for queries ranging from very local to global, see Figure 7: We choose 1 000 random sample points s and for each power of two $r = 2^k$, we determine the node t with *Dijkstra rank* $\mathrm{rk}_s(t) = r$, where (for a fixed source s) the Dijkstra rank of a node t is the rank w.r.t. the order which Dijkstra's algorithm settles the nodes in. We then use our algorithm to make an s-t-query. By plotting the resulting statistics for each value $r = 2^k$, we can see how the performance scales with a natural measure of difficulty of the query. We represent the distributions as box-and-whisker plots [**20**]: each box spreads from the lower to the upper quartile and contains the median, the whiskers extend to the minimum and maximum value omitting outliers, which are plotted individually.

Note that even the median query times for the largest Dijkstra rank (which is the best case) are higher than the average query times given in Table 5. This is due to the fact that logging the statistics required to create the depicted plot causes a certain overhead.

For the generous approach, we can easily recognise the three levels of transit-node routing with small transition zones in between: For ranks 2^{18}–2^{24} we usually have $\neg\mathcal{L}_3(s,t)$ and thus only require cheap distance table accesses in level 3. For ranks 2^{12}–2^{16}, we need additional lookups in the table of level 2 so that the queries

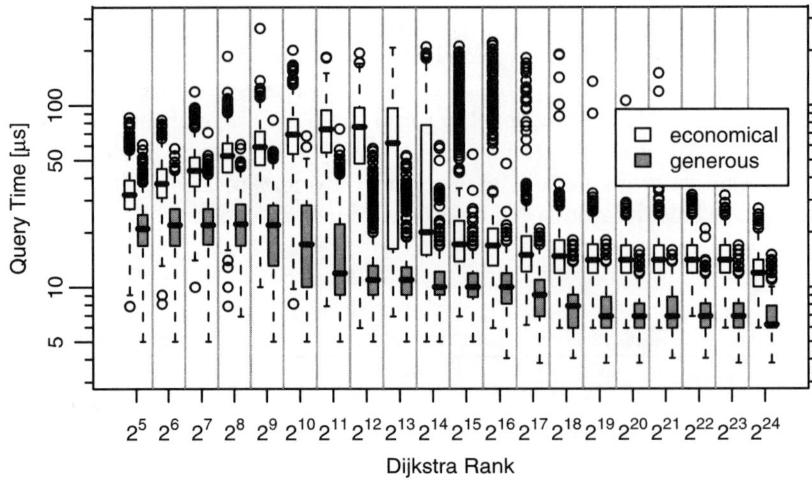

FIGURE 7. Local queries on the European network with the travel time metric using the economical and the generous variant.

get somewhat more expensive. In this range, outliers can be considerably more costly, indicating that occasional local searches are needed. For small ranks we usually need local searches and additional lookups in the level-1 table. Still, the combination of a local search in a very small area and table lookups in all three levels usually results in query times of less than 30 μs.

In case of the economical approach, we observe a high variance in query times for ranks 2^{13}–2^{14}. In this range, all types of queries occur and the difference between the level-3 queries and the local queries is rather big since the economical variant does not make use of level 1. For small Dijkstra ranks, we see a growth of the query times that is typical for highway-node routing (or highway hierarchies).

5.2.5. *Distance Table Accesses.* Figure 8 represents a histogram of the number of topmost distance table accesses during an s-t-query. For Europe, we observe an average number of table accesses of 75 and a maximum of $37 \cdot 40 = 1\,480$.

5.3. Outputting Complete Path Descriptions. So far, we have reported only the times needed to compute the shortest-path length between two nodes. Now, we determine a complete description of the shortest path. In Table 6 we give the additional preprocessing time and the additional disk space for the unpacking data structures. Furthermore, we report the additional time that is needed to determine a complete description of the shortest path and to traverse it[11] summing up the weights of all edges as a sanity check—assuming that the query to determine the shortest-path length has already been performed. We restrict ourselves to the travel time metric and the generous variant. Currently, we provide an efficient implementation only for the case that the path goes through the top level. In all other cases, we just perform a normal highway-node query and use the path unpacking routines of highway-node routing. The effect on the average query times is very small since most queries are correctly answered using only the top search. In order to unpack edges of the overlay graphs, we use two different variants that

[11]Note that we do *not* traverse the path in the original graph, but we directly scan the assembled description of the path.

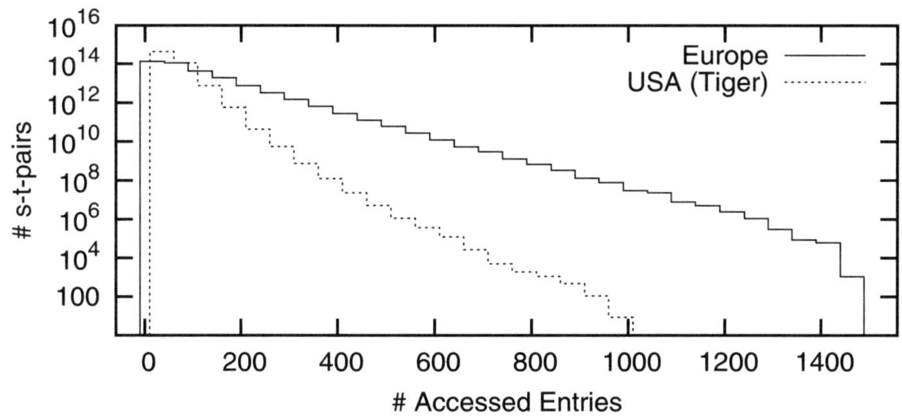

FIGURE 8. Histogram of the number of entries in the topmost distance table that have to be accessed during an s-t-query.

TABLE 6. Additional preprocessing (pp) time, additional disk space and query time that is needed to determine a complete description of the shortest path and to traverse it summing up the weights of all edges—assuming that the query to determine its lengths has already been performed. Moreover, the average number of hops—i.e., the average path length in terms of number of nodes—is given.

	Europe				USA (Tiger)			
	pp [s]	space [MB]	query [μs]	# hops (avg.)	pp [s]	space [MB]	query [μs]	# hops (avg.)
Variant 2	18	91	314	1 370	29	124	869	4 551
Variant 3	505	227	153	1 370	277	221	264	4 551

are introduced as 'Variant 2' and 'Variant 3' in [**7**]. Variant 2 consists of a purely recursive data structure where the description of a shortcut edge may contain shortcuts of a lower level, while Variant 3 also stores non-recursive (i.e., fully unpacked) representations of the most important edges.

Note that the figures for Variant 3 have been obtained using an older implementation of transit-node routing based on highway hierarchies and a different set of parameters since the current implementation of highway-node routing does not support this variant.

5.4. Further Experiments. In Table 7, we also give some results for subgraphs of the US road network, which support our claim that we achieve almost constant query times irrespective of the size of the road network: while the sizes range between 264 346 and 23 947 347 nodes, the query times vary only from 3.7 to 5.0 μs for the generous variant. Note that these results have been obtained using an older implementation of transit-node routing based on highway hierarchies [**23, 4, 3**].

TABLE 7. Results for US subgraphs with travel time metric using the generous variant.

graph	#nodes	preproc. time [min]	total disk space [MB]	query time [μs]
NY	264 346	4	147	4.6
BAY	321 270	2	105	4.2
COL	435 666	3	156	4.7
FLA	1 070 376	7	418	3.8
NW	1 207 945	7	325	3.7
NE	1 524 453	16	578	4.1
CAL	1 890 815	15	554	3.8
LKS	2 758 119	26	890	4.2
E	3 598 623	30	1 159	4.4
W	6 262 104	47	1 801	4.2
CTR	14 081 816	148	4 169	5.0
USA	23 947 347	205	6 108	4.9

6. Discussion

Transit-node routing provides the fastest available query times for large static real-world road networks. Speedups compared to Dijkstra's algorithm exceed factor one million. The extremely good query performance does not imply prohibitive preprocessing times or memory consumption. In fact, the preprocessing is still clearly faster than many other route planning techniques that achieve considerably smaller speedups. Moreover, transit-node routing is not only optimised for long-range queries, but also answers local queries very efficiently.

6.1. Alternative Instantiations. There seem to be two basic approaches to transit-node routing. One that starts with a locality filter $\mathcal{L}$ and then has to find a good set of transit nodes $\mathcal{T}$ for which $\mathcal{L}$ works (e.g., the grid-based implementation [**2**]). The other approach starts with $\mathcal{T}$ and then has to find a locality filter that can be efficiently evaluated and detects as accurately as possible whether local search is needed (e.g., our abstract and concrete instantiations, Sections 3–4). Both basic approaches fit in the generic framework introduced in Section 2. In [**3, 23**], we describe a few additional general preprocessing techniques that might be useful for instantiations that differ from the one specified in Section 3.

6.2. Future Work. Although conceptually simple, an efficient implementation of transit-node routing has so many ingredients that there are many further optimisation opportunities and a large spectrum of trade-offs between query time, preprocessing time, and space usage. For example, in order to reduce the latter, we could store the access-node information only at level 1; this would reduce the size of the access-node data by a factor of six. For reducing the average query time, we could try to precompute information analogous to edge flags or geometric containers [**17, 18, 29**] that tells us which access nodes lead to which regions of the graph [**5**].

There are many interesting ways to choose transit nodes. For example nodes with high node reach [**13, 10**] could be a good starting point. Here, we can directly influence $|\mathcal{T}|$, and the resulting reach bound might help defining a simple locality filter. However, it seems that geometric reach or travel time reach do not reflect the inhomogeneous density of real-world road networks. Hence, it would be interesting if we could efficiently approximate reach based on the Dijkstra rank.

Another interesting approach might be to start with some locality filter that guarantees uniformly small local searches and then to view it as an optimisation problem to choose a small set of transit nodes that cover all the local search spaces.

Parallel processing can easily be used to accelerate preprocessing, or to execute many queries in parallel. With very fine grained multi-core parallelism it might even be possible to accelerate an individual query. Forward local search, backward local search, and each table lookup are largely independent of each other.

Acknowledgements. We would like to thank Holger Bast, Stefan Funke, Kirill Müller, and Dorothea Wagner for interesting discussions on transit-node routing and Timo Bingmann for work on visualisation tools. Two anonymous reviewers provided valuable suggestions.

References

1. 9th DIMACS Implementation Challenge, *Shortest Paths*, `http://www.dis.uniroma1.it/~challenge9/`, 2006.
2. H. Bast, S. Funke, and D. Matijevic, *TRANSIT—ultrafast shortest-path queries with linear-time preprocessing*, 9th DIMACS Implementation Challenge [**1**], 2006, an updated version of the paper appears in this book.
3. H. Bast, S. Funke, D. Matijevic, P. Sanders, and D. Schultes, *In transit to constant time shortest-path queries in road networks*, Workshop on Algorithm Engineering and Experiments (ALENEX), 2007, pp. 46–59.
4. H. Bast, S. Funke, P. Sanders, and D. Schultes, *Fast routing in road networks with transit nodes*, Science **316** (2007), no. 5824, 566.
5. R. Bauer, D. Delling, P. Sanders, D. Schieferdecker, D. Schultes, and D. Wagner, *Combining hierarchical and goal-directed speed-up techniques for Dijkstra's algorithm*, 7th Workshop on Experimental Algorithms (WEA), LNCS, vol. 5038, Springer, 2008, to appear.
6. D. Delling, M. Holzer, K. Müller, F. Schulz, and D. Wagner, *High-performance multi-level graphs*, 9th DIMACS Implementation Challenge [**1**], 2006, an updated version of the paper appears in this book.
7. D. Delling, P. Sanders, D. Schultes, and D. Wagner, *Highway hierarchies star*, 9th DIMACS Implementation Challenge [**1**], 2006, an updated version of the paper appears in this book.
8. E. W. Dijkstra, *A note on two problems in connexion with graphs.*, Numerische Mathematik **1** (1959), 269–271.
9. J. Fakcharoenphol and S. Rao, *Planar graphs, negative weight edges, shortest paths, and near linear time*, 42nd IEEE Symposium on Foundations of Computer Science, 2001, pp. 232–241.
10. A. Goldberg, H. Kaplan, and R. F. Werneck, *Reach for A*: Efficient point-to-point shortest path algorithms*, Workshop on Algorithm Engineering and Experiments (ALENEX) (Miami), 2006, pp. 129–143.
11. A. V. Goldberg and C. Harrelson, *Computing the shortest path: A* meets graph theory*, 16th ACM-SIAM Symposium on Discrete Algorithms, 2005, pp. 156–165.
12. A. V. Goldberg and R. F. Werneck, *Computing point-to-point shortest paths from external memory*, Workshop on Algorithm Engineering and Experiments (ALENEX), 2005, pp. 26–40.
13. R. Gutman, *Reach-based routing: A new approach to shortest path algorithms optimized for road networks*, Workshop on Algorithm Engineering and Experiments (ALENEX), 2004, pp. 100–111.
14. P. E. Hart, N. J. Nilsson, and B. Raphael, *A formal basis for the heuristic determination of minimum cost paths*, IEEE Transactions on System Science and Cybernetics **4** (1968), no. 2, 100–107.
15. P. N. Klein, *Multiple-source shortest paths in planar graphs*, 16th ACM-SIAM Symposium on Discrete Algorithms, SIAM, 2005, pp. 146–155.

16. S. Knopp, P. Sanders, D. Schultes, F. Schulz, and D. Wagner, *Computing many-to-many shortest paths using highway hierarchies*, Workshop on Algorithm Engineering and Experiments (ALENEX), 2007.
17. U. Lauther, *An extremely fast, exact algorithm for finding shortest paths in static networks with geographical background*, Geoinformation und Mobilität – von der Forschung zur praktischen Anwendung, vol. 22, IfGI prints, Institut für Geoinformatik, Münster, 2004, pp. 219–230.
18. R. Möhring, H. Schilling, B. Schütz, D. Wagner, and T. Willhalm, *Partitioning graphs to speed up Dijkstra's algorithm*, 4th International Workshop on Efficient and Experimental Algorithms (WEA), 2005, pp. 189–202.
19. K. Müller, *Design and implementation of an efficient hierarchical speed-up technique for computation of exact shortest paths in graphs*, Diploma Thesis, Universität Karlsruhe (TH), 2006.
20. R Development Core Team, *R: A Language and Environment for Statistical Computing*, `http://www.r-project.org`, 2004.
21. P. Sanders and D. Schultes, *Highway hierarchies hasten exact shortest path queries*, 13th European Symposium on Algorithms (ESA), LNCS, vol. 3669, Springer, 2005, pp. 568–579.
22. P. Sanders and D. Schultes, *Engineering highway hierarchies*, 14th European Symposium on Algorithms (ESA), LNCS, vol. 4168, Springer, 2006, pp. 804–816.
23. ———, *Robust, almost constant time shortest-path queries in road networks*, 9th DIMACS Implementation Challenge [**1**], 2006, workshop version.
24. D. Schultes, *Route planning in road networks*, Ph.D. thesis, Universität Karlsruhe (TH), 2008.
25. D. Schultes and P. Sanders, *Dynamic highway-node routing*, 6th Workshop on Experimental Algorithms (WEA), LNCS, vol. 4525, Springer, 2007, pp. 66–79.
26. F. Schulz, D. Wagner, and C. D. Zaroliagis, *Using multi-level graphs for timetable information*, Workshop on Algorithm Engineering and Experiments (ALENEX), LNCS, vol. 2409, Springer, 2002, pp. 43–59.
27. M. Thorup, *Compact oracles for reachability and approximate distances in planar digraphs*, 42nd IEEE Symposium on Foundations of Computer Science, 2001, pp. 242–251.
28. U.S. Census Bureau, Washington, DC, *UA Census 2000 TIGER/Line Files*, `http://www.census.gov/geo/www/tiger/tigerua/ua_tgr2k.html`, 2002.
29. D. Wagner and T. Willhalm, *Geometric speed-up techniques for finding shortest paths in large sparse graphs*, 11th European Symposium on Algorithms (ESA), LNCS, vol. 2832, Springer, 2003, pp. 776–787.
30. ———, *Drawing graphs to speed up shortest-path computations*, Workshop on Algorithm Engineering and Experiments (ALENEX), 2005.

Peter Sanders, Universität Karlsruhe (TH), Fakultät für Informatik, Postfach 69 80, 76128 Karlsruhe, Germany
E-mail address: `sanders@ira.uka.de`

Dominik Schultes, Universität Karlsruhe (TH), Fakultät für Informatik, Postfach 69 80, 76128 Karlsruhe, Germany
E-mail address: `schultes@ira.uka.de`

DIMACS Series in Discrete Mathematics
and Theoretical Computer Science
Volume **74**, 2009

Single-Source Shortest Paths with the Parallel Boost Graph Library

Nick Edmonds, Alex Breuer, Douglas Gregor, and Andrew Lumsdaine

ABSTRACT. The Parallel Boost Graph Library (Parallel BGL) is a library of graph algorithms and data structures for distributed-memory computation on large graphs. Developed with the Generic Programming paradigm, the Parallel BGL is highly customizable, supporting various graph data structures, arbitrary vertex and edge properties, and different communication media. In this paper, we describe the implementation of three parallel variants of Dijkstra's single-source shortest paths algorithm in the Parallel BGL. We also provide an experimental evaluation of these implementations using synthetic and real-world benchmark graphs from the 9th DIMACS Implementation Challenge and present performance results of solving the single-source shortest path problem for graphs with over a billion vertices on a modest size cluster.

1. Introduction

Large-scale graph problems arise in many application areas. As with other large-scale computations, large-scale graph computations can potentially benefit from the use of high-performance parallel computing systems. Parallel graph algorithms have been well-studied in the literature [**29, 36**] and selected algorithms have been implemented for shared memory [**16, 19, 26, 37, 22**], distributed memory [**10, 35**], and highly multithreaded architectures [**6, 7**]. While such implementations are important to demonstrate proof of concept for parallel graph algorithms, they tend to be of limited use in practice because such implementations are not typically reusable outside of their test environments. As a result, new uses of a given parallel graph algorithm will almost surely have to be implemented from scratch.

In mainstream software development, software libraries provide a common infrastructure that amortizes the costs of implementing widely-used algorithms and data structures. In computational sciences, software libraries can be extremely valuable to an entire research community. Libraries provide a language that improves dissemination of research results and simplifies comparison of alternatives. In addition, a single widely-used (and thus widely-studied) implementation is more likely to be reliable, correct, and efficient. While there are several high-quality

1991 *Mathematics Subject Classification.* Primary 68R10, 68W15; Secondary 05C12, 05C85.
This work was supported by a grant from the Lilly Endowment and by NSF grants EIA-0131354 and EIA-0202048.

sequential graph libraries available, such as LEDA [**47**], Stanford GraphBase [**41**], and JUNG [**60**] there are relatively few attempts at parallel graph libraries. Of the parallel graph libraries that have been reported in the literature [**4, 14, 34**], none provide the functionality or flexibility needed in a general-purpose library.

In addition to flexibility, performance is an important concern for software libraries. Libraries such as BLAS [**42**] in the scientific computing community exist specifically to provide high levels of performance to applications using them. However, tensions may arise between the needs for flexibility in a library and the needs for performance. Recently, the generic programming paradigm has emerged as an approach for library development that simultaneously meets the needs of flexibility and performance.

The sequential Boost Graph Library (BGL) [**53**], (formerly the Generic Graph Component Library [**44**]), is a high-performance generic graph library that is part of the Boost library collection [**11**]. The Boost libraries are a collection of open-source, peer-reviewed C++ libraries that have driven the evolution of library development in the C++ community [**2**] and ANSI/ISO C++ standard committee [**5**]. Following the principles of generic programming [**50, 56**] and written in a style similar to the C++ Standard Template Library (STL) [**49, 57**], the algorithms provided by the Parallel BGL are parameterized by the data types on which they operate. Arbitrary graph data types can be used with BGL algorithms; in particular, independently-developed third-party graph types can be used without the need to modify the BGL algorithms themselves. The BGL does provide its own data types, which are parameterized by the underlying storage types. This parameterization allows extensive customization of the BGL, from storing user-defined data types with the vertices and edges of a graph to completely replacing the Parallel BGL graph types with application-specific data structures, without incurring additional overhead. Table 1 lists some of the algorithms that are currently implemented in the BGL.

Following our philosophy of software libraries and reuse, we have recently developed the Parallel Boost Graph Library (Parallel BGL) [**30**] on top of the sequential BGL. The Parallel BGL provides data structures and algorithms for parallel computation on graphs. The Parallel BGL retains much of the interface of the (sequential) BGL upon which it is built, greatly simplifying the task of porting programs from the sequential BGL to the Parallel BGL. Because it is built on top of the sequential BGL, the Parallel BGL also retains the performance and flexibility of the underlying BGL. Table 2 lists some of the algorithms that are currently implemented within the Parallel BGL.

2. The (Parallel) Boost Graph Library

The Parallel BGL is a generic graph library written for high performance and maximum reusability, and is itself built upon the generic Boost Graph Library. The core of the BGL—sequential or parallel—is a set of generic graph algorithms, that are polymorphic with respect to the underlying graph types. BGL graph algorithms can be applied to any graph data type, even to types that are not included with the library, provided that type supplies all the functionality required by the algorithm. Additionally, BGL graph algorithms are often customizable in other ways, through visitor objects and abstract representations of vertex and edge properties (e.g., edge weights). The genericity of the BGL is such that it can (for example) operate on

Breadth-first search [**15**]	Dijkstra's shortest paths [**15**]
Depth-first search [**15**]	Bellman-Ford shortest paths [**15**]
King ordering [**40**]	Dynamic connected components [**21**]
Transpose [**15**]	Kruskal's minimum spanning tree [**15**]
Topological sort [**15**]	Prim's minimum spanning tree [**15**]
Sloan ordering [**54**]	Floyd-Warshall all-pairs shortest paths [**15**]
Gursoy-Atun layout [**33**]	Kamada-Kawai spring layout [**38**]
Transitive closure [**15**]	Strongly connected components [**15**]
Connected components [**15**]	Reverse Cuthill-McKee ordering [**17**]
Articulation points [**3**]	Smallest last vertex ordering [**46**]
Push-Relabel max flow [**15**]	Johnson's all-pairs shortest paths [**15**]
Sequential vertex coloring [**55**]	Brandes betweenness centrality [**12**]
Edmonds-Karp max flow [**15**]	Incremental connected components [**21**]
Biconnected components [**3**]	Fruchterman-Reingold force directed layout [**23**]
Minimum degree ordering [**25**]	

TABLE 1. Algorithms currently implemented in the sequential BGL.

Breadth-first search [**15**]	Biconnected components [**35**]
Dijkstra shortest path [**15**]	Boruvka Minimum spanning tree [**15**]
Depth-first search [**15**]	Strongly-connected components [**15**]
Dinic Max-flow [**20**]	Fructerman-Rheingold layout [**23**]
PageRank [**51**]	Δ-stepping shortest path [**48**]
Graph coloring [**10**]	Dehne-Götz Minimum spanning tree [**19**]
Connected components [**15**]	Crauser *et al.* single-source shortest path [**16**]

TABLE 2. Algorithms currently implemented in the Parallel BGL.

a LEDA [**47**] graph just as efficiently as it can operate on its own graph types, without requiring the user to perform any data conversion.

The Parallel BGL employs a unique, modular architecture built on the sequential BGL. Figure 1 illustrates the components in the sequential and Parallel BGL and their interactions. There are three primary kinds of interfaces in the Parallel BGL: graphs, which describe the structure of a graph that may either be stored in (distributed) memory or generated on-the-fly; property maps, which associate additional information with the vertices and edges of a graph; and process groups, which facilitate communication among distributed data structures and algorithms. Two of these interfaces, graphs and property maps, are inherited from—and are therefore compatible with—the sequential Boost Graph Library. These are illustrated by the dark shaded blocks in Figure 1.

The Parallel BGL components (represented by lightly-shaded blocks in Figure 1) typically wrap their sequential BGL counterparts, building distributed-memory parallel processing functionality on top of efficient sequential code. Layering a parallel library on top of a sequential library has many benefits: one need not reimplement core data structures or algorithms; potential users may have some prior knowledge of the interfaces that can be transferred from the sequential to the parallel library; and improvements to the sequential library will immediately improve the parallel library. However, the performance of the parallel library then becomes

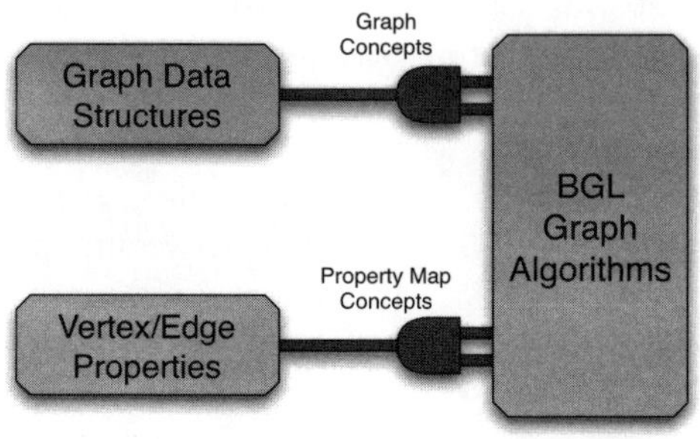

(a) Boost Graph Library

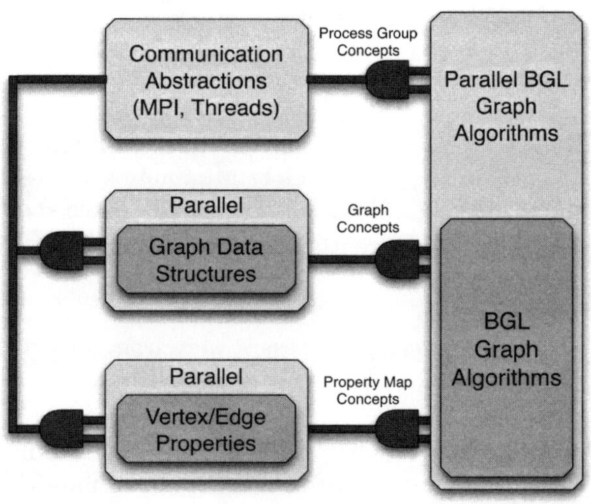

(b) Parallel Boost Graph Library

FIGURE 1. Architectures of the sequential and parallel Boost Graph Libraries, illustrating how various abstractions (communication medium, graph data structures, graph properties) "plug into" the generic graph algorithms via concepts.

dependent on the sequential library, making it extremely important that the sequential library be both efficient and customizable. We have found that generic libraries such as the (sequential) BGL can be layered in this fashion without performance degradation, and they even allow reuse of algorithm implementations in the parallel context. The implementation of breadth-first search, Dijkstra's shortest paths, and Fruchterman-Reingold force-directed layout in the Parallel BGL, for instance, are merely invocations of the generic implementations of the sequential BGL using appropriate distributed data structures [**31**].

Components in the (Parallel) BGL are loosely connected via *concepts*, which provide abstraction without unnecessary performance penalties. A *concept* is essentially an interface description. Generic algorithms are written using concepts, and any data type that meets the requirements of the concept can be used with the generic algorithm. Concepts differ from abstract base classes or "interfaces" in object-oriented languages in several ways, two of which are important in the context of the BGL. First, concepts are purely compile-time entities, and unlike virtual functions their use incurs no run-time overhead. Second, concepts permit *retroactive modeling*, which means that one can provide the appropriate concept interface for a data type *without changing the data type*; this allows external data structures to be used with generic libraries such as the (Parallel) BGL.

2.1. Generic Algorithms. The generic algorithms in the (Parallel) BGL are implemented using C++ templates, which provide compile-time polymorphism with no run-time overhead. Figure 2 contains the complete implementation of the sequential breadth-first search algorithm in the BGL. The four template parameters for this algorithm correspond to the graph type itself (IncidenceGraph), the queue that will be used to store vertices (Buffer), the visitor that will be notified for various events during the breadth-first search (BFSVisitor), and the *property map* that will be used to keep track of which vertices have been seen (ColorMap).

The most important feature of breadth_first_visit() is that it can operate on an arbitrary graph type, as long as the type meets certain (minimal) requirements. In particular, to be used as a graph with breadth_first_visit(), the functions out_edges(), source(), and target() must exist for the given type. These requirements are part of the Incidence Graph concept, which is documented elsewhere in greater detail [**53**]. For types that provide the required functionality, but do not have these required functions, a simple adaptation layer can easily map from the provided functionality to the required function names. This kind of adaptation is provided in BGL and allows LEDA graphs (for example) to be used directly with BGL algorithms.

The ColorMap template parameter is also interesting because it separates the notion of the "color" of a vertex from the storage of the vertex itself. In BGL terminology, ColorMap is a *property map*, because it provides a particular property for each vertex of the graph. When initiating a breadth-first search, all vertices will be white. When a vertex is seen, it is colored gray. Once all of its outgoing edges have been visited, it is colored black. This information can be stored either inside the graph or in an external data structure, such as an array or hash table.

The remaining two template parameters, Buffer and BFSVisitor, allow further customization of the behavior of the breadth-first search algorithm. We will revisit these parameters when we discuss the implementation of (parallel) Dijkstra's algorithm in the Parallel BGL in Section 3.

2.2. Graph Data Structures. In addition to generic algorithms, the (Parallel) BGL provides several configurable graph data structures. The sequential BGL provides an adjacency list representation, a more compact and efficient (but less versatile) compressed sparse row representation, and an adjacency matrix representation, any of which can be used for the vast majority of the generic algorithms in

```
template<class IncidenceGraph, class Buffer, class BFSVisitor, class ColorMap>
void
breadth_first_visit   (const IncidenceGraph& g,
                       typename graph_traits<VertexListGraph>::vertex_descriptor s,
                       Buffer& Q, BFSVisitor vis, ColorMap color)
{
  put(color, s, Color::gray());                          vis.discover_vertex (s, g);
  Q.push(s);
  while (! Q.empty()) {
    Vertex u = Q.top(); Q.pop();                         vis.examine_vertex(u, g);
    for (tie (ei, ei_end) = out_edges(u, g); ei != ei_end; ++ei) {
      Vertex v = target(*ei, g);                         vis.examine_edge(*ei, g);
      ColorValue v_color = get(color, v);
      if (v_color == Color::white()) {                   vis.tree_edge(*ei, g);
        put(color, v, Color::gray());                    vis.discover_vertex (v, g);
        Q.push(v);
      } else {                                           vis.non_tree_edge(*ei, g);
        if (v_color == Color::gray())                    vis.gray_target (*ei, g);
        else                                             vis.black_target (*ei, g);
      }
    } // end for
    put(color, u, Color::black());                       vis.finish_vertex (u, g);
  } // end while
}
```

FIGURE 2. Generic implementation of the (sequential) breadth-first search algorithm in the BGL. The algorithm resides in the left column and the associated event points are written in the right column.

the BGL. In addition, the BGL provides an adaptation layer for various third-party graph types.

The Parallel BGL provides distributed counterparts to the adjacency list and compressed sparse row graphs of the BGL. Graphs in the Parallel BGL are distributed using a row-wise decomposition: each processor stores a disjoint set of vertices and all edges outgoing from those vertices. Figure 3 (a) shows a small graph consisting of nine vertices, distributed across three processors (indicated by the shaded rectangles). Figure 3 (b) illustrates how this graph can be represented using a distributed adjacency list in the Parallel BGL. Each processor contains several vertices. Attached to those vertices is a list of edges. For instance, vertex a on the leftmost processor has two outgoing edges, (a, b) and (a, d), so b and d are stored in its list. We note that we use the term "distributed" in a general sense. The distributed types in Parallel BGL can be used in shared-memory environments, in which case parallelism can be effected via MPI or threads.

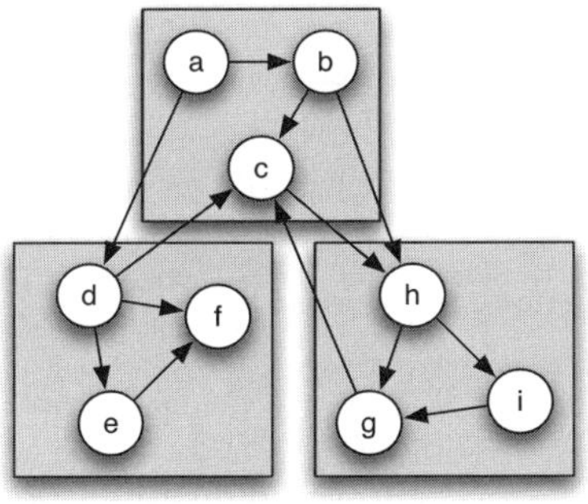

(a) Distributed graph

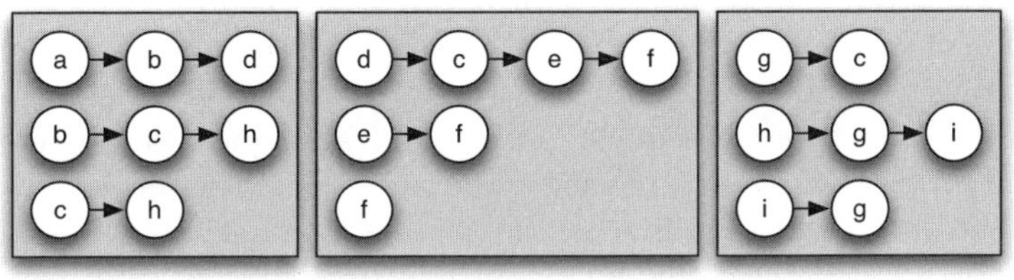

(b) Distributed adjacency list representation

FIGURE 3. A distributed directed graph represented as an adjacency list across three processors.

The Parallel BGL algorithms can operate on any distributed graph type that meets the distributed graph requirements, allowing the user to choose the best data structure for her task. The distributed compressed sparse row graph uses the same distribution scheme as the distributed adjacency list shown in Figure 3 (b), but instead of maintaining separate lists for the out-edges of each vertex, the out-edge lists are packed into a single, contiguous array. The distributed compressed sparse row graph therefore requires far less memory than the distributed adjacency list and exhibits better locality, resulting in better performance. However, the distributed compressed sparse row graph requires significantly more effort to use. For example, because insertion into the middle of the edge list is $\mathcal{O}(|E|)$, the edges must be ordered by source vertex before they are inserted. We typically use distributed compressed sparse row for benchmarking (see Section 4.5 for a performance comparison between these two graph data structures).

3. Implementing Single-Source Shortest Paths

We implemented three single-source shortest paths algorithms in the Parallel BGL, all of which are roughly based on Dijkstra's sequential algorithm. Dijkstra's algorithm computes shortest paths by incrementally growing a tree of shortest paths for a weighted graph $G = (V, E)$ from the source vertex s out to the most distant vertices. The primary data structure used in Dijkstra's algorithm is a priority queue of vertices that have been seen but not yet processed and ordered based on their distance $d(u)$ from the source vertex s. Prior to execution of Dijkstra's algorithm, $d(u) = \infty$ for all $u \neq s$, $d(s) = 0$, and the priority queue contains the vertex s. At each step in the computation, the algorithm removes the vertex u with the smallest $d(u)$ from the priority queue, then *relaxes* each outgoing edge (u, v). The *relax* step determines whether there is a better path from s to v via u, i.e., if $d(u) + w(u, v) < d(v)$ (where $w(u, v)$ is the non-negative weight of edge (u, v)). When a better path is found, $d(v)$ is updated with the value of $d(u) + w(u, v)$ and it is either inserted into the priority queue (if v had not previously been seen) or its position in the queue is updated. Dijkstra's algorithm terminates when the priority queue is empty.

Dijkstra's algorithm can be readily adapted for distributed memory. The graph itself is distributed using a row-wise decomposition, with each processor owning both a subset of the vertices as well as all edges outgoing from those vertices (as in the adjacency list of Figure 3). Likewise, the priority queue is distributed, with each processor's priority queue storing only those vertices owned by that processor. The processors will only remove vertices from their local priority queue, so each processor only relaxes the edges outgoing from vertices that it owns. When an edge is relaxed, the owner of the source vertex sends a message containing the target vertex, its new distance, and (optionally) the name of the source vertex to the target vertex's owner. To ensure that only the vertex u with the smallest $d(u)$ (globally) is selected, the computation is divided into supersteps: at the beginning of each superstep, the processors coordinate to determine the global minimum distance $\mu = min\{d(u) : u \text{ is queued}\}$. The processors then select a vertex u with $d(u) = \mu$, and the owner of vertex u removes it from the local priority queue and relaxes its outgoing edges. All processors then receive the messages produced by u's owner as edges are relaxed, update their local priority queues with new vertices and new distances, and the superstep completes. Successive supersteps process the remaining vertices in the shortest paths tree, one vertex per superstep. The algorithm terminates when all local priority queues are empty.

There are two obvious opportunities for parallelizing the naïvely distributed Dijkstra's algorithm. The first opportunity is to relax all of the edges outgoing from the active vertex u in parallel, effectively parallelizing the inner loop of Dijkstra's algorithm. However, the speedup that we can attain by parallelizing this loop is limited by the number of outgoing edges from a given vertex. With a row-wise distribution of the graph data structure, only a single processor has direct access to the outgoing edges of u. Therefore, parallelizing in this manner requires replication or distribution of the outgoing edges for the active vertex u, Even within a shared-memory system, the effect of parallelizing only this inner loop is limited. In many real-world graphs, the average out-degree is a relatively small constant, so parallelizing the relaxation of outgoing edges is not likely to yield good scalability except in the case of extremely dense graphs. For this reason, the current implementation of Parallel BGL therefore does not include parallel relaxation of edges. We are studying the inclusion of this parallel edge relaxation in future versions of the Parallel BGL.

The second opportunity for parallelizing Dijkstra's algorithm is to remove several vertices from the priority queue simultaneously, relaxing their edges in parallel. At the beginning of each superstep, the distributed-memory formulation of Dijkstra's algorithm determines the global minimum distance, μ, and selects a single vertex u to relax. Instead, we could allow every vertex u with $d(u) = \mu$ to be removed from the priority queue (and its outgoing edges relaxed) within the superstep. The scalability of the algorithm is then tied to the number of vertices that can be removed from the distributed priority queue within a single superstep and how well those vertices are distributed among the processors. Ideally, each superstep would remove a large number of vertices, evenly distributed among the processors. Unfortunately, real-world graphs rarely have a large number of vertices with the same distance from the source vertex, so the degree of parallelism we can extract from this direct parallelization is limited.

To expose more parallelism in Dijkstra's algorithm we need to remove more vertices from the priority queue in each superstep. This is accomplished by allowing the removal of vertices whose distances exceed μ. When removing a vertex u with $d(u) > \mu$ and relaxing its outgoing edges, it is possible that another processor will find a better route from the source s to u. When a better route is found, u will need to be *re*-inserted into the priority queue so that its edges will be relaxed again, but with a smaller value of $d(u)$. Thus, there is a trade-off between exposing more parallelism (by removing more vertices in each superstep) and avoiding unnecessary work (by limiting how many re-insertions will occur). We have implemented three variations on Dijkstra's algorithm that use different strategies to decide which vertices u with $d(u) > \mu$ should be considered.

3.1. Implementation Strategy. Dijkstra's algorithm can be viewed as a modified breadth-first search. A breadth-first search is typically implemented using a first-in first-out (FIFO) queue. Breadth-first search initially places the start vertex s into the queue. At each step, it extracts a vertex from the head of the queue, visits its outgoing edges, and places all new target vertices into the tail of the queue.

Dijkstra's algorithm changes breadth-first search in two ways. First, the FIFO queue is replaced with the priority queue. Second, when visiting the outgoing edges for the active vertex, Dijkstra's algorithm relaxes those edges and updates the ordering in the priority queue. Within the (sequential) Boost Graph Library, Dijkstra's algorithm is implemented as a call to breadth-first search that replaces the FIFO queue with a relaxed heap and provides a visitor that relaxes edges. A simplified version of the visitor is shown in Figure 4. It uses two events to update the queue: tree_edge() is invoked when breadth-first search traverses an edge whose target has not yet been seen, hence the edge is part of the breadth-first spanning tree, and gray_target(), which is invoked when the target of an edge has been seen but not processed. The visitor functions for both events "relax" edges, although only the latter needs to update the priority queue directly. Using this visitor, Dijkstra's algorithm is implemented as a simple call to breadth_first_visit():

```
dijkstra_shortest_paths
  (Graph &g, Vertex source)
{
    relaxed_heap<Vertex> Q; // Priority queue
    dijkstra_bfs_visitor bfs_vis(Q); // Visitor that updates the priority queue
    breadth_first_visit(graph, source, Q, bfs_vis);
}
```

The Parallel BGL contains a distributed-memory parallel breadth-first search implementation, upon which we have built two of the parallel Dijkstra variants. The eager and Crauser *et al.* variants use the visitor shown in Figure 4, but they provide

```
struct dijkstra_bfs_visitor {
  template<typename Edge, typename Graph>
  void tree_edge(Edge e, Graph& g) {
    if (distance(source(e, g)) + weight(e) < distance(target(e, g)))
        distance(target(e, g)) = distance(source(e, g)) + weight(e);
  }

  template<typename Edge, typename Graph>
  void gray_target(Edge e, Graph& g) {
    if (distance(source(e, g)) + weight(e) < distance(target(e, g)))
      Q.update(target(e, g), distance(source(e, g)) + weight(e));
  }
};
```

FIGURE 4. Breadth-first search visitor that relaxes each outgoing edge and updates the queue appropriately. This visitor is used by both the sequential and two of the parallel formulations of Dijkstra's algorithm in the (Parallel) BGL.

different distributed priority queue implementations, each using a different heuristic to determine which vertices should be removed in a superstep. Implementing other distributed priority queue heuristics for Dijkstra's algorithm is relatively simple with the Parallel BGL: one need only implement a queue that models the new heuristics and then call breadth_first_visit() with an instance of the new queue and the Dijkstra visitor from Figure 4, as shown above.

3.2. Eager Dijkstra's Algorithm. The "eager" Dijkstra's algorithm uses a simple heuristic to determine which vertices should be removed in a given superstep. The eager algorithm uses a constant lookahead factor λ, and in each superstep the processors remove every vertex u such that $d(u) \leq \mu + \lambda$, ordered by increasing values of $d(u)$. Keeping vertices sorted by increasing values of $d(u)$ allows the closest vertices to be relaxed first which minimizes the total number of edge relaxations. It also simplifies finding all vertices such that $d(u) \leq \mu + \lambda$.

When $\lambda = 0$, the eager algorithm is equivalent to the naïve parallelization of Dijkstra's algorithm, and exposes very little parallelism. Larger values of λ can expose more parallelism, but might result in a work-inefficient algorithm if too many vertices need to be re-inserted into the priority queue. When $\lambda = min\{weight(e)|e \in E\}$, we can expose additional parallelism without introducing any re-insertions. The optimal value for λ depends on the graph density, shape, and weight distribution, among other factors. We provide an experimental evaluation of the effect of λ on performance in Section 4.3.

The eager Dijkstra distributed queue from the Parallel BGL is responsible for implementing both the eager lookahead behavior of the algorithm and for managing synchronization among the processors. In addition to push() and pop(), it implements empty() and update() operations. Whenever the push() or update() operation is invoked, a message is sent to the owning process. These messages are only processed at the end of each superstep, which occurs inside the empty() method. empty() returns false so long as the local queue contains at least one vertex u such that $d(u) \leq \mu + \lambda$. When no such vertex exists, the processor synchronizes with all of the other processors, receiving "push" messages and finally recomputing the global minimum value, μ. Note that empty() only returns false when *all* priority queues on all processors are empty, signaling termination of the algorithm. The design and implementation process used to arrive at this formulation of a distributed queue, and its use with the sequential breadth_first_visit() implementation to effect a parallel algorithm, is further described in [**31**].

3.3. Crauser *et al.*'s Algorithm. The parallel Dijkstra variant due to Crauser *et al.* [**16**] uses more precise heuristics to increase the number of vertices removed in each superstep without causing any re-insertions. The algorithm uses two separate criteria, the OUT-criterion and the IN-criterion, which can be combined to determine which vertices should be removed in a given superstep. Unlike the eager algorithm, there are no parameters that need to be tuned.

The OUT-criterion computes a threshold L based on the weights of the outgoing edges in the graph. L is given the value $min\{d(u)+weight(u,w) : u$ is queued and $(u,w) \in E\}$. Any vertex v with $d(v) < L$ can safely be removed from the queue, because L bounds the smallest distance value $d(v)$ that can be achieved by relaxing the outgoing edges of any queued vertex.

The IN-criterion computes a threshold based on the incoming edges. If $d(v) - min\{weight(u,v) : (u,v) \in E\} \leq \mu$ (where μ is the global minimum) for a queued vertex v, then v can safely be removed from the queue, because there is no vertex in the queue with an outgoing edge to v that could be relaxed.

The OUT- and IN-criteria can be used in conjunction, so that each superstep removes all vertices that meet either criterion. On random graphs with uniform edge weights, each superstep will remove on average $\mathcal{O}(n^{2/3})$ vertices with high probability [**16**].

In the Parallel BGL, we have implemented Crauser *et al.*'s algorithm by creating a new distributed priority queue that applies the OUT- and IN-criteria. This distributed priority queue is very similar to the eager Dijkstra queue. However, the Crauser *et al.* priority queue contains three relaxed heaps for each processor, ordered by $d(v)$, $d(u) + weight(u,w) : u$ is queued and $(u,w) \in E$ (for the OUT-criterion), and $d(v) - min\{weight(u,v) : (u,v) \in E\}$ (for the IN-criterion). The three relaxed heaps are maintained simultaneously, so that both the IN- and OUT-criteria can be used together. The implementation of Crauser *et al.*'s algorithm is a single call to breadth_first_visit(), using the new distributed priority queue and the Dijkstra visitor from Figure 4.

3.4. Δ-stepping. The Δ-stepping [**48**], algorithm is a label-correcting algorithm that is parameterized by a lookahead parameter Δ, much like the λ lookahead factor in the Eager Dijkstra algorithm. As with the Eager Dijkstra algorithm, all vertices whose tentative distances are within Δ of the current global minimum distance are processed in parallel.

Unlike the Eager Dijkstra algorithm, however, Δ-stepping uses an approximate bucket data structure in lieu of a priority queue. The bucket data structure contains an array of buckets, B, each of which has width Δ. Thus, the ith bucket in B will store all vertices whose tentative distance falls in the range $[i\Delta, (i+1)\Delta)$, i.e., those vertices that will be processed in parallel in a superstep. By maintaining only an approximate ordering of elements in its central data structure, Δ-stepping performs less work ordering its bucket data structure than Eager Dijkstra does ordering its priority queue.

Δ-stepping also reduces the amount of work performed when relaxing the outgoing edges of each vertex in the current bucket by classifying edges into *light edges*—those whose weights are $\leq \Delta$, and whose relaxation could cause re-insertion of vertices into the current bucket—and *heavy edges*—those whose weights are $> \Delta$, and therefore cannot cause re-insertions of vertices into the current bucket. Δ-stepping repeatedly relaxes only the light edges outgoing from the vertices in the current bucket, until it has determined that no processor has performed any re-insertions into the current bucket. At this point, the heavy edges outgoing from every vertex that was in the current bucket will be relaxed. By delaying the relaxation of heavy edges, Δ-stepping ensures that the heavy edges are relaxed only once, whereas the light edges may need to be relaxed several times.

3.5. Using Dijkstra's Algorithm. The eager and Crauser *et al.* implementations of Dijkstra's algorithm in the (Parallel) BGL are provided by the function template dijkstra_shortest_paths(), which can operate on both distributed and non-distributed graphs. The algorithm is polymorphic based on the graph type, and can be invoked for any suitable graph from source vertex source and with the specified edge weights:

```
dijkstra_shortest_paths(graph, source, weight_map(edge_weights));
```

The actual implementation of Dijkstra's algorithm selected at compile time depends on what kind of graph is provided in the call. For instance, graph could be a non-distributed adjacency list, in which case the sequential Dijkstra's algorithm will be used:

```
adjacency_list<vecS, vecS, directedS> graph; // non-distributed adjacency list
dijkstra_shortest_paths(graph, source, weight_map(edge_weights));
  // sequential Dijkstra's
```

The same sequential Dijkstra's algorithm can instead be used with a (non-distributed) compressed sparse row graph, providing more compact storage and potentially improving algorithm performance:

```
compressed_sparse_row_graph<directedS> graph; // non-distributed CSR graph
dijkstra_shortest_paths(graph, source, weight_map(edge_weights));
  // sequential Dijkstra's
```

On the other hand, if graph were a distributed graph, dijkstra_shortest_paths() would instead apply Crauser *et al.*'s algorithm for distributed-memory parallel shortest paths. The graph in this case is an instance of adjacency_list that uses a distributedS selector, indicating that the vertices should be distributed across the processors. The distributedS selector is parameterized by the process group type, which indicates how parallel communication will be performed. In this case, we have used the bsp_process_group implemented over MPI. The fact that the graph is distributed is encoded within the type of the graph itself, allowing the Parallel BGL to perform a compile-time dispatch to select a distributed algorithm. The call to dijkstra_shortest_paths(), and the majority of the code leading up to the call, remains unchanged when one moves from the non-distributed graph types of the (sequential) BGL to the distributed graph types of the Parallel BGL.

```
adjacency_list<vecS, distributedS<vecS, mpi::bsp_process_group>, directedS> graph;
  // distributed adjacency list
dijkstra_shortest_paths(graph, source, weight_map(edge_weights));
  // Crauser et al.'s for distributed memory
```

To use the eager Dijkstra algorithm in lieu of Crauser *et al.*'s algorithm for a distributed graph, the user need only supply a lookahead value λ. Note that in the following example, the period separating the weight_map parameter from the lookahead parameter is not an error; rather, it is a form of named parameters used within both Boost Graph Libraries. Here, we illustrate the application of the eager Dijkstra algorithm to a distributed graph stored in compressed sparse row format:

```
compressed_sparse_row_graph<directedS, void, void, no_property,
  distributedS<mpi::bsp_process_group> > graph; // distributed CSR graph
dijkstra_shortest_paths(graph, source, weight_map(edge_weights).lookahead(15));
  // Eager Dijkstra's
```

The third implementation of Dijkstra's algorithm in the (Parallel) BGL is provided by the function template delta_stepping_shortest_paths(), this algorithm has no analog in the sequential BGL but work on unifying the implementations of the single-source shortest path algorithms is underway. In the same fashion as dijkstra_shortest_paths(), the Δ-stepping algorithm is polymorphic based on graph type, and can be invoked for any suitable graph from source vertex source and with the specified edge weights:

```
delta_stepping_shortest_paths(graph, source, weight_map(edge_weights));
```

An optional lookahead value Δ can also be supplied in a similar fashion to the eager Dijkstra algorithm:

```
delta_stepping_shortest_paths(graph, source, weight_map(edge_weights).lookahead(10));
```

4. Evaluation

For this paper we evaluated the performance and scalability of our single-source shortest paths implementations using various synthetic and real-world graphs; the sequential performance of the BGL has been demonstrated previously [**43, 44**]. All performance evaluations were performed on the Indiana University Computer Science Department's "Odin" research cluster. Odin consists of 128 compute nodes connected via Infiniband. Each node contains 4GB of dual-channel PC3200 (400 MHz) DDR-DRAM with two 2.0GHz AMD Opteron 270 processors (1MB Cache) running Red Hat Enterprise Linux WS Release 4. Each node contains a Mellanox InfiniHost HCA on a 133MHz PCI-X bus running OFED 1.1, connected to a 144 port Voltaire SDR switch. For our tests we left one processor idle on each node. The Parallel BGL tests were compiled using a pre-release version of Boost 1.34.0 [**11**] (containing the sequential BGL) and the latest development version of the Parallel BGL [**30**]. All programs were compiled with version 9.0 of the Intel C++ compiler with optimization flags −O3 −tpp7 −ipo −fno−alias. All MPI tests use version 1.2b3 of Open MPI [**24**] with the openib module.

We performed our experiments with real-world and synthetic data from the 9th DIMACS Implementation Challenge [**1**]. We used several different kinds of synthetic graphs generated using GTgraph [**9**], each of which exhibits different

	Parallel BGL Δ-stepping (1 processor)	BGL	SQ Ref. Solver
Time (s)	26.9801	9.26762	9.33730

TABLE 3. Performance of the sequential BGL and reference implementations on the USA road network (average over 50 source vertices).

graph properties. Additionally, we use real-world graph data for the network of roads in the United States. The graphs we have used in this evaluation are:

Random: Graphs as produced with the GTgraph random generator, using random (n,m) graphs. These graphs tend to have very little structure.

RMAT: Graphs as produced by the GTgraph implementation of the RMAT [**13**] power-law graph algorithm.

SSCA: Graphs as generated by the GTgraph implementation of the HPCS SSCA # 2 benchmark [**8**]. This algorithm begins by producing cliques of size uniformly distributed between 1 and n (we set n to 8 for weak scaling tests and 5 for comparison against the USA Roads data) and then adds inter-clique edges with probability p (we set p to 0.5 for weak scaling tests and 0.25 for comparison against the USA Roads data). These graphs also tend to have a large number of multiple edges.

USA Roads: Complete U.S. Road network from the UA 2000 Census TIGER/-Line data [**58**]. The graph is very sparse; it has $\sim$ 24 million vertices and $\sim$ 58 million edges.

European Roads: Road networks of 17 European countries from the PTV Europe data [**52**]. The graph is very sparse; it has $\sim$ 19 million vertices and $\sim$ 23 million edges.

Erdős-Rényi: Random graphs generated using the Parallel BGL random graph generator.

The Parallel BGL offers a variety of graph distribution options. All graphs use the default block distribution except where otherwise noted. The block distribution assigns $P/|V|$ vertices to each of P processors where each block of $P/|V|$ vertices is contiguous with regard to vertex order.

4.1. Sequential Performance. Although the focus of this paper is the Parallel BGL and it's single-source shortest paths algorithms, a brief discussion of the performance of the sequential BGL is in order to provide a baseline. Dijkstra's single-source shortest path algorithm is implemented using the sequential BGL's breadth-first search and a multi-level bucket queue [**27**] implementing Goldberg's caliber heuristic [**28**].

Table 3 demonstrates that the performance of the sequential BGL is comparable to the reference solver [**1**]. Also shown is the performance of the Δ-stepping parallel algorithm on a single processor. On a single processor the parallel implementation is noticeably slower due to the additional overhead imposed by the distributed data structures and communication code needed to support parallel computation. All single processor numbers beyond this table are the results of running the parallel algorithm with the required parallel data structures on a single processor.

4.2. Strong Scaling. To understand how well the parallel implementations of Dijkstra's algorithm in the Parallel BGL scale as more computational resources are provided, we evaluated the performance of each algorithm on fixed-size graphs. We generated synthetic graphs that are comparable in size to the USA road network, with $\sim 24M$ vertices and $\sim 58M$ edges. Both the random and RMAT graphs were created by specifying the number of vertices and edges parameters, with all other parameters set to defaults. The SSCA graph was generated by specifying the number of vertices, setting the maximum clique size to five, setting the probability of inter-clique edges to 0.25 and setting the maximum edge weight to 100. Note that with the SSCA graph, we were unable to generate graphs as sparse as the USA road network; to do so would require an unrealistically small maximum clique size. Thus, the SSCA graph contains about $148M$ edges. For the Eager Dijkstra algorithm, we have selected a lookahead value $\lambda = 8$ based on experimental evidence gathered for random graphs (Figure 8).

Figure 5 illustrates the strong scalability of the three single-source shortest path implementations. The random data appear to scale linearly with all algorithms. This is the result of a relatively uniform distribution of work due to the uniform nature of the random graph. The RMAT data also scale very linearly, though they exhibit less speedup than the random data. We speculate this is likely due to a less balanced work distribution caused by large variances in the degree of the vertices. The USA road data begin to scale inversely at 8 processors using the eager Dijkstra algorithm for reasons that are examined in section 4.3. The poor scalability of the USA road data using the Crauser *et al.* algorithm is most likely due to the conservative nature of the algorithm preventing it from removing a sufficient number of vertices in each superstep. When insufficient numbers of vertices are removed, the result is more communication rounds which increase runtime. Removing insufficient numbers of vertices in a superstep can also lead to load imbalances, further decreasing performance. The USA road data scale up to approximately 16 processors with the Δ-stepping algorithm but beyond that there is insufficient work available due to the small size of the graph to scale further. The SSCA data scale linearly with the eager Dijkstra algorithm, though the speedup as not as significant as on the other synthetic data. With the Crauser *et al.* algorithm and the Δ-stepping algorithm the SSCA data begins to scale inversely around 32 processors as the more complex communication and data-structure overhead of these algorithms begins to dominate the increasingly smaller amount of work available on each processor. The SSCA data contain more edges than the other synthetic graph types, but a significant portion of these edges are duplicates in the sense that they have the same source and target. The shortest path algorithms still have to consider the duplicate edges; moreover, the larger number of edges overall leads to a larger memory footprint. Without duplicate edges, the SSCA graph is only slightly denser than the random and RMAT graphs, yet it exhibits the same scaling behavior. This illustrates that the poor scaling of the SSCA graphs is indeed a product of their structure, not their density.

In order to determine if the performance observed was typical of structured data sets such as road networks, both variants of parallel Dijkstra's algorithm were also run on the European road network data. Figure 7 shows similar scalability results to the USA road network data. This supports our theory that the structure of the road network data limits available parallelism.

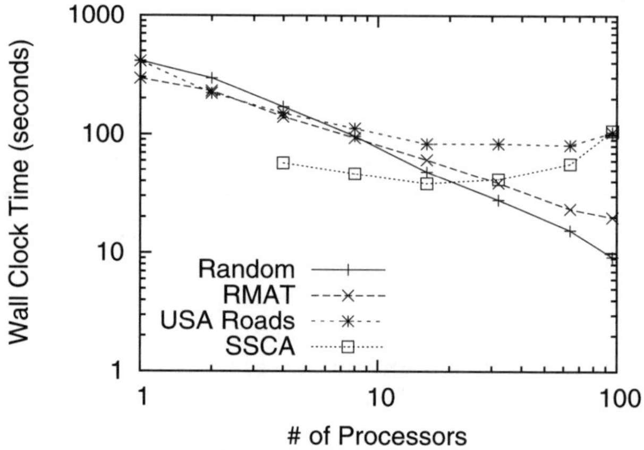

(a) Crauser *et al.*'s Algorithm

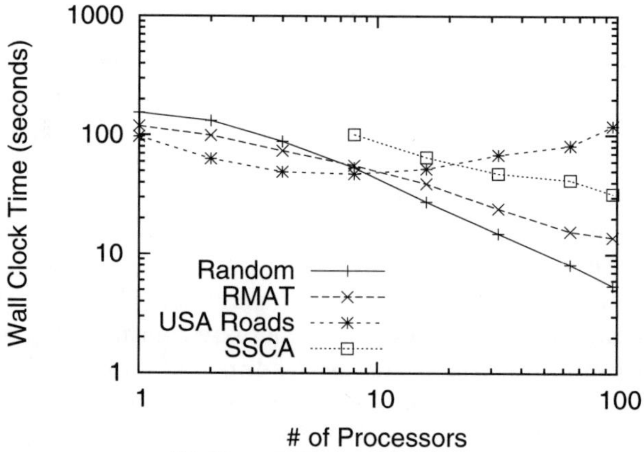

(b) Eager Dijkstra Algorithm

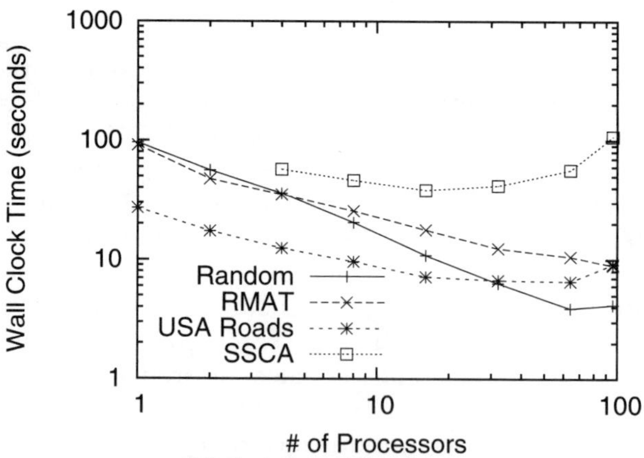

(c) Δ-stepping Algorithm

FIGURE 5. Strong scalability for the three single-source shortest paths algorithms, using fixed-size graphs with $\sim 24M$ vertices and $\sim 58M$ edges.

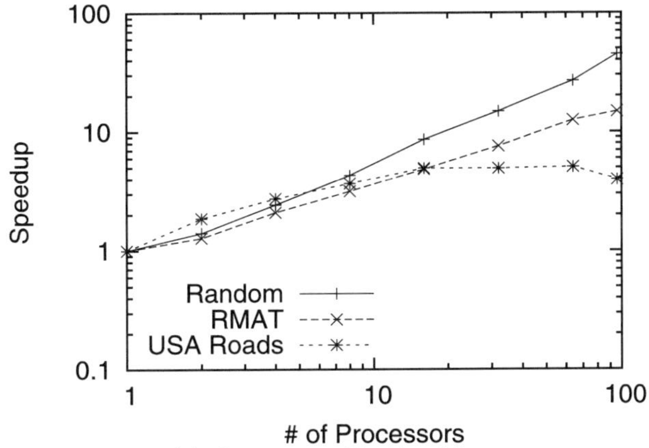

(a) Crauser *et al.*'s Algorithm

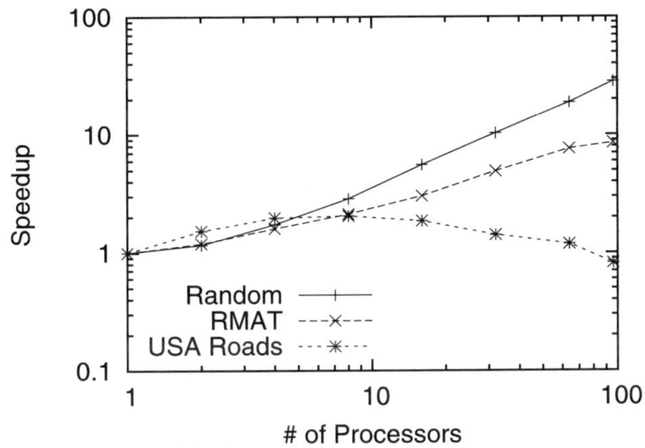

(b) Eager Dijkstra Algorithm

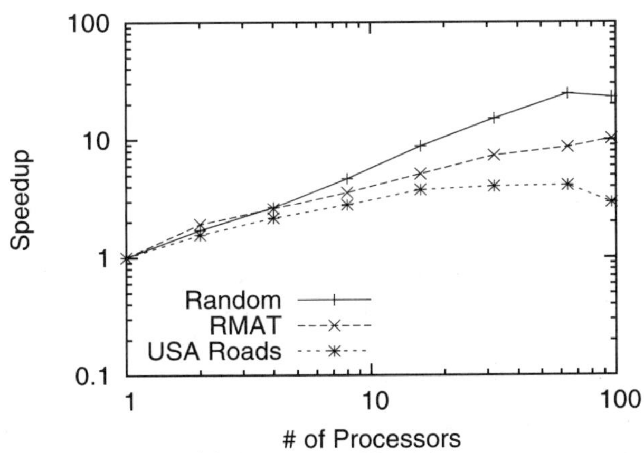

(c) Δ-stepping Algorithm

FIGURE 6. Parallel speedup for the three single-source shortest paths algorithms, using fixed-size graphs with $\sim 24M$ vertices and $\sim 58M$ edges.

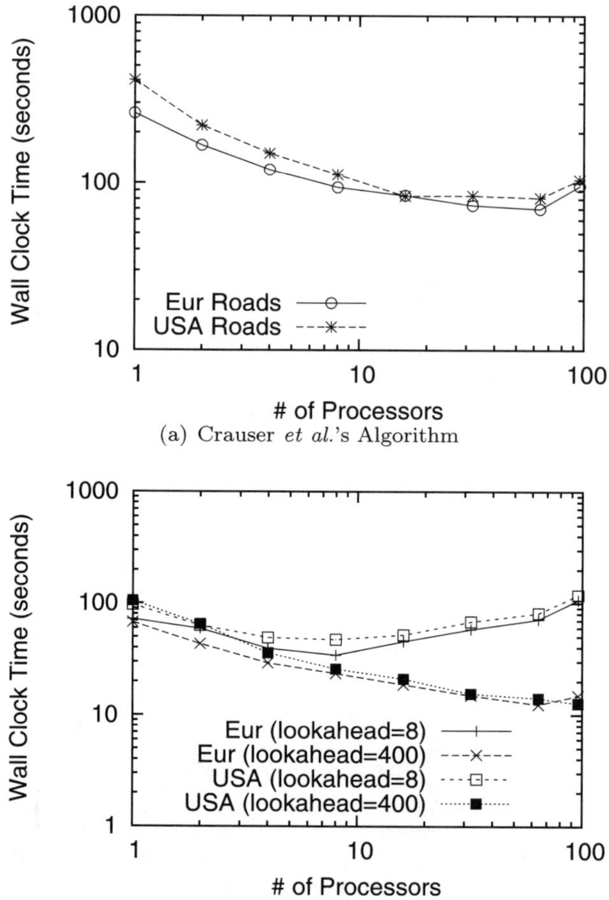

(a) Crauser *et al.*'s Algorithm

(b) Eager Dijkstra Algorithm

FIGURE 7. Strong scalability for two variants of parallel Dijkstra's algorithm, using the USA and European road networks

It should be noted that all of these graphs are relatively small compared to the size of problems the Parallel BGL is capable of solving. The Parallel BGL does introduce some communication overhead in order to manage the distributed data structures, therefore for small problem sizes the sequential BGL may be a more appropriate choice. However for problem sizes too large to fit in core on a single machine, the Parallel BGL is a fast, efficient, and scalable alternative. For problems that do fit in core on a single machine the Parallel BGL may still be able to provide a faster solution using small numbers of processors.

4.3. Eager Dijkstra Lookahead Factor. To determine an appropriate lookahead value to use in our scalability tests we evaluated a range of options on a random graph generated using the GTgraph [**9**] generator. This graph is comparable in size to the USA road network. We chose a random graph in order to reduce any bias the particular structures of the other graphs may have had on the lookahead value. Previous tests using the Parallel BGL [**32**] have indicated that optimal lookahead

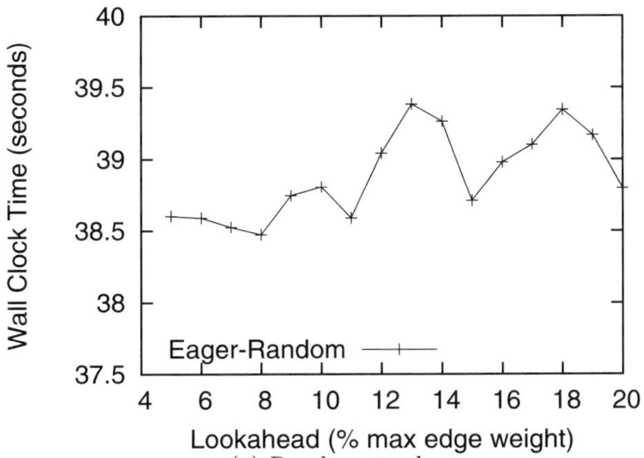

(a) Random graph

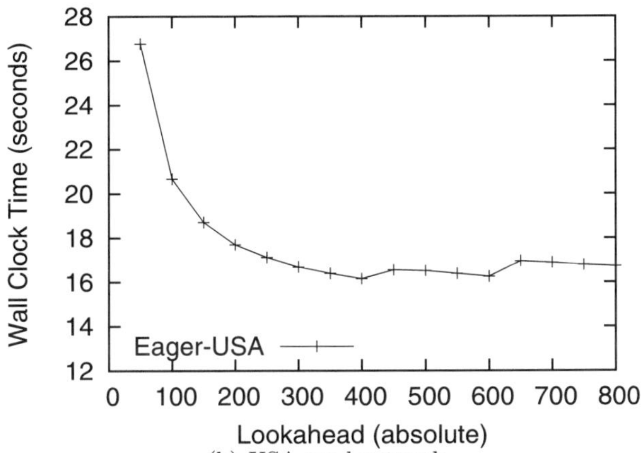

(b) USA road network

FIGURE 8. Effect of lookahead value on the performance of the eager Dijkstra's algorithm, using fixed-size graphs with $\sim 24M$ vertices and $\sim 58M$ edges.

values for random graphs tend to be around 10% of the maximum edge weight in the graph so we examined values around 10%.

Figure 8 shows a minimum at a lookahead value of 8, thus this value was chosen for our scalability results. Examining the strong scalability results in Figure 5 obtained using this lookahead value indicates that the USA road network data scale very poorly. We speculated that this was likely the result of a poor choice of lookahead value for this particular graph, so we tried a variety of alternate lookahead values. Figure 8 shows that the optimal lookahead value for the USA road network was 400, much larger than our original approximation of 8.

Figure 9 illustrates the performance difference that the choice of a lookahead value can have on the eager Dijkstra's algorithm. Optimal lookahead values vary not only with graph size and structure, but also with graph shape, density, and edge weight distribution. We theorize that the poor scaling behavior of the USA

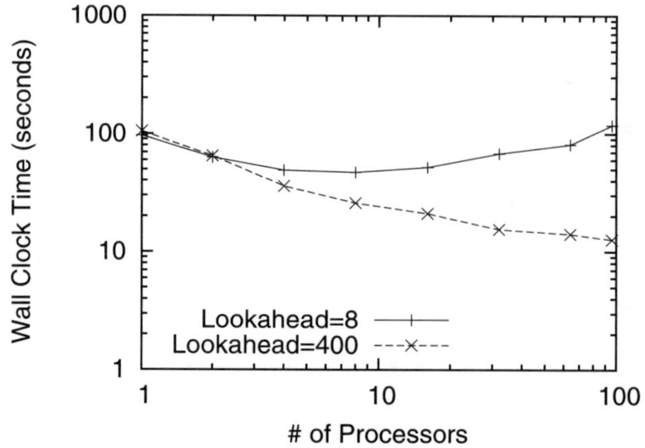

(a) Wall clock time

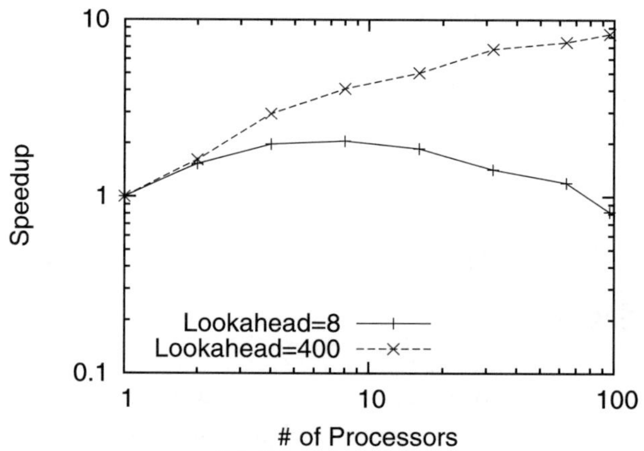

(b) Parallel speedup

FIGURE 9. Results of running the eager Dijkstra's algorithm with two different lookahead values on the USA road network

road network was due to sparse regions containing many edges with large weights. Traversing high-weight edges in these sparse regions may require several algorithm iterations if the lookahead value is small. Each iteration is relatively expensive due to the all-to-all communication that needs to occur between each BSP superstep and thus increasing the number of iterations causes severe performance degradation. Conversely if the lookahead value is large enough to cause these edges to be explored in a single iteration, then many iterations may be saved and runtime significantly reduced.

Figure 9 shows that using an appropriate lookahead value yields much better scalability on the USA road network. The parallel speedup begins to taper off around 16 processors because the data set is too small to provide adequate local work to overcome the communication overhead. Given a larger data set the shortest paths algorithms in the Parallel BGL should scale well up to hundreds of processors.

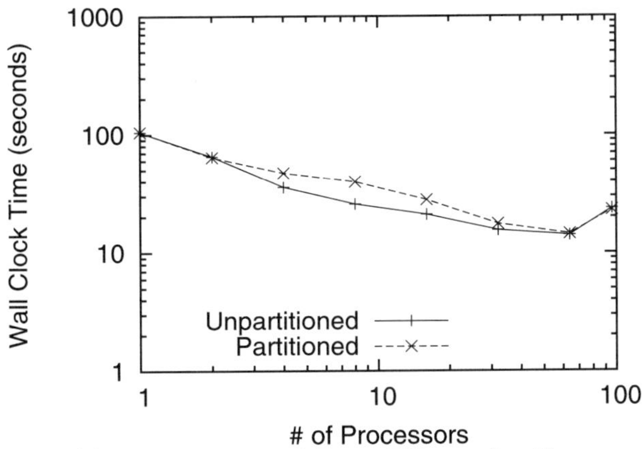

(a) USA road network Eager Dijkstra algorithm

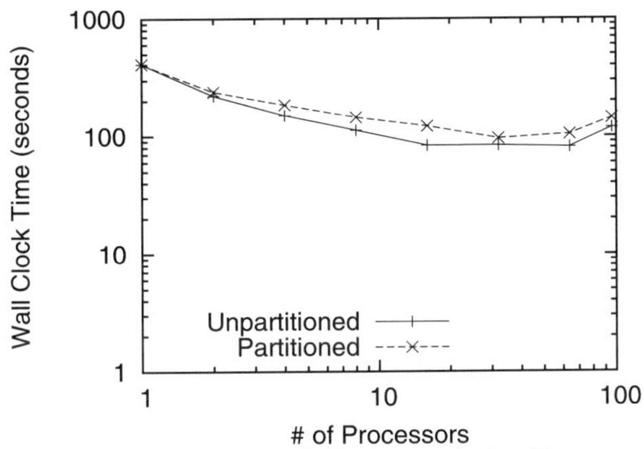

(b) USA road network Crauser *et al.* algorithm

FIGURE 10. Parallel Dijkstra run times on METIS partitioned and unpartitioned USA road network

4.4. Graph Partitioning. To evaluate the effects of data distribution on algorithm performance, we applied a data partition to the USA road network. We computed this partition using the k-way METIS [**39**] partitioning program, but memory pressure prevented us from using edge weights or vertex coordinates in our partitioning. We ran both the Crauser *et al.* algorithm and Eager Dijkstra algorithm on the partitioned graphs and compared the results to the unpartitioned version. Figure 10 provides the comparison between the partitioned and unpartitioned USA road network. Partitioning did not prove useful, and in some cases was detrimental to performance. The most likely cause of this is that the TIGER/Line data already exhibit good locality properties in their file representation while METIS was hampered by the inability to consider edge weights or vertex coordinates due to memory pressure making it unlikely to improve on the existing partitioning and in some cases choose poorer partitionings.

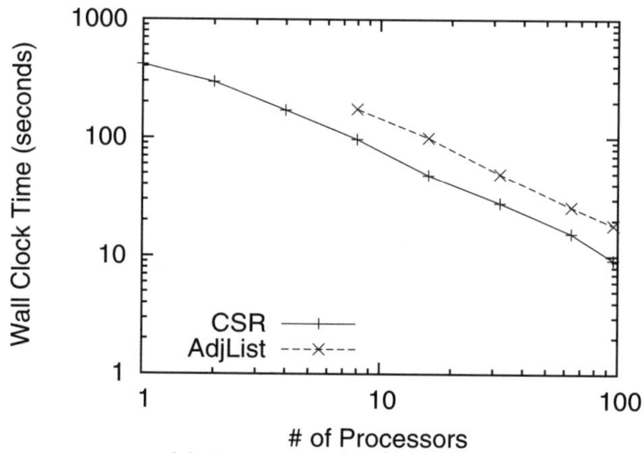

(a) Crauser *et al.*'s Algorithm

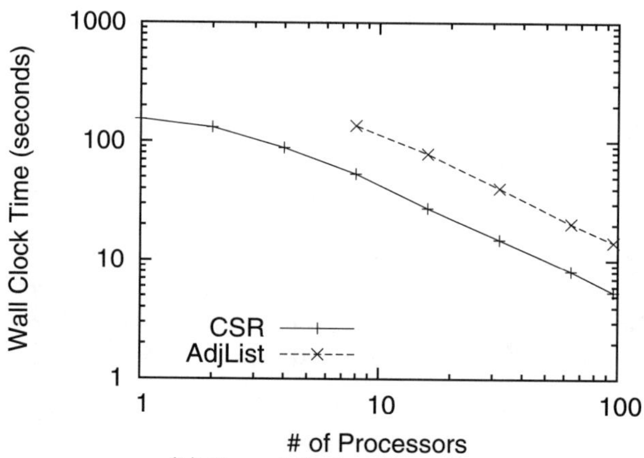

(b) Eager Dijkstra Algorithm

FIGURE 11. Effect of graph data type on the performance of two variants of parallel Dijkstra's algorithm on the USA road network.

4.5. Graph Data Structures. The Parallel BGL offers several graph data structures including an adjacency list and compressed sparse row (CSR) graph. Changing the data structures used by an algorithm can mean the difference between fitting the algorithm's working set in core and paging. Additionally, more compact data structures often gain a performance advantage from cache reuse. Because more data elements can be kept in cache the likelihood of finding a data element there is higher. Compact data structures such as CSR also have drawbacks, in this case $\mathcal{O}(|E|)$ cost for edge insertion, which may require the use of more verbose data structures in some cases.

Figure 11 shows that the CSR graph representation out-performs the adjacency list representation; this can be attributed to its more compact size and efficient access methods. The data for the Adjacency List graph type are missing for fewer processors because of the higher memory footprint of that graph type. This figure also shows that there is a constant performance overhead for the less-compact

Adjacency List which may be explained by cache effects and indirection. These results demonstrate the importance of choosing appropriate data structures when implementing an algorithm. Fortunately, the concepts in the Parallel BGL simplify making these choices by explicitly specifying the requirements that data structures must model.

4.6. Weak Scaling. To understand how the parallel implementations of Dijkstra's algorithm scale as the problem size scales, we evaluated the performance of each algorithm on graphs where $|V| \propto |E| \propto p$ (where p is the number of processors). We generated random and RMAT graphs with $\sim 1M$ vertices and $\sim 10M$ edges per processor and SSCA graphs with $\sim 1M$ vertices per processor and as close to $\sim 10M$ edges per processor as possible. Figure 12 illustrates the run times on the GTgraph graphs. We also wanted to generate weak scaling results for the largest possible graph we could fit in core. Reading graphs from disk and generating them using the sequential generator was very expensive so we used the Parallel BGL Erdős-Rényi generator to generate graphs similar to the GTgraph random graphs in core. These graphs had $\sim 2.5M$ vertices and $\sim 12.5M$ edges per processor which results in a maximum graph size of $\sim 240M$ vertices and $\sim 1.2B$ edges on 96 processors. Figure 13 shows the run times for the weak scaling tests on these graphs produced with the Parallel BGL Erdős-Rényi generator. All weak scaling results use a lookahead value $\lambda = 8$.

These experiments show that the runtime increases even though the amount of data per processor remains constant. This is because the amount of work performed by the Crauser *et al.* algorithm is $\mathcal{O}(|V|\log|V| + |E|)$ [**16**]. As we vary $|V|$ linearly with the number of processors the amount of work increases faster than the number of processors. This yields more work per processor which gives rise to the sub-linear scaling exhibited in Figure 12. In fact, the amount of work per processor is equal to $\frac{|V|}{p} \times \log \frac{|V|}{p}$ where p is the number of processors. In our weak scaling experiment $|V| \propto p$ so the work per processor is proportional to $\log(|V|)$. The eager Dijkstra's algorithm has a similar $\log(|V|)$ factor in the amount of work performed due to the priority queue operations and thus it is reasonable to presume that it will scale sub-linearly as well. Some curve-fitting to the weak scaling data showed that a logarithmic curve does fit the data better than any other function (radical, linear, etc.). This suggests that our weak scaling performance closely approximates the shape of the theoretical maximum though the constant factors may differ.

4.7. Very Large Graphs. We also used Parallel BGL Erdős-Rényi generator to generate larger graphs than those provided in the challenge data sets which demonstrate the capability of the Parallel BGL to solve very large graph problems. Figure 14 shows run times for strong scaling tests on graphs with between 2^{28} and 2^{30} vertices and average degree ~ 4. The largest of these graphs had 2^{30} vertices and 2^{32} edges.

5. Related Work

The CGM*graph* library [**14**] implements several graph algorithms, including Euler tour, connected components, spanning tree, and bipartite graph detection. It uses its own communication layer built on top of MPI and based on the Course Grained Multicomputer (CGM) [**18**] model, a variant of the BSP model. From

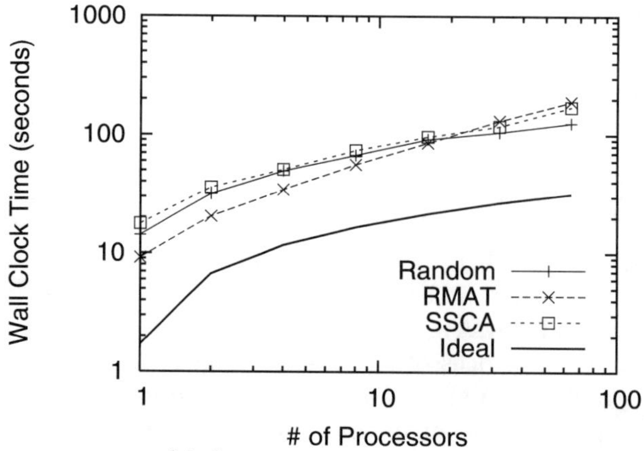

(a) Crauser *et al.*'s Algorithm

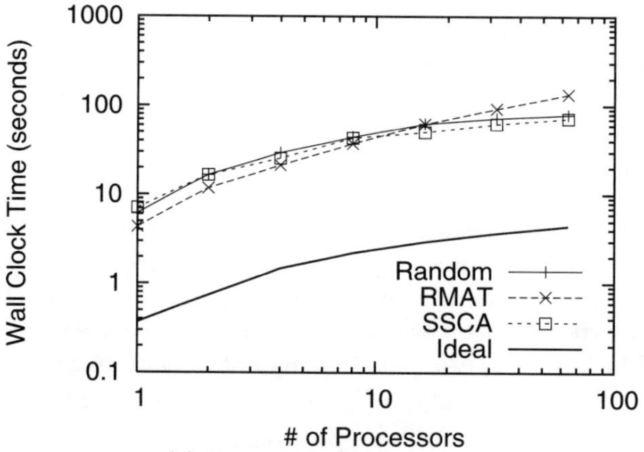

(b) Eager Dijkstra Algorithm

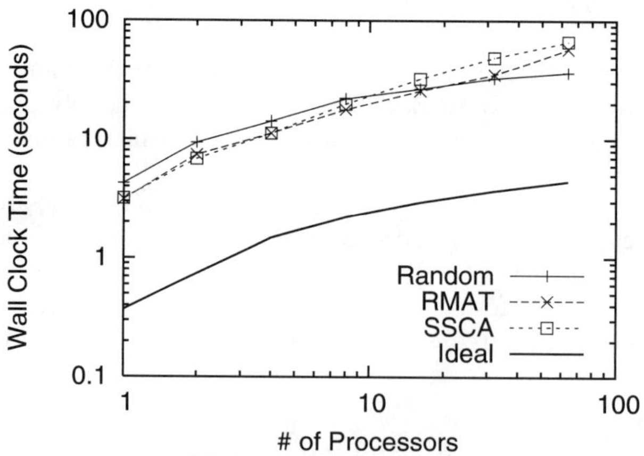

(c) Δ-stepping Algorithm

FIGURE 12. Weak scalability for the three parallel single-source shortest path algorithms , using graphs with an average of $\sim 1M$ vertices and $\sim 10M$ edges per processor.

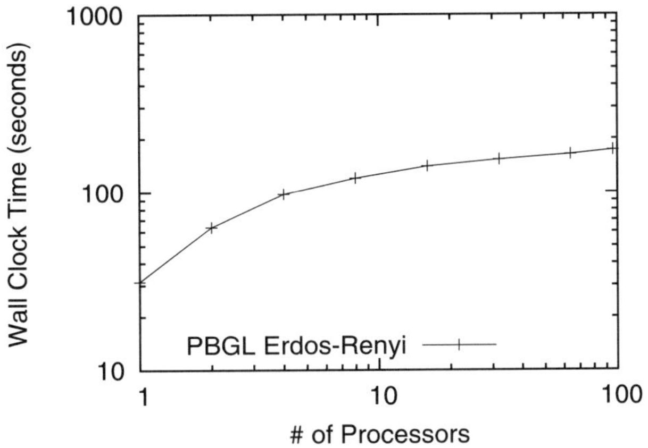

(a) Crauser *et al.*'s Algorithm

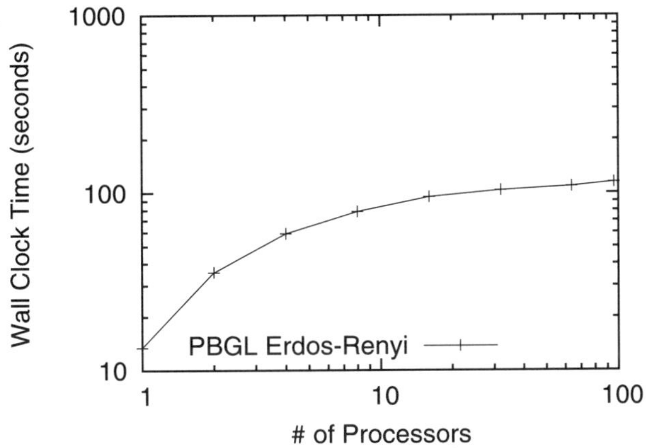

(b) Eager Dijkstra Algorithm

FIGURE 13. Weak scalability for two variants of parallel Dijkstra's algorithm, using BGL-generated graphs with an average of $\sim 2.5M$ vertices and $\sim 12.5M$ edges per processor.

an architectural standpoint, CGM*graph* and the Parallel BGL adopt different programming paradigms: the former adheres to Object-Oriented principles whereas the latter follows the principles of generic programming. We have previously shown that the Parallel BGL's implementation of connected components is at least an order of magnitude faster than CGM*graph*'s implementation [**31**], which we believe is due to generic programming's dual focus on genericity and efficiency. However, CGM*graph* does not provide any implementations of parallel shortest paths.

The ParGraph library [**34**] shares many goals with the Parallel BGL. Both libraries are based on the sequential BGL and aim to provide flexible, efficient parallel graph algorithms. The libraries differ in their approach to parallelism: the Parallel BGL stresses source-code compatibility with the sequential BGL, eliminating most explicit communication. ParGraph, on the other hand, represents communication

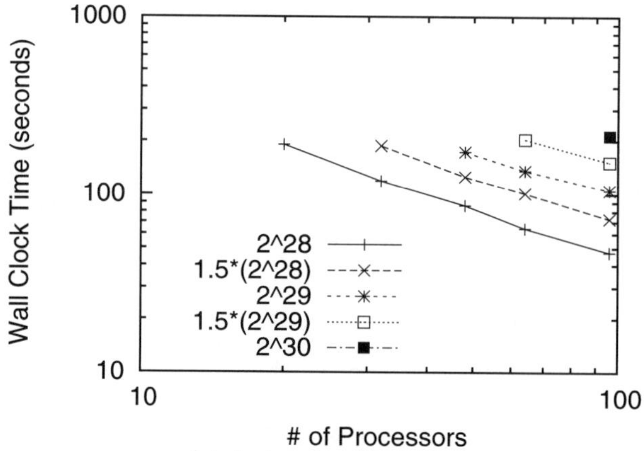

(a) Δ-stepping Algorithm

FIGURE 14. Weak scalability for the Delta-Stepping algorithm, using BGL-generated graphs with between 2^{28} and 2^{30} vertices and average degree ~ 4. Chart labels indicate the number of vertices in the graph being tested.

explicitly, which may permit additional optimizations. ParGraph does not provide any implementations of parallel shortest paths.

The Standard Template Adaptive Parallel Library (STAPL) [**4**] is a generic parallel library providing data structures and algorithms whose interfaces closely match those of the C++ Standard Template Library. STAPL and Parallel BGL both share the explicit goal of parallelizing an existing generic library, but their approach to parallelization is quite different. STAPL is an adaptive library, that will determine at *run time* how best to distribute a data structure or parallelize an algorithm, whereas the Parallel BGL encodes distribution information (i.e., the process group) into the data structure types and makes parallelization decisions at *compile time*. Run time decisions potentially offer a more convenient interface, but compile time decisions permit the library to optimize itself to particular features of the task or communication model (an *active library* [**59**]), effectively eliminating the cost of any abstractions we have introduced. STAPL includes a distributed graph container with several algorithms, but it is unclear whether any parallel shortest paths algorithms have been implemented.

Performance results for the Δ-stepping algorithm on large graphs using the Cray MTA-2 have been presented [**45**]. The MTA-2 is a large, shared-memory machine with a massively multithreading architecture. The MTA-2 processors have hardware support for a large number of threads, and can efficiently switch from one hardware thread to another when an outstanding memory request is serviced. Thus, the MTA-2 is able to tolerate latency better than most architectures, particularly the commodity clusters for with the Parallel BGL was designed. Unfortunately, the largest MTA-2 ever built contains only 40 processors, limiting the practicality of graph algorithm implementations on this machine.

6. Conclusion

The Parallel Boost Graph Library provides flexible distributed graph data structures and generic algorithms. Using the facilities of the Parallel BGL, we implemented three parallel, distributed-memory variants of Dijkstra's algorithm for single-source shortest paths. By building on the pre-existing distributed-memory breadth-first search, the sequential implementation of Dijkstra's algorithm in the (sequential) BGL, and reusing the Parallel BGL's data structures, we were able to implement parallel Dijkstra's algorithm with a relatively small amount of effort and evaluate performance with several different graph types. The results showed that our solutions are computationally efficient and scalable.

Naturally, the scalability of a graph algorithm depends on the structure of the graph on which it operates. Our generic, flexible library scales well on unstructured graphs both in terms of parallel speedup as well as problem size. Problem size scaling is limited by the superlinear complexity of the Crauser and Eager Dijkstra algorithms. Graphs with grid-like structures do not scale as well as unstructured graphs.

We expect to extend the work in this paper in a number of ways. In the short term, we intend to continue to refine the implementations we already have in order to capture the state of the art in distributed-memory parallel algorithms. In the longer term, we will extend the Parallel BGL to shared-memory parallel processing and take advantage of hybrid models for clusters of SMPs. To facilitate rapid prototyping and development of new sequential and parallel graph algorithms, we are also refining our Python interfaces to sequential and Paralell BGL. Ultimately, our goal is to continue to develop the (Parallel) BGL as a robust platform for parallel graph algorithm and data structure research.

References

[1] 9th DIMACS Implementation Challenge - Shortest Paths. `http://www.dis.uniroma1.it/~challenge9/`, March 2006.

[2] D. Abrahams and A. Gurtovoy. *C++ Template Metaprogramming: Concepts, Tools, and Techniques from Boost and Beyond.* Addison-Wesley, 2004.

[3] A. V. Aho, J. E. Hopcroft, and J. D. Ullman. *Data Structures and Algorithms.* Addison-Wesley, 1983.

[4] P. An, A. Jula, S. Rus, S. Saunders, T. Smith, G. Tanase, N. Thomas, N. Amato, and L. Rauchwerger. STAPL: An adaptive, generic parallel programming library for C++. In *Workshop on Languages and Compilers for Parallel Computing*, pages 193–208, August 2001.

[5] M. Austern. (Draft) Technical report on standard library extensions. Technical Report N1711=04-0151, ISO/IEC JTC 1, Information Technology, Subcommittee SC 22, Programming Language C++, 2004.

[6] D. Bader, G. Cong, and J. Feo. A comparison of the performance of list ranking and connected components algorithms on SMP and MTA shared-memory systems. Technical report, October 2004.

[7] D. Bader and K. Madduri. Designing multithreaded algorithms for breadth-first search and st-connectivity on the cray MTA-2. In *Proceedings of 35th International Conference on Parallel Processing*, pages 523–530, August 2006.

[8] D. A. Bader and K. Madduri. Design and implementation of the HPCS graph analysis benchmark on symmetric multiprocessors. Technical report, 2005.

[9] D. A. Bader and K. Madduri. GTgraph: A suite of synthetic graph generators. `http://www-static.cc.gatech.edu/~kamesh/GTgraph/`, February 2006.

[10] E. G. Boman, D. Bozdağ, U. Catalyurek, A. H. Gebremedhin, and F. Manne. A scalable parallel graph coloring algorithm for distributed memory computers. In *Lecture Notes in Computer Science*, volume 3648, pages 241–251, August 2005.

[11] Boost. *Boost C++ Libraries.* `http://www.boost.org/`.

[12] U. Brandes. A faster algorithm for betweenness centrality. *Journal of Mathematical Sociology*, 25(2):163–177, 2001.

[13] D. Chakrabarti, Y. Zhan, and C. Faloutsos. Rmat: A recursive model for graph mining. In *Proceedings of 4th International Conference on Data Mining*, pages 442–446, April 2004.

[14] A. Chan and F. Dehne. CGM*graph*/CGM*lib*: Implementing and testing CGM graph algorithms on PC clusters. In *PVM/MPI*, pages 117–125, 2003.

[15] T. Cormen, C. Leiserson, and R. Rivest. *Introduction to Algorithms.* McGraw-Hill, 1990.

[16] A. Crauser, K. Mehlhorn, U. Meyer, and P. Sanders. A parallelization of dijkstra's shortest path algorithm. In L. Brim, J. Gruska, and J. Zlatuska, editors, *Mathematical Foundations of Computer Science*, volume 1450 of *Lecture Notes in Computer Science*, pages 722–731. Springer, 1998.

[17] E. H. Cuthill and J. McKee. Reducing the bandwidth of sparse symmetric matrices. In *Proc. 24th National Conference of the ACM*, pages 157–172. ACM Press, 1969.

[18] F. Dehne, A. Fabri, and A. Rau-Chaplin. Scalable parallel geometric algorithms for coarse grained multicomputers. In *Proceedings of the Ninth Annual Symposium on Computational Geometry*, pages 298–307. ACM Press, 1993.

[19] F. Dehne and S. Götz. Practical parallel algorithms for minimum spanning trees. In *Symposium on Reliable Distributed Systems*, pages 366–371, 1998.

[20] E. A. Dinic. Algorithm for solution of a problem of maximum flow in networks with power estimation. *Soviet Mathematics Doklady*, 11:1277–1280, 1970.

[21] D. Eppstein, Z. Galil, and G. Italiano. Dynamic graph algorithms. In *In CRC Handbook of Algorithms and Theory of Computation, Chapter 22. CRC Press*, page 95, 1997.

[22] L. Fleischer, B. Hendrickson, and A. Pinar. On identifying strongly connected components in parallel. In *Parallel and Distributed Processing (IPDPS)*, volume 1800 of *Lecture Notes in Computer Science*, pages 505–511. Springer, 2000.

[23] T. Fruchterman and E. Reingold. Graph drawing by force-directed placement. *Software–Practice and Experience*, 21(11):1129–1164, 1991.

[24] E. Gabriel, G. E. Fagg, G. Bosilca, T. Angskun, J. J. Dongarra, J. M. Squyres, V. Sahay, P. Kambadur, B. Barrett, A. Lumsdaine, R. H. Castain, D. J. Daniel, R. L. Graham, and T. S. Woodall. Open MPI: Goals, concept, and design of a next generation MPI implementation. In *Proceedings, 11th European PVM/MPI Users' Group Meeting*, pages 97–104, Budapest, Hungary, September 2004.

[25] A. George and J. W. H. Liu. The evolution of the minimum degree ordering algorithm. volume 31, pages 1–19, March 1989.

[26] S. Goddard, S. Kumar, and J. F. Prins. Connected components algorithms for mesh connected parallel computers. In S. N. Bhatt, editor, *Parallel Algorithms*, volume 30 of *DIMACS Series in Discrete Mathematics and Theoretical Computer Science*, pages 43–58. American Mathematical Society, 1997.

[27] A. Goldberg. Shortest path algorithms: Engineering aspects. In *Proceedings of 12th International Symposium, ISAAC*, pages 502–512. Springer, 2001.

[28] A. Goldberg. A simple shortest path algorithm with linear average time. Technical report, STAR Lab., InterTrust Tech., Inc., 2001.

[29] A. Grama, A. Gupta, G. Karypis, and V. Kumar. *Introduction to Parallel Computing, Second Edition.* Addison-Wesley, 2003.

[30] D. Gregor, N. Edmonds, A. Breuer, P. Gottschling, B. Barrett, and A. Lumsdaine. The Parallel Boost Graph Library. `http://www.osl.iu.edu/research/pbgl`, 2005.

[31] D. Gregor and A. Lumsdaine. Lifting sequential graph algorithms for distributed-memory parallel computation. In *Proceedings of the 2005 ACM SIGPLAN conference on Object-oriented programming, systems, languages, and applications*, pages 423–437, October 2005.

[32] D. Gregor and A. Lumsdaine. The Parallel BGL: A generic library for distributed graph computations. In *In Proceedings of the Fourth Workshop on Parallel Object-Oriented Scientific Computing*, July 2005.

[33] A. Gürsoy and M. Atun. Neighborhood preserving load balancing: A self-organizing approach. In *Euro-Par Parallel Processing*, volume 1900 of *Lecture Notes in Computer Science*, pages 324–341, 2000.

[34] F. Hielscher and P. Gottschling. ParGraph. `http://pargraph.sourceforge.net/`, 2004.

[35] W. Hohberg. How to find biconnected components in distributed networks. *Journal of Parallel and Distributed Computint*, 9(4):374–386, 1990.

[36] J. Jaja. *An Introduction to Parallel Algorithms*. Addison-Wesley Professional, 1992.

[37] D. B. Johnson and P. T. Metaxas. A parallel algorithm for computing minimum spanning trees. In *ACM Symposium on Parallel Algorithms and Architectures*, pages 363–372, 1992.

[38] T. Kamada and S. Kawai. An algorithm for drawing general undirected graphs. *Information Processing Letters*, 31:7–15, 1989.

[39] G. Karypis and V. Kumar. Multilevel k-way partitioning scheme for irregular graphs. *Journal of Parallel and Distributed Computing*, 48(1):96–129, 1998.

[40] I. King. An automatic reordering scheme for simultaneous equations derived from network analysis. *International Journal for Numerical Methods in Engineering*, 2:523–533, 1970.

[41] D. E. Knuth. *Stanford GraphBase: A Platform for Combinatorial Computing.* ACM Press, 1994.

[42] C. L. Lawson, R. J. Hanson, D. R. Kincaid, and F. T. Krogh. Basic Linear Algebra Subprograms for Fortran usage. *ACM Transactions on Mathematical Software*, 5(3):308–323, September 1979.

[43] L.-Q. Lee, J. Siek, and A. Lumsdaine. Generic graph algorithms for sparse matrix ordering. In *International Symposium on Computing in Object-Oriented Parallel Environments*, volume 1732 of *Lecture Notes in Computer Science*, pages 120–129, 1999.

[44] L.-Q. Lee, J. Siek, and A. Lumsdaine. The Generic Graph Component Library. In *Proceedings of the 14th ACM SIGPLAN conference on Object-oriented programming, systems, languages, and applications*, pages 339–414, 1999.

[45] K. Madduri, D. Bader, J. W. Berry, and J. R. Crobak. Parallel shortest path algorithms for solving large-scale instances. 9th DIMACS Implementation Challenge - Shortest Paths, 2006. An updated version of the paper appears in this book.

[46] D. W. Matula and L. L. Beck. Smallest-last ordering and clustering and graph coloring algorithms. *Journal of the ACM*, 30(3):417–427, 1983.

[47] K. Mehlhorn and S. Näher. *The LEDA Platform of Combinatorial and Geometric Computing.* Cambridge University Press, 1999.

[48] U. Meyer and P. Sanders. Delta-stepping: A parallel single source shortest path algorithm. In *Proceedings of the 6th Annual European Symposium on Algorithms*, pages 393–404. Springer-Verlag, 1998.

[49] D. R. Musser, G. J. Derge, and A. Saini. *STL Tutorial and Reference Guide.* Addison-Wesley, 2nd edition, 2001.

[50] D. R. Musser and A. A. Stepanov. Generic programming. In P. P. Gianni, editor, *Symbolic and algebraic computation: ISSAC '88, Rome, Italy, July 4–8, 1988: Proceedings*, volume 358 of *Lecture Notes in Computer Science*, pages 13–25, Berlin, 1989. Springer Verlag.

[51] L. Page, S. Brin, R. Motwani, and T. Winograd. The PageRank citation ranking: Bringing order to the Web. Technical report, Stanford Digital Library Technologies Project, November 1998.

[52] PTV Europe. European road graphs. `http://i11www.iti.uni-karlsruhe.de/resources/roadgraphs.php`.

[53] J. Siek, L.-Q. Lee, and A. Lumsdaine. *The Boost Graph Library: User Guide and Reference Manual.* Addison-Wesley Longman Publishing Co., Inc., Boston, MA, USA, 2002.

[54] S. W. Sloan. An algorithm for profile and wavefront reduction of sparse matrices. *International Journal for Numerical Methods in Engineering*, 23:239–251, 1986.

[55] T. Steihaug and A. Hossain. Graph coloring and the estimation of sparse jacobian matrices using row and column partitioning. Technical Report Report 72, Department of Informatics, University of Bergen, 1992.

[56] A. A. Stepanov. Generic programming. *Lecture Notes in Computer Science*, 1181:40, 1996.

[57] A. A. Stepanov and M. Lee. The Standard Template Library. Technical Report X3J16/94-0095, WG21/N0482, ISO Programming Language C++ Project, May 1994.

[58] U.S. Census Bureau. UA census 2000 TIGER/Line files. `http://www.census.gov/geo/www/tiger/tigerua/ua_tgr2k.html`.

[59] T. L. Veldhuizen. *Active Libraries and Universal Languages.* PhD thesis, Indiana University Computer Science, May 2004.

[60] S. White, J. O'Madadhain, D. Fisher, and Y.-B. Boey. Java Universal Network/Graph framework. `http://jung.sourceforge.net/`, 2004.

Open Systems Laboratory, Indiana University, Bloomington, IN 47405
E-mail address: ngedmond@osl.iu.edu

Open Systems Laboratory, Indiana University, Bloomington, IN 47405
E-mail address: abreuer@osl.iu.edu

Open Systems Laboratory, Indiana University, Bloomington, IN 47405
E-mail address: dgregor@osl.iu.edu

Open Systems Laboratory, Indiana University, Bloomington, IN 47405
E-mail address: lums@osl.iu.edu

DIMACS Series in Discrete Mathematics
and Theoretical Computer Science
Volume **74**, 2009

Parallel Shortest Path Algorithms for Solving Large-Scale Instances

Kamesh Madduri, David A. Bader, Jonathan W. Berry, and Joseph R. Crobak

ABSTRACT. We present an experimental study of the single source shortest path problem with non-negative edge weights (NSSP) on large-scale graphs using the Δ-stepping parallel algorithm. We report performance results on the Cray MTA-2, a multithreaded parallel computer. The MTA-2 is a high-end shared memory system offering two unique features that aid the efficient parallel implementation of irregular algorithms: the ability to exploit fine-grained parallelism, and low-overhead synchronization primitives. Our implementation exhibits remarkable parallel speedup when compared with competitive sequential algorithms, for low-diameter sparse graphs. For instance, Δ-stepping on a directed scale-free graph of 100 million vertices and 1 billion edges takes less than ten seconds on 40 processors of the MTA-2, with a relative speedup of close to 30. To our knowledge, these are the first performance results of a shortest path problem on realistic graph instances in the order of billions of vertices and edges.

1. Introduction

We present an experimental study of the Δ-stepping parallel algorithm [**53**] for solving the single source shortest path problem on large-scale graph instances. In addition to applications in combinatorial optimization problems, shortest path algorithms are finding increasing relevance in the domain of complex network analysis. Popular graph theoretic analysis metrics such as betweenness centrality [**10, 29, 36, 43, 45**] are based on shortest path algorithms. Our parallel implementation targets graph families that are representative of real-world, large-scale networks [**8, 13, 26, 54, 55**]. Real-world graphs are typically characterized by a low diameter, heavy-tailed degree distributions modeled by power laws, and self-similarity. They are often very large, with the number of vertices and edges ranging from several hundreds of thousands to billions. On current workstations, it is not possible to do exact in-core computations on these graphs due to the limited physical memory. In such cases, parallel computing techniques can be applied to obtain

1991 *Mathematics Subject Classification.* Primary 05C38, 68W10; Secondary 68W40.
Key words and phrases. shortest paths, parallel algorithms, multithreaded architectures.

exact solutions for memory and compute-intensive graph problems quickly. For instance, recent experimental studies on Breadth-First Search for large-scale graphs show that a parallel in-core implementation is two orders of magnitude faster than an optimized external memory implementation [**2, 5**]. In this paper, we present an efficient parallel implementation for the single source shortest paths problem that can handle scale-free instances in the order of billions of edges. In addition, we conduct an experimental study of performance on several other graph families, and this work is our submission to the 9th DIMACS Implementation Challenge [**20**] on Shortest Paths. Preliminary results from this work are discussed in [**47**].

Sequential algorithms for the single source shortest path problem with non-negative edge weights (NSSP) are studied extensively, both theoretically [**22, 24, 27, 28, 34, 37, 50, 59, 61**] and experimentally [**16, 23, 31, 32, 33, 64**]. Let m and n denote the number of edges and vertices in the graph respectively. Nearly all NSSP algorithms are based on the classical Dijkstra's [**24**] algorithm. Using Fibonacci heaps [**27**], Dijkstra's algorithm can be implemented in $O(m + n \log n)$ time. Thorup [**61**] presents an $O(m+n)$ RAM algorithm for undirected graphs that differs significantly different from Dijkstra's approach. Instead of visiting vertices in the order of increasing distance, it traverses a *component tree*. Meyer [**51**] and Goldberg [**33**] propose simple algorithms with linear average time for uniformly distributed edge weights.

Parallel algorithms for solving NSSP are reviewed in detail by Meyer and Sanders [**50, 53**]. There are no known PRAM algorithms that run in sub-linear time and $O(m + n \log n)$ work. Parallel priority queues [**12, 25**] for implementing Dijkstra's algorithm have been developed, but these linear work algorithms have a worst-case time bound of $\Omega(n)$, as they only perform edge relaxations in parallel. Several matrix-multiplication based algorithms [**30, 38**], proposed for the parallel All-Pairs Shortest Paths (APSP), involve running time and efficiency trade-offs. Parallel approximate NSSP algorithms [**17, 44, 60**] based on the randomized Breadth-First search algorithm of Ullman and Yannakakis [**63**] run in sub-linear time. However, it is not known how to use the Ullman-Yannakakis randomized approach for exact NSSP computations in sub-linear time.

Meyer and Sanders give the Δ-stepping [**53**] NSSP algorithm that divides Dijkstra's algorithm into a number of *phases*, each of which can be executed in parallel. For random graphs with uniformly distributed edge weights, this algorithm runs in sub-linear time with linear average case work. Several theoretical improvements [**48, 49, 52**] are given for Δ-stepping (for instance, finding shortcut edges, adaptive bucket-splitting), but it is unlikely that they would be faster than the simple Δ-stepping algorithm in practice, as the improvements involve sophisticated data structures that are hard to implement efficiently. On a random d-regular graph instance (2^{19} vertices and $d = 3$), Meyer and Sanders report a speedup of 9.2 on 16 processors of an Intel Paragon machine, for a distributed memory implementation of the simple Δ-stepping algorithm. For the same graph family, we are able to solve problems three orders of magnitude larger with near-linear speedup on the Cray MTA-2. For instance, we achieve a speedup of 14.82 on 16 processors and 29.75 on 40 processors for a random d-regular graph of size 2^{29} vertices and d set to 3.

The literature contains few experimental studies on parallel NSSP algorithms [**39, 41, 56, 62**]. Prior implementation results on distributed memory machines resorted to graph partitioning [**1, 15, 35**], and running a sequential NSSP algorithm on the sub-graph. Heuristics are used for load balancing and termination detection [**40, 42**]. The implementations perform well for certain graph families and problem sizes, but in the worst case, there is no speedup.

Implementations of PRAM graph algorithms for arbitrary sparse graphs are typically memory intensive, and the memory accesses are fine-grained and highly irregular. This often leads to poor performance on cache-based systems. On distributed memory clusters, few parallel graph algorithms outperform the best sequential implementations due to long memory latencies and high synchronization costs [**3, 4**]. Parallel shared memory systems are a more supportive platform. They offer higher memory bandwidth and lower latency than clusters, and the global shared memory greatly improves developer productivity. However, parallelism is dependent on the cache performance of the algorithm [**57**] and scalability is limited in most cases.

We present our shortest path implementation results on the Cray MTA-2, a massively multithreaded parallel machine. The MTA-2 is a high-end shared memory system offering two unique features that aid considerably in the design of irregular algorithms: fine-grained parallelism and low-overhead word-level synchronization. The MTA-2 has no data cache; rather than using a memory hierarchy to reduce latency, the MTA-2 processors use hardware multithreading to tolerate the latency. The word-level synchronization support complements multithreading and makes performance primarily a function of parallelism. Since graph algorithms have an abundance of parallelism, yet often are not amenable to partitioning, the MTA-2 architectural features lead to superior performance and scalability.

Our recent results highlight the exceptional performance of the MTA-2 for implementations of key combinatorial optimization and graph theoretic problems such as list ranking [**3**], connected components [**3, 9**], subgraph isomorphism [**9**], Breadth-First Search and *st*-connectivity [**5**]. We recently studied multithreaded implementations of Thorup's algorithm for solving SSSP on undirected graphs and report preliminary results in [**19**]. Thorup's algorithm constructs and traverses the component hierarchy data structure in order to identify all vertices that can be settled at a given time. This strategy is well suited to a shared-memory environment since the component hierarchy can be constructed only once, then shared by multiple concurrent SSSP computations. On the MTA-2, Δ-Stepping is faster than this implementation for a single source, but Thorup's implementation beats Δ-stepping for simultaneous SSSP runs on 40 processors. We refer the interested reader to [**19**] for more details.

The main contributions of this paper are as follows:

- *An experimental study of solving the single-source shortest paths problem in parallel using the Δ-stepping algorithm.* Prior studies have predominantly focused on running sequential NSSP algorithms on graph families that can be easily partitioned, whereas we also consider several arbitrary, sparse graph instances. We also analyze performance using machine-independent algorithmic operation counts.
- *Demonstration of the power of massive multithreading for graph algorithms on highly unstructured instances.* We achieve impressive performance on low-diameter random and scale-free graphs.
- *Solving NSSP for large-scale realistic graph instances in the order of billions of edges.* Δ-stepping on a synthetic directed scale-free graph of 100

million vertices and 1 billion edges takes 9.73 seconds on 40 processors of the MTA-2, with a relative speedup of approximately 31. These are the first results that we are aware of, for solving instances of this scale and also achieving near-linear speedup. Also, the sequential performance of our implementation is comparable to competitive NSSP implementations.

This paper is organized as follows. Section 2 provides a brief overview of Δ-stepping. Our parallel implementation of Δ-stepping is discussed in Section 3. Section 4 and 5 describe our experimental setup, performance results and analysis. We conclude with a discussion on implementation improvements and future plans in Section 6. Appendix A describes the MTA-2 architecture.

2. Review of the Δ-stepping Algorithm

2.1. Preliminaries. Let $G = (V, E)$ be a graph with n vertices and m edges. Let $s \in V$ denote the source vertex. Each edge $e \in E$ is assigned a non-negative real weight by the length function $l : E \rightarrow \mathbb{R}$. Define the *weight of a path* as the sum of the weights of its edges. The single source shortest paths problem with non-negative edge weights (NSSP) computes $\delta(v)$, the weight of the *shortest* (minimum-weighted) path from s to v. $\delta(v) = \infty$ if v is unreachable from s. We set $\delta(s) = 0$.

Most shortest path algorithms maintain a *tentative distance* value for each vertex, which are updated by *edge relaxations.* Let $d(v)$ denote the tentative distance of a vertex v. $d(v)$ is initially set to ∞, and is an upper bound on $\delta(v)$. *Relaxing* an edge $\langle v, w\rangle \in E$ sets $d(w)$ to the minimum of $d(w)$ and $d(v) + l(v, w)$. Based on the manner in which the tentative distance values are updated, most shortest path algorithms can be classified into two types: *label-setting* or *label-correcting.* Label-setting algorithms (for instance, Dijkstra's algorithm) perform relaxations only from *settled* ($d(v) = \delta(v)$) vertices, and compute the shortest path from s to all vertices in exactly m edge relaxations. Based on the values of $d(v)$ and $\delta(v)$, at each iteration of a shortest path algorithm, vertices can be classified into *unreached* ($d(v) = \infty$), *queued* ($d(v)$ is finite, but v is not settled) or *settled.* Label-correcting algorithms (e.g., Bellman-Ford) relax edges from unsettled vertices also, and may perform more than m relaxations. Also, all vertices remain in a *queued* state until the final step of the algorithm. Δ-stepping belongs to the label-correcting type of shortest path algorithms.

2.2. Algorithmic Details. The Δ-stepping algorithm (see Algorithm 1) is an "approximate bucket implementation of Dijkstra's algorithm" [**53**]. It maintains an array of buckets B such that $B[i]$ stores the set of vertices $\{v \in V : v$ is queued and $d(v) \in [i\Delta, (i+1)\Delta]\}$. Δ is a positive real number that denotes the "bucket width".

In each *phase* of the algorithm (the inner *while* loop in Algorithm 1, lines 9–14, when bucket $B[i]$ is not empty), all vertices are removed from the current bucket, added to the set S, and *light* edges ($l(e) \leq \Delta$, $e \in E$) adjacent to these vertices are relaxed (see Algorithm 2). This may result in new vertices being added to the current bucket, which are deleted in the next phase. It is also possible that vertices previously deleted from the current bucket may be reinserted, if their tentative distance is improved. *Heavy* edges ($l(e) > \Delta$, $e \in E$) are not relaxed in a phase, as they result in tentative values outside the current bucket. Once the current bucket

Algorithm 1: Δ-stepping algorithm.

Input: $G(V, E)$, source vertex s, length function $l : E \rightarrow \mathbb{R}$
Output: $\delta(v), v \in V$, the weight of the shortest path from s to v

1 **foreach** $v \in V$ **do**
2     $heavy(v) \longleftarrow \{\langle v, w\rangle \in E : l(v, w) > \Delta\}$;
3     $light(v) \longleftarrow \{\langle v, w\rangle \in E : l(v, w) \leq \Delta\}$;
4     $d(v) \longleftarrow \infty$;
5 relax(s, 0);
6 $i \longleftarrow 0$;
7 **while** B *is not empty* **do**
8     $S \longleftarrow \phi$;
9   **while** $B[i] \neq \phi$ **do**
10         $Req \longleftarrow \{(w, d(v) + l(v, w)) : v \in B[i] \wedge \langle v, w\rangle \in light(v)\}$;
11         $S \longleftarrow S \cup B[i]$;
12         $B[i] \longleftarrow \phi$;
13   **foreach** $(v, x) \in Req$ **do**
14   relax(v, x);
15     $Req \longleftarrow \{(w, d(v) + l(v, w)) : v \in S \wedge \langle v, w\rangle \in heavy(v)\}$;
16   **foreach** $(v, x) \in Req$ **do**
17   relax(v, x);
18     $i \longleftarrow i + 1$;
19 **foreach** $v \in V$ **do**
20     $\delta(v) \longleftarrow d(v)$;

Algorithm 2: The *relax* routine in the Δ-stepping algorithm.

Input: v, weight request x
Output: Assignment of v to appropriate bucket

1 **if** $x < d(v)$ **then**
2     $B[\lfloor d(v)/\Delta \rfloor] \leftarrow B[\lfloor d(v)/\Delta \rfloor] \setminus \{v\}$;
3     $B[\lfloor x/\Delta \rfloor] \leftarrow B[\lfloor x/\Delta \rfloor] \cup \{v\}$;
4     $d(v) \leftarrow x$;

remains empty after relaxations, all heavy edges out of the vertices in S are relaxed at once (lines 15–17 in Algorithm 1). The algorithm continues until all the buckets are empty.

Observe that edge relaxations in each phase can be done in parallel, as long as individual tentative distance values are updated atomically. The number of phases bounds the parallel running time, and the number of *reinsertions* (insertions of vertices previously deleted) and *rerelaxations* (relaxation of their out-going edges) costs an overhead over Dijkstra's algorithm. The performance of the algorithm also depends on the value of the bucket-width Δ. For $\Delta = \infty$, the algorithm is similar to the Bellman-Ford algorithm. It has a high degree of parallelism, but is inefficient compared to Dijkstra's algorithm. Δ-stepping tries to find a good compromise between the number of parallel phases and the number of re-insertions. Theoretical bounds on the number of phases and re-insertions, and the average case analysis of the parallel algorithm are presented in [**53**]. We summarize the salient results.

Let d_c denote the maximum shortest path weight, and P_Δ denote the set of paths with weight at most Δ. Define a parameter l_{max}, an upper bound on the maximum number of edges in any path in P_Δ. The following results hold true for any graph family.

- The number of buckets in B is $\lceil d_c/\Delta \rceil$.
- The total number of reinsertions is bounded by $|P_\Delta|$, and the total number of rerelaxations is bounded by $|P_{2\Delta}|$.
- The number of phases is bounded by $\frac{d_c}{\Delta} l_{max}$, i.e., no bucket is expanded more than l_{max} times.

For graph families with random edge weights and a maximum degree of d, Meyer and Sanders [**53**] theoretically show that $\Delta = \theta(1/d)$ is a good compromise between work efficiency and parallelism. The sequential algorithm performs $O(dn)$ expected work divided between $O(\frac{d_c}{\Delta} \cdot \frac{\log n}{\log \log n})$ phases *with high probability*. In practice, in case of graph families for which d_c is $O(\log n)$ or $O(1)$, the parallel implementation of Δ-stepping yields sufficient parallelism for our parallel system.

3. Parallel Implementation of Δ-stepping

See Appendix A for details of the MTA-2 architecture and parallelization primitives.

The bucket array B is the primary data structure used by the parallel Δ-stepping algorithm. We implement individual buckets as *dynamic arrays* that can be resized when needed and iterated over easily. To support constant time insertions and deletions, we maintain two auxiliary arrays of size n: a mapping of the vertex ID to its current bucket, and a mapping from the vertex ID to the position of the vertex in the current bucket (see Figure 1 for an illustration). All new vertices are added to the end of the array, and deletions of vertices are done by setting the corresponding locations in the bucket and the mapping arrays to -1. Note that once bucket i is finally empty after a light edge relaxation phase, there will be no more insertions into the bucket in subsequent phases. Thus, the memory can be reused once we are done relaxing the light edges in the current bucket. Also observe that all the insertions are done in the relax routine, which is called once in each phase, and once for relaxing the heavy edges.

We implement a timed pre-processing step to *semi-sort* the edges based on the value of Δ. All the light edges adjacent to a vertex are identified in parallel and stored in contiguous virtual locations, and so we visit only light edges in a phase. The $O(n)$ work pre-processing step scales well in parallel on the MTA-2.

We also support fast parallel insertions into the request set R. R stores $\langle v, x \rangle$ pairs, where $v \in V$ and x is the requested tentative distance for v. We only add a vertex v to R if it satisfies the condition $x < d(v)$. We do not store duplicates in R. We use a sparse set representation similar to one used by Briggs and Torczon [**11**] for storing vertices in R. This sparse data structure uses two arrays of size n: a *dense* array that contiguously stores the elements of the set, and a *sparse* array that indicates whether the vertex is a member of the set. Thus, it is easy to iterate over the request set, and membership queries and insertions are constant time. Unlike other Dijkstra-based algorithms, we do not relax edges in one step.

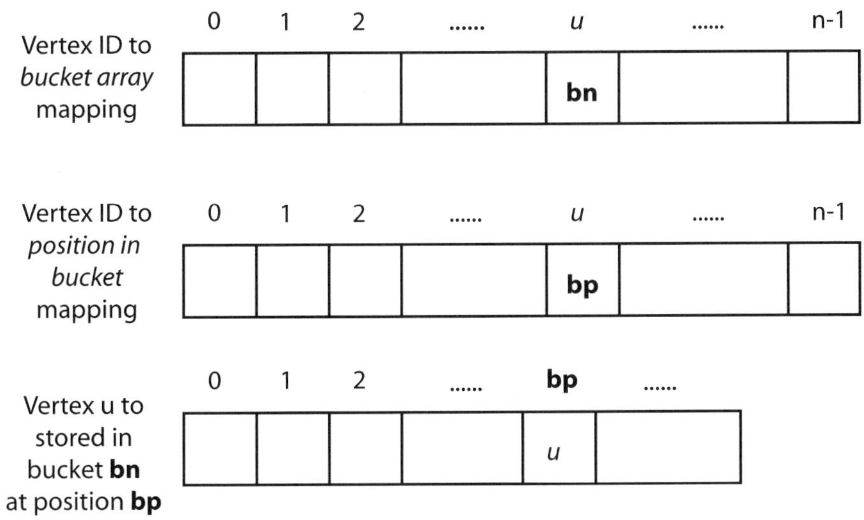

FIGURE 1. Bucket array and auxiliary data structures.

Instead, we inspect adjacencies (light edges) in each phase, construct a request set of vertices, and then relax *vertices* in the relax step.

All vertices in the request set R are relaxed in parallel in the relax routine. In this step, we first delete a vertex from the old bucket, and then insert it into the new bucket. Instead of performing individual insertions, we first determine the expansion factor of each bucket, expand the buckets, and then add all vertices into their new buckets in one step. Since there are no duplicates in the request set, no synchronization is involved for updating the tentative distance values.

To saturate the MTA-2 processors with work and to obtain high system utilization, we need to minimize the number of phases and non-empty buckets, and maximize the request set sizes. Entering and exiting a parallel phase involves a negligible running time overhead in practice. However, if the number of phases is $O(n)$, this overhead dominates the actual running time of the implementation. Also, we enter the relax routine once every phase. There are several implicit barrier synchronizations in the algorithm that are proportional to the number of phases. Our implementation reduces the number of barriers. Our source code for the Δ-stepping implementation, along with the MTA-2 graph generator ports, is freely available online [**46**].

4. Experimental Setup

4.1. Platforms. We report parallel performance results on a 40-processor Cray MTA-2 system with 160 GB uniform shared memory. Each processor has a clock speed of 220 MHz and support for 128 hardware threads. The Δ-stepping code is written in C with MTA-2 specific pragmas and directives for parallelization. We compile it using the MTA-2 C compiler (Cray Programming Environment (PE) 2.0.3) with -O3 and -par flags.

The MTA-2 code also compiles and runs on sequential processors without any modifications. Our test platform for the sequential performance results is one processor of a dual-core 3.2 GHz 64-bit Intel Xeon machine with 6GB memory, 1MB cache and running RedHat Enterprise Linux 4 (Linux kernel 2.6.9). We compare the sequential performance of our implementation with the DIMACS reference solver [**21**]. Both the codes are compiled with the Intel C compiler (icc) Version 9.0, with the flags `-O3`. The source code is freely available online [**46**].

4.2. Problem Instances. We evaluate sequential and parallel performance on several graph families. Some of the generators and graph instances are part of the DIMACS Shortest Path Implementation Challenge benchmark package [**20**]:

- *Random graphs*: Random graphs are generated by first constructing a Hamiltonian cycle, and then adding $m - n$ edges to the graph at random. The generator may produce parallel edges as well as self-loops. We define the random graph family *Random4-n* such that n is varied, $\frac{m}{n} = 4$, and the edge weights are chosen from a uniform random distribution.
- *Grid graphs*: This synthetic generator produces two-dimensional meshes with grid dimensions x and y. *Long-n* ($x = \frac{n}{16}$, $y = 16$) and *Square-n* grid ($x = y = \sqrt{n}$) families are defined, similar to random graphs.
- *Road graphs*: Road graph families with transit time (*USA-road-t*) and distance (*USA-road-d*) as the length function.

In addition, we also study the following families:

- *Scale-free graphs*: We use the R-MAT graph model [**14**] for real-world networks to generate scale-free graphs. We define the family *ScaleFree4-n* similar to random graphs.
- *Log-uniform weight distribution*: The above graph generators assume randomly distributed edge weights. We report results for an additional *log-uniform* distribution also. The generated integer edge weights are of the form 2^i, where i is chosen from the uniform random distribution $[1, \log C]$ (C denotes the maximum integer edge weight). We define *Random4logUnif-n* and *ScaleFree4logUnif-n* families for this weight distribution.

4.3. Methodology. For sequential runs, we report the execution time of the reference DIMACS NSSP solver (an efficient implementation of Goldberg's algorithm [**34**], which has expected-case linear case for some inputs) and the baseline Breadth-First Search (BFS) on every graph family. The BFS running time is a natural lower bound for NSSP codes and is a good indicator of how optimized the shortest path implementations are. It is reasonable to directly compare the execution times of the reference code and our implementation: both use a similar adjacency array representation for the graph, are written in C, and compiled and run in identical experimental settings. Note that our implementation is optimized for the MTA-2 and we make no modifications to the code before running on a sequential machine. The time taken for semi-sorting and mechanisms to reduce memory contention on the MTA-2 both constitute overhead on a sequential processor. Also, our implementation assumes real-weighted edges, and we cannot use fast bitwise operations. By default, we set the value of Δ to $\frac{n}{m}$ for all graph instances. We will show that this choice of Δ may not be optimal for all graph classes and weight distributions.

On the MTA-2, we compare our implementation running time with the execution time of a multithreaded level-synchronized breadth-first search [**6**], optimized for low-diameter graphs. The multithreaded BFS scales as well as δ-stepping for the core graph families, and the execution time serves as a lower bound for the shortest path running time.

On a sequential processor, we execute the BFS and shortest path codes on all the core graph families, for the recommended problem sizes. However, for parallel runs, we only report results for sufficiently large graph instances in case of the synthetic graph families. We parallelize the synthetic core graph generators and port them to run on the MTA-2.

Our implementations accept both directed and undirected graphs. For all the synthetic graph instances, we report execution times on directed graphs in this paper. The road networks are undirected graphs. We also assume the edge weights to be distributed in $[0, 1]$ in the Δ-stepping implementation. So we have a pre-processing step to scale the integer edge weights in the core problem families to the interval [0, 1], dividing the integer weights by the maximum edge weight.

The first run on the MTA-2 is usually slower than subsequent ones (by about 10% for a typical Δ-stepping run). So we report the average running time for 10 successive runs. We run the code from three randomly chosen source vertices and average the running time. We found that using three sources consistently gave us execution time results with little variation on both the MTA-2 and the reference sequential platform. We tabulate the sequential and parallel performance metrics in Appendix B, and report execution time in seconds. If the execution time is less than 1 millisecond, we round the time to four decimal digits. If it is less than 100 milliseconds, we round it to three digits. In all other cases, the reported running time is rounded to two decimal digits.

5. Results and Analysis

5.1. Sequential Performance. First we present the performance results of our implementation on the reference sequential platform for the core graph families. The BFS, Δ-stepping, and reference DIMACS implementation execution times on the recommended core graph instances are given in Appendix B.1. We observe that the ratio of the Δ-stepping execution time to the Breadth-First Search time varies between 3 and 10 across different problem instances. Also, the DIMACS reference code is about 1.5 to 2 times faster than our implementation for large problem instances in each family. As noted previously, we design an optimized multithreaded implementation of the shortest path algorithm, and some of the mechanisms specific to the MTA-2 may be an overhead on the reference sequential platform. Thus, the sequential execution times quantify the additional work due to parallelization.

Table 1 summarizes the performance for random graph instances. For the Random4-n family, n is varied from 2^{11} to 2^{21}, the maximum edge weight is set to n, and the graph density is constant. For the largest instance, Δ-stepping execution time is 1.7 times slower than the reference implementation and 5.4 times the BFS execution time. For the Random4-C family, we normalize the weights to the maximum integer weight. We do not observe any trend similar to the reference implementation, where the execution time gradually rises as the maximum weight increases. This suggests that the Δ-stepping algorithm performance is independent of maximum integer edge weight, provided the edge weights follow a uniform random distribution and Δ is set appropriately.

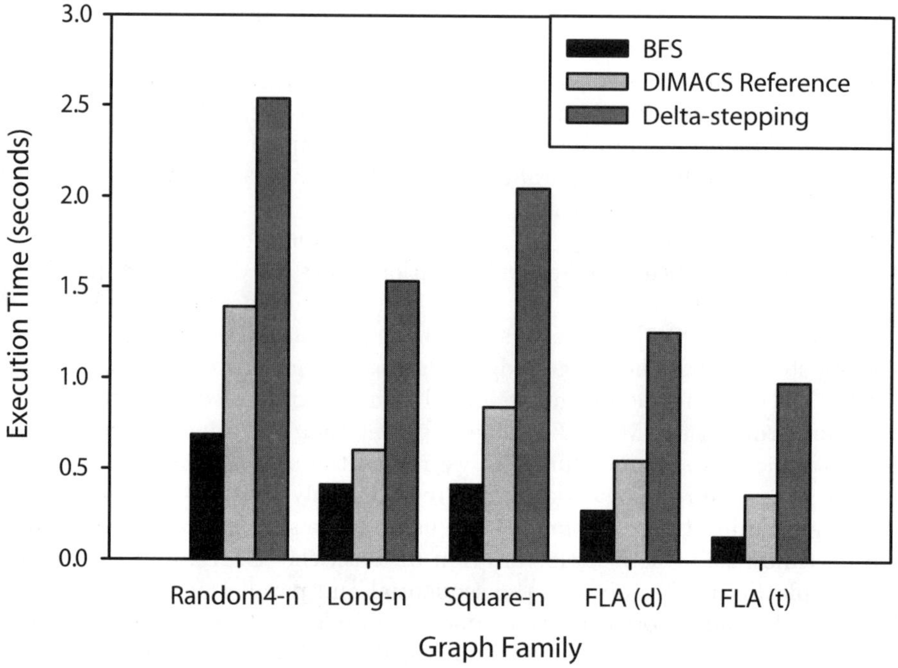

FIGURE 2. Sequential performance of our Δ-stepping implementation, on the core graph families. All the synthetic graphs are directed, with 2^{20} vertices and $\frac{m}{n} \approx 4$. FLA(d) and FLA(t) are road networks corresponding to Florida, with 1070376 vertices and 2712768 edges.

The sequential performance of Δ-stepping on Long grid graphs (Table 2) is similar to that on Random graphs. However, the reference implementation is slightly faster on long grids. For square grids and road networks, the Δ-stepping to BFS ratio is comparatively higher (e.g., BFS to Δ-stepping ratio is 4.71 for the largest Square-n graph, and 3.74 for the largest Random4-n graph) than the Random and Long grid families.

Figure 2 and Figure 3 summarize the key observations from the tables in Appendix B.1. Comparing execution time across graphs of the same size in Figure 2, we find that the Δ-stepping running time for the Random4-n graph instance is slightly higher than the rest of the families. The Δ-stepping running time is also comparable to the execution time of the reference implementation for all graph families. Figure 3 plots the execution time normalized to the problem size (or the running time per edge) for Random4-n and Long-n families. Observe that the Δ-stepping implementation execution time scales with problem size at a faster rate compared to BFS or the DIMACS reference implementation. This suggests a slight increase in additional computation as the problem size is scaled up.

5.2. Δ-stepping analysis. To better understand the algorithm performance across graph families, we study machine-independent algorithm operation counts. The parallel performance is dependent on the value of Δ, the number of phases, the size of the request set in each phase.

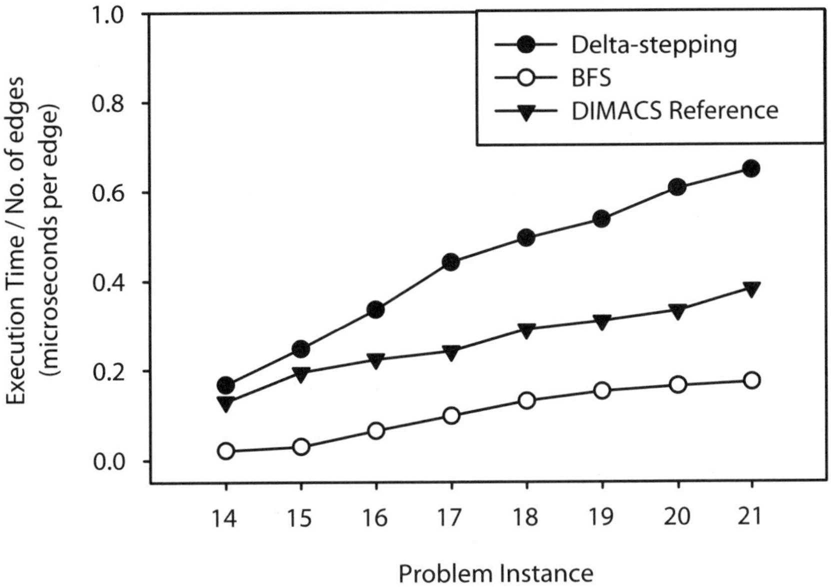

(a) Random4-n family. Problem instance i denotes a directed graph of 2^i vertices, $m = 4n$ edges, and maximum weight $C = n$.

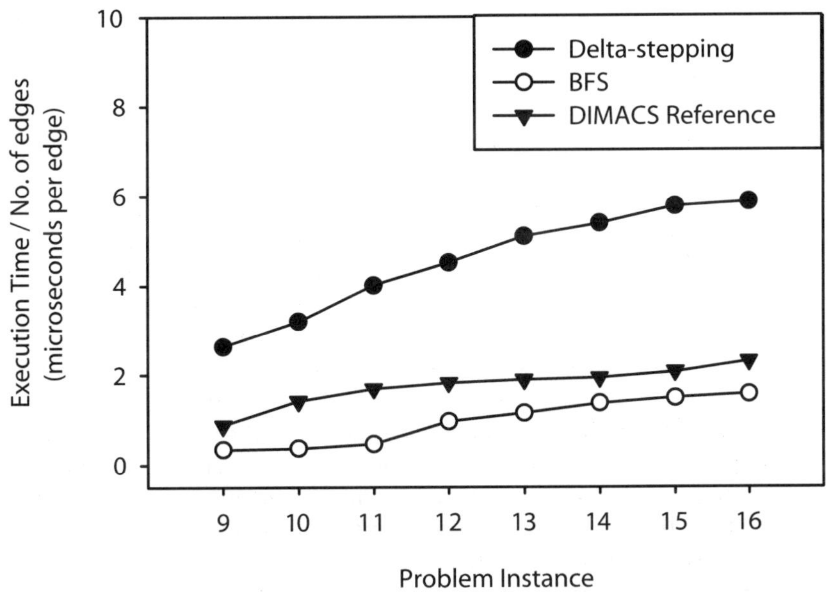

(b) Long-n family. Problem instance i denotes a grid with $x = 2^i$ and $y = 16$. $n = xy$ and $\frac{m}{n} \approx 4$.

FIGURE 3. Δ-stepping sequential execution time as a function of problem size.

Size of request sets. Figure 4 and Figure 5 plot the size of the light request set in each phase, for each core graph family. The choice of Δ in these experiments is motivated by the observation of Meyer and Sanders [**53**] that for graph families with random edge weights and a maximum degree of d, $\Delta = \theta(1/d)$ would be a good compromise between work efficiency and parallelism. Since $d \approx m/n(\approx 4)$ for most of the test graph instances, Δ is set to 0.25 by default for all runs. We also evaluate the performance of the algorithm as the value of Δ is varied (see Figure 5.2 and Figure 5.2). If the request set size is less than 10, it is not plotted.

Consider the random graph family (Figure 4(a)). It executes in 84 phases, and the request set sizes vary from 0 to 27,000. Observe the recurring pattern of nearly three bars stacked together in the plot. This indicates that all the light edges in a bucket are relaxed in roughly three phases, and the bucket then becomes empty. The size of the relax set is relatively high for several phases, which provides scope for exploiting multithreaded parallelism. The relax set size plot of a similar problem instance from the Long grid family (Figure 4(b)) stands in stark contrast to the random graph plot. It takes about 200,000 phases to execute (compared to the 84 phases for the random graph), and the maximum request size is only 15. Both of these values indicate that our implementation performance is significantly dependent on the graph diameter, and that the parallel performance would be poor on long grid graphs (e.g. meshes with a very high aspect ratio). On square grids (Figure 5(a)), Δ-stepping takes fewer phases, and the request set sizes go up to 500. For a road network instance (NE USA-road-d, Figure 5(b)), the algorithm takes 23,000 phases to execute, and only a few phases (about 30) have request set counts greater than 1000. As expected, the number of phases are proportional to the graph diameter in all the cases.

Algorithm operation counts. Figure 6 and Figure 7 plot several key Δ-stepping operation counts for various graph classes. Along with the core graph families, we include ScaleFree4-n, RandomlogUnif4-n, and LonglogUnif4-n graph classes. All synthetic graphs are roughly of the same size. Figure 6(a) plots the average shortest path weight for various graph classes. Scale-free and Long grid graphs are on the two extremes, with the graph diameter again being the determining factor. A log-uniform edge weight distribution also results in low average edge weight. The number of phases (see Figure 6(b)) is highest for Long grid graphs. The number of buckets shows a similar trend as the average shortest path weight. Figure 7(b) plots the total number of insertions for each graph family. The number of vertices is 2^{20} for all graph families (slightly higher for the road network), and so Δ-stepping results in roughly 20% overhead in insertions for all the graph families with random edge weights. Note the number of insertions for graphs with log-uniform weight distributions. Δ-stepping performs a lot of excess work for these families, because the value of Δ is quite high for this particular distribution.

Influence of Δ. We next evaluate the performance of the algorithm as Δ is varied (tables in Appendix B.2). Figure 5.2 and Figure 5.2 plot the execution time of various graph instances on a sequential machine, and one processor of the MTA-2. Δ is varied from 0.1 to 10 in each case. We find that the absolute running times on a 3.2 GHz Xeon processor and the MTA-2 are comparable for random, square grid and road network instances. However, on long grid graphs (Figure 8(b)), the MTA-2 execution time is two orders of magnitude greater than the sequential time. The number of phases and the total number of relaxations vary as Δ is varied

(Tables 5, 6, and 7). On the MTA-2, the running time is not only dependent on the work done, but also on the number of phases and the average number of relax requests in a phase. For instance, in the case of long grids (see Figure 8(b), with

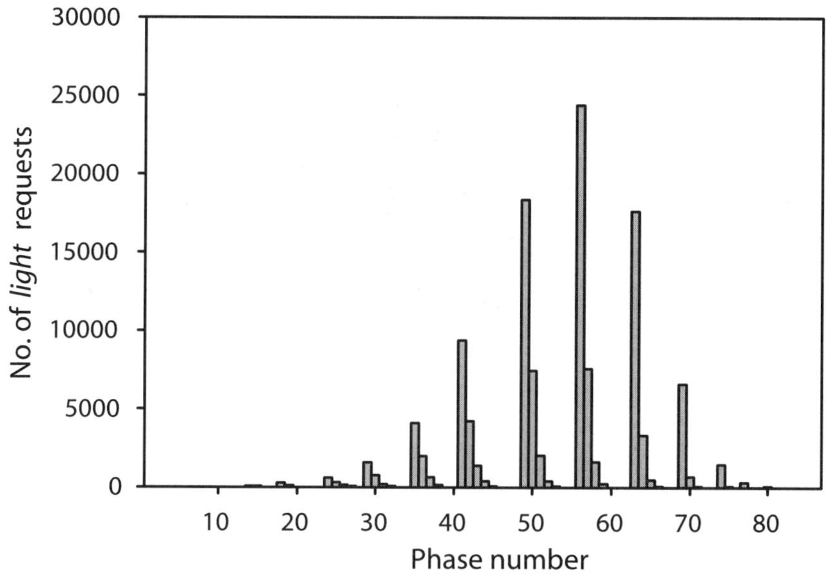

(a) Random4-n family, $n = 2^{20}$.

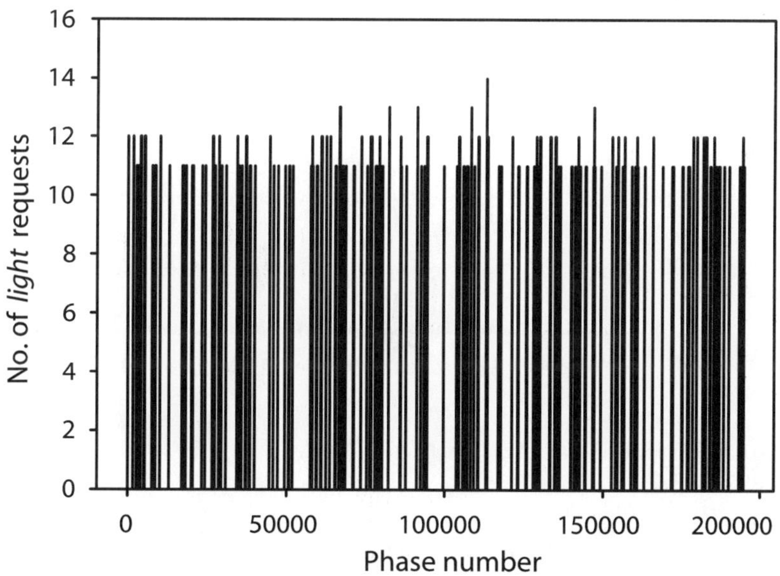

(b) Long-n family, $n = 2^{20}$.

FIGURE 4. Δ-stepping algorithm: Size of the light request set at the end of each phase, for the core graph families. Request set sizes less than 10 are not plotted.

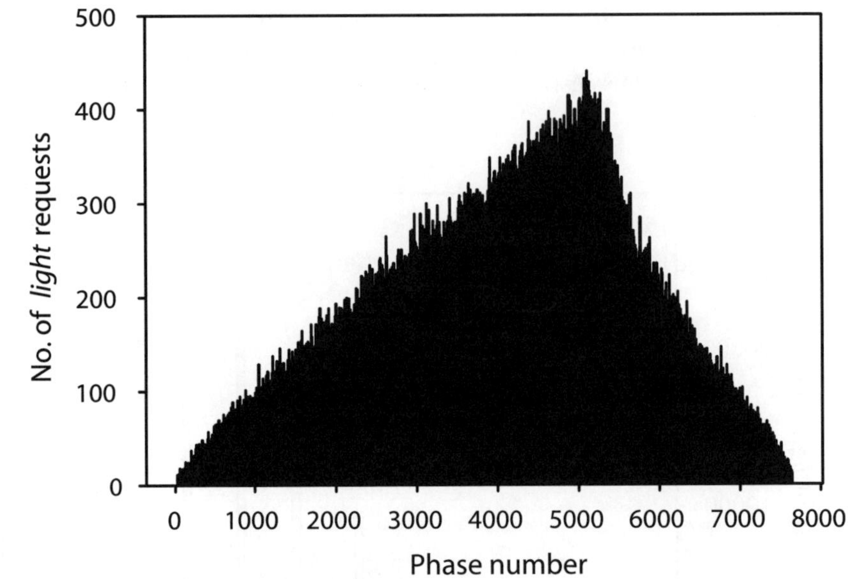

(a) Square-n family, $n = 2^{20}$.

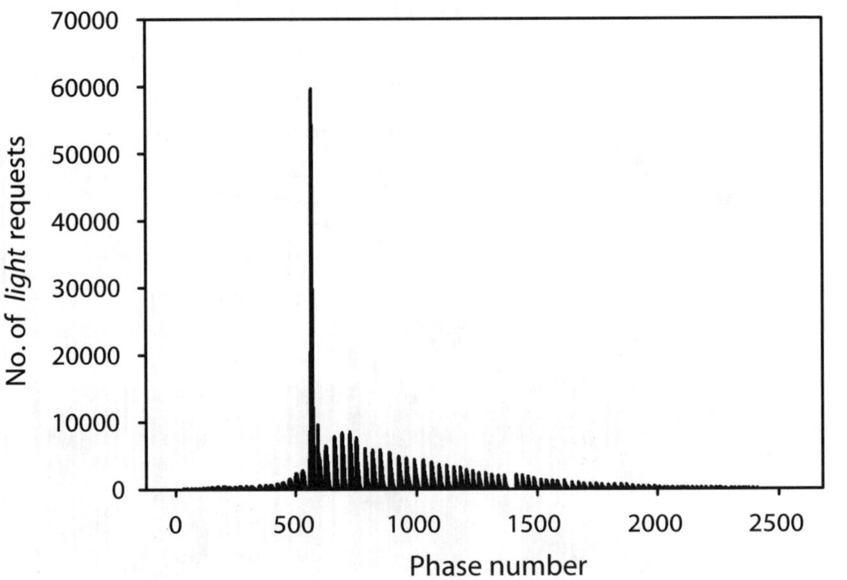

(b) USA-road-d family, Northeast USA (NE). $n = 1524452$, $m = 3897634$.

FIGURE 5. Δ-stepping algorithm: Size of the light request set at the end of each phase, for the core graph families. Request set sizes less than 10 are not plotted.

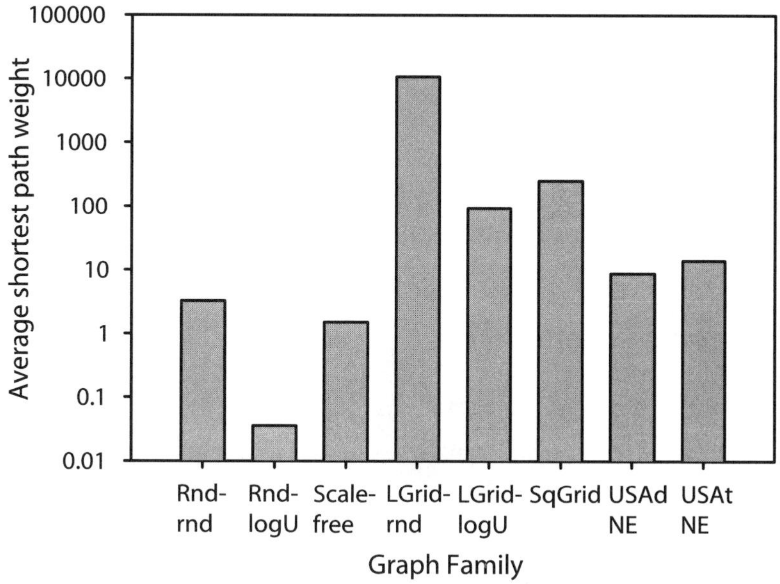

(a) Average shortest path weight ($\frac{1}{n} * \sum_{v \in V} \delta(v)$).

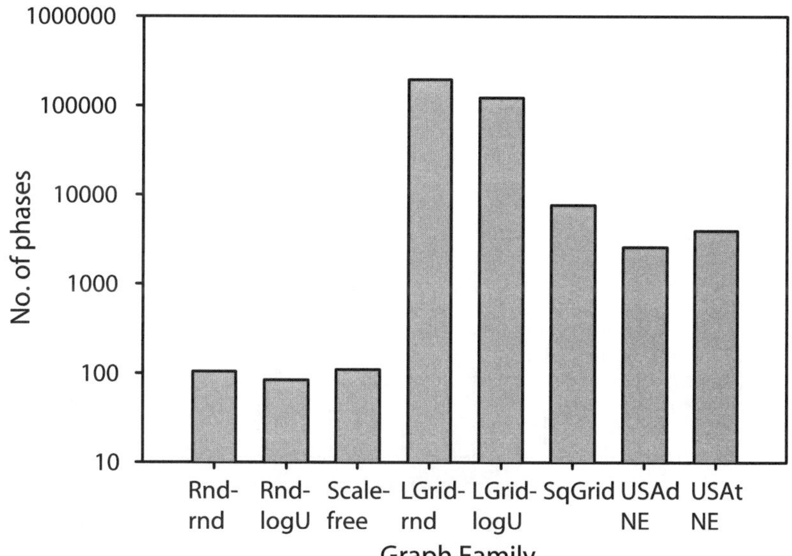

(b) No. of phases.

FIGURE 6. Δ-stepping algorithm performance statistics for various graph classes. All synthetic graph instances have n set to 2^{20} and $m \approx 4n$. Rnd-rnd: Random graph with random edge weights, Rnd-logU: Random graph with log-uniform edge weights, Scale-free: Scale-free graph with random edge weights, LGrid: Long grid, SqGrid: Square grid, USA NE: 1524452 vertices, 3897634 edges.

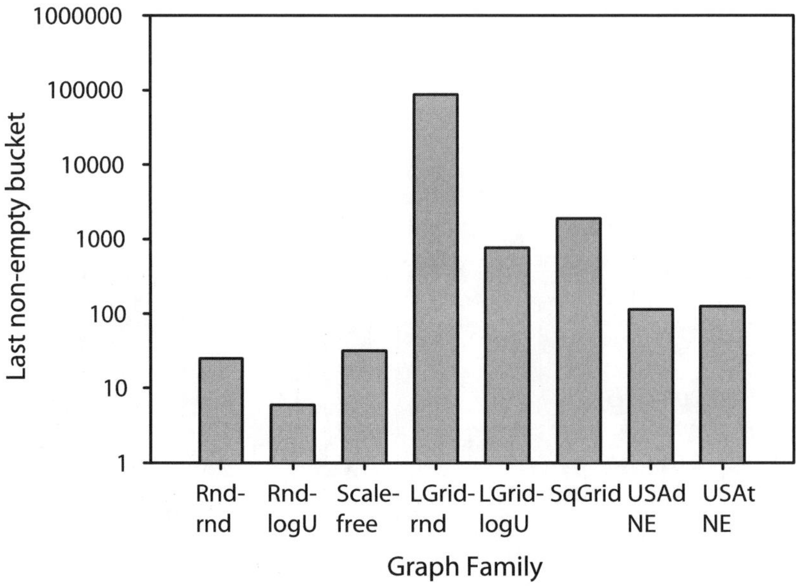

(a) Last non-empty bucket.

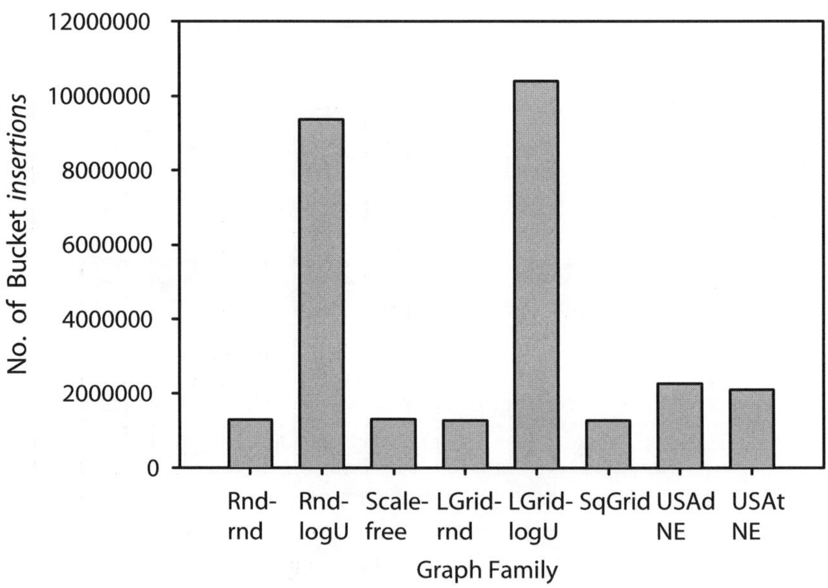

(b) Number of relax requests.

FIGURE 7. Δ-stepping algorithm performance statistics for various graph classes. All synthetic graph instances have n set to 2^{20} and $m \approx 4n$. Rnd-rnd: Random graph with random edge weights, Rnd-logU: Random graph with log-uniform edge weights, Scale-free: Scale-free graph with random edge weights, LGrid: Long grid, SqGrid: Square grid, USA NE: 1524452 vertices, 3897634 edges. Plot (b) uses a linear scale.

execution time plotted on a log scale), the running time decreases significantly as the value of Δ is decreased, as the number of phases reduce. On a sequential processor, however, the running time is only dependent on the work done (number of insertions). If the value of Δ is greater than the average shortest path weight, we perform excess work and the running time noticeably increases (observe the execution time for $\Delta = 5, 10$ on the random graph and the road network). The optimal value of Δ (and the execution time on the MTA-2) is also dependent on the number of processors. For a particular Δ, it may be possible to saturate a single processor of the MTA-2 with the right balance of work and phases. The execution time on a 40-processor run may not be minimal with this value of Δ.

5.3. Parallel Performance. In this section, we discuss the parallel scaling of Δ-stepping in detail (see tables in Appendix B.3). We ran Δ-stepping and the level-synchronous parallel BFS on graph instances from the core families, scale-free graphs, and graphs with log-uniform edge weight distributions. Define the speedup on p processors of the MTA-2 as the ratio of the execution time on 1 processor to the execution time on p processors. Since the computation on the MTA-2 is thread-centric rather than processor-centric, note that the single processor run is also parallel. In all graph classes except long grids, there is sufficient parallelism to saturate a single processor of the MTA-2 for reasonably large problem instances.

5.3.1. *Unstructured Instances.* As expected from the discussion in the previous section, Δ-stepping performs best for low-diameter random and scale-free graphs with randomly distributed edge weights (see Figure 10 and Figure 11). We attain a speedup of approximately 31 on 40 processors for a directed random graph of nearly a billion edges, and the ratio of the BFS and Δ-stepping execution time is a constant factor (about 3-5) throughout. The implementation performs equally well for scale-free graphs, that are more difficult to handle due to the irregular degree distribution. The execution time on 40 processors of the MTA-2 for the scale-free graph instance is only 1 second slower than the running time for a random graph and the speedup is approximately 30 on 40 processors. We have already shown that the execution time for smaller graph instances on a sequential machine is comparable to the DIMACS reference implementation, a competitive NSSP algorithm. Thus, attaining a speedup of 30 for a realistic scale-free graph instance of one billion edges (Figure 11) is a remarkable result.

Table 8 gives the execution time of Δ-stepping on the Random4-n family, as the number of vertices is increased from 2^{21} to 2^{28}, and the number of processors is varied from 1 to 40. Observe that the relative speedup increases as the problem size is increased (for e.g., on 40 processors, the speedup for $n = 2^{21}$ is just 3.96, whereas it is 31.04 for 2^{28} vertices). This is because there is insufficient parallelism in a problem instance of size 2^{21} to saturate 40 processors of the MTA-2. As the problem size increases, the ratio of Δ-stepping execution time to multithreaded BFS running time decreases. On an average, Δ-stepping is 5 times slower than BFS for this graph family.

Table 9 gives the execution time for random graphs with a log-uniform weight distribution. With Δ set to $\frac{n}{m}$, we do a lot of additional work. The Δ-stepping to BFS ratio is typically 40 in this case, about 8 times higher than the corresponding ratio for random graphs with random edge weights. However, the execution time scales well with the number of processors for large problem sizes.

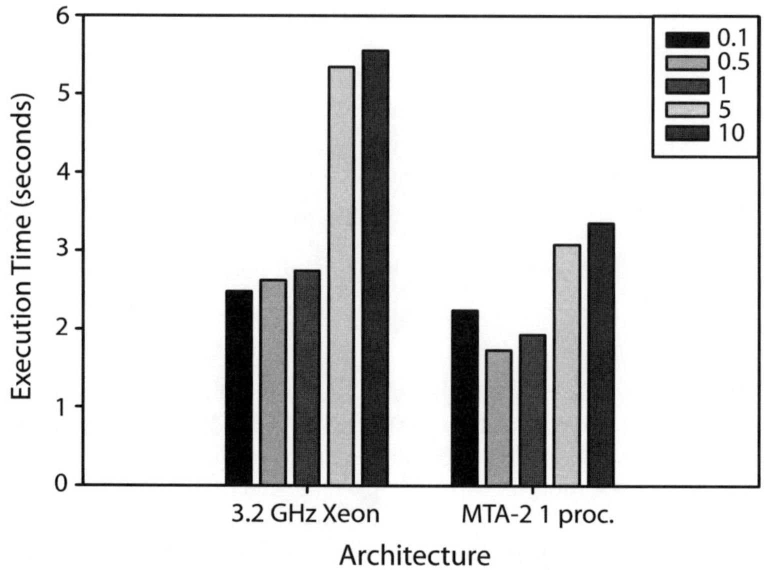

(a) Random4-n family. 2^{20} vertices.

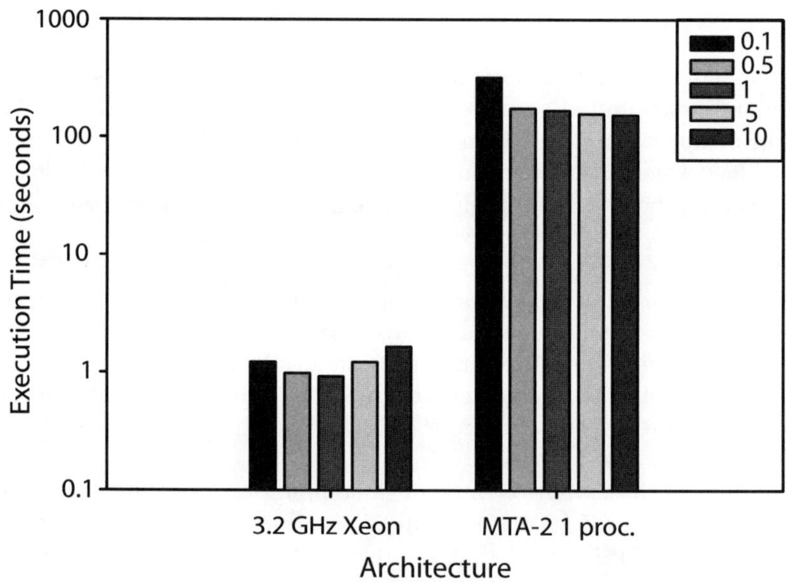

(b) Long-n family. 2^{20} vertices.

FIGURE 8. A comparison of the execution time on the reference sequential platform and a single MTA-2 processor, as the bucket-width Δ is varied.

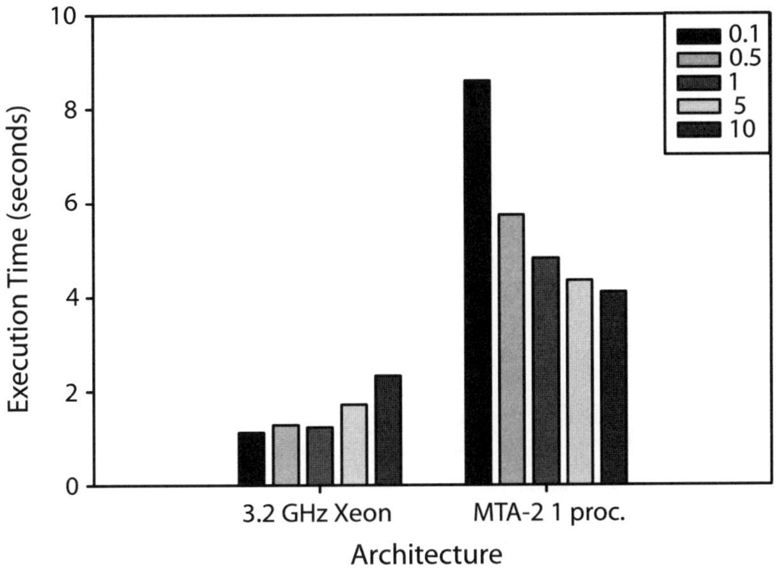

(a) Square-n family. 2^{20} vertices.

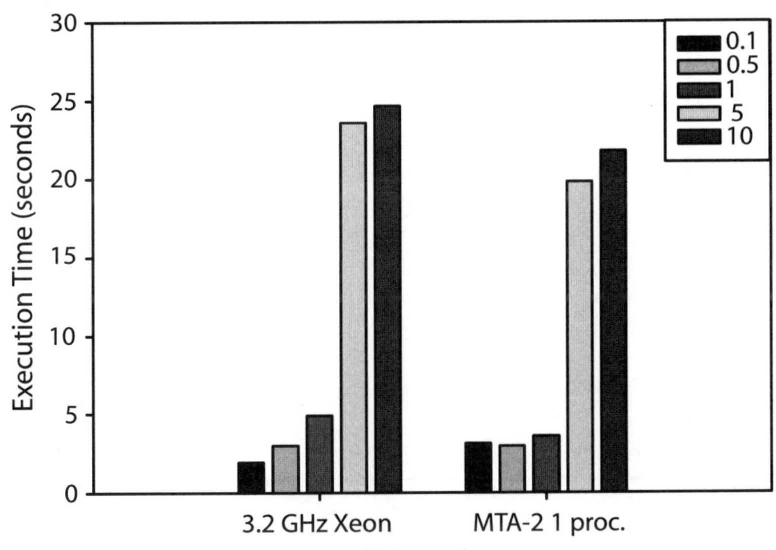

(b) USA-road-d family, Florida (FLA). 1070376 vertices, 2712798 edges.

FIGURE 9. A comparison of the execution time on the reference sequential platform and a single MTA-2 processor, as the bucket-width Δ is varied.

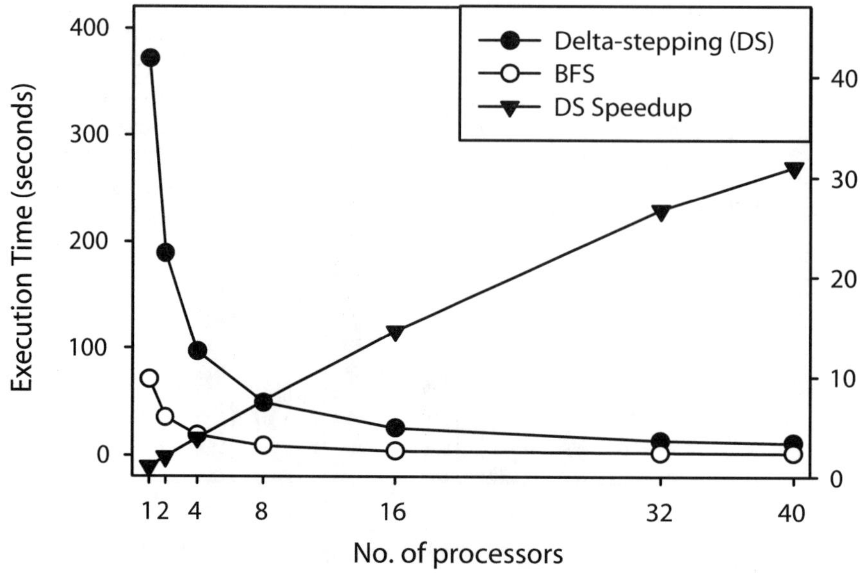

(a) Execution time and Relative Speedup (linear scale).

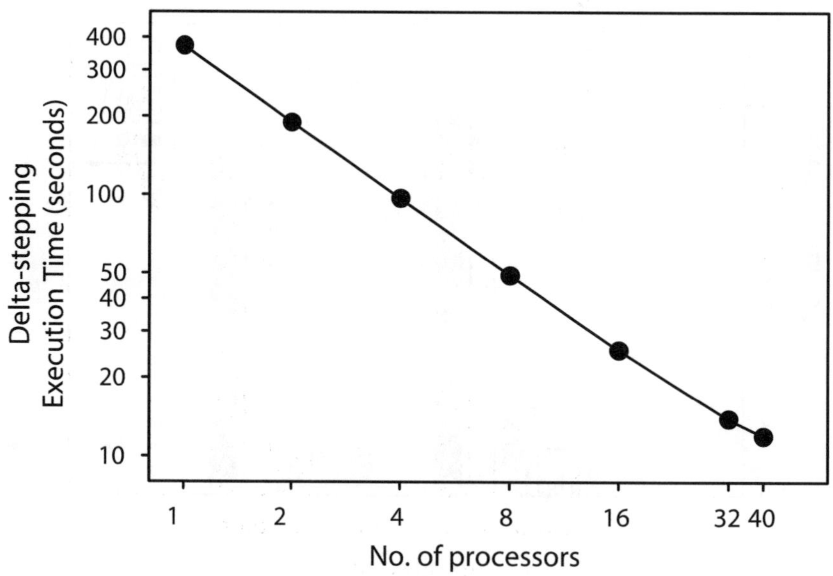

(b) Execution time vs. No. of processors (log-log scale).

FIGURE 10. Δ-stepping execution time and relative speedup on the MTA-2 for a Random4-n graph instance (directed graph, n=2^{28} vertices and $m = 4n$ edges, random edge weights).

Table 10 summarizes the execution time for the Random4-C family. The maximum edge weight is varied from 4^0 to 4^{15} while keeping m and n constant. We do not notice any trend in the execution time in this case, as we normalize the edge

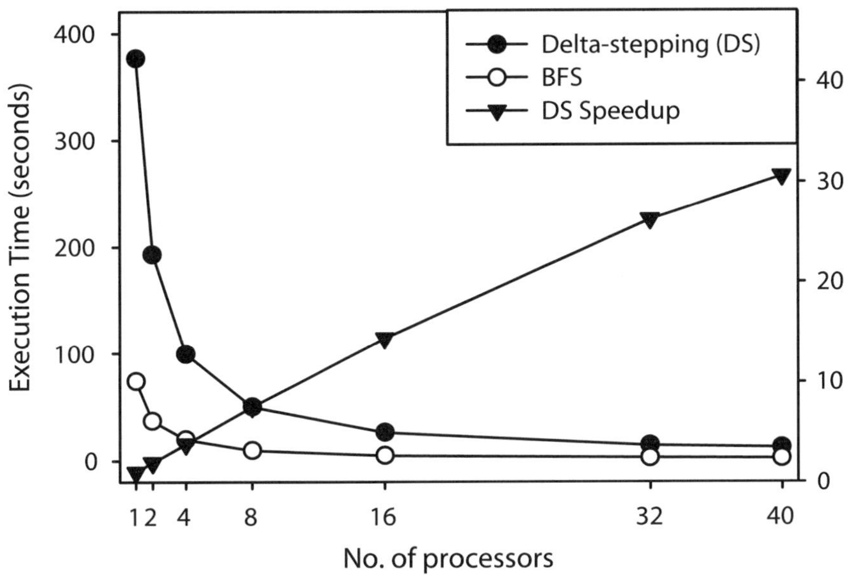

(a) Execution time and Relative Speedup (linear scale).

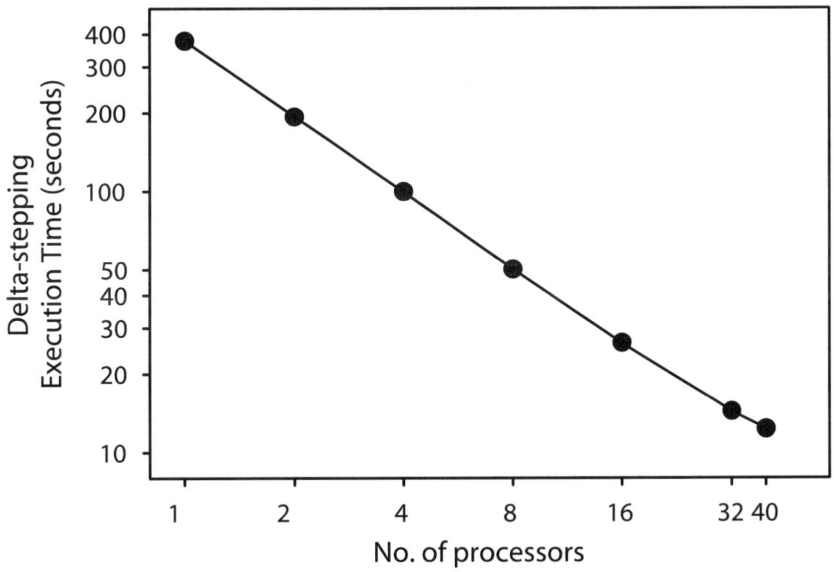

(b) Execution time vs. No. of processors (log-log scale).

FIGURE 11. Δ-stepping execution time and relative speedup on the MTA-2 for a ScaleFree4-n graph instance (directed graph, n=2^{28} vertices and $m = 4n$ edges, random edge weights).

weights to fall in the interval [0, 1]. Similarly, there is no noticeable trend in case of the Long-C family (Table 12).

5.3.2. *Long and Square Mesh Instances.* Tables 11 and 13 give the execution times for Δ-stepping on the long and square grid graphs respectively, as the problem size and number of processors are varied. For Long-n graphs with Δ set to $\frac{n}{m}$, there is insufficient parallelism to fully utilize even a single processor of the MTA-2. The execution time of the level-synchronous BFS also does not scale with the number of processors. In fact, as we see in Figure 12(a), the running time goes up in case of multiprocessor runs, as the parallelization overhead becomes significant. Also, note that the execution time on a single processor of the MTA-2 is two orders of magnitude slower than the reference sequential processor (Figure 8(b)). In case of square grid graphs (Figure 12(b)), there is sufficient parallelism to utilize up to 4 processors for a graph instance of 2^{24} vertices. For all other instances, the running time does not scale for multiprocessor runs. The ratio of the running time to BFS is about 5 in this case, and the Δ-stepping MTA-2 single processor time is comparable to the sequential reference platform running time for smaller instances.

5.3.3. *Road networks.* Table 14, Table 15 and Figure 13 summarize the running times on the USA and PTV Europe [**58**] road networks. The execution time and parallel performance is highly dependent on the value of Δ, as the normalized edge weights do not have a uniform random distribution. The behavior is best exemplified by the Europe network instance with transit time as the length function. For this graph, the maximum edge weight is 44.8458 million and the mean is 16.78 million, whereas the median weight is only 166. For $\Delta = 0.4 \approx \frac{n}{m}$, the algorithm performance tends to the worst case behavior. Now, consider the performance for Δ values of 10^{-4} and 10^{-3}. The total number of relax requests for $\Delta = 10^{-4}$ is nearly 31 million (73% more than the optimal 18 million requests), whereas for $\Delta = 10^{-3}$, the number of relax requests is 137 million (627% more than optimal). Although we do significantly more work for $\Delta = 10^{-3}$, the running time of a single MTA-2 processor is about 60 seconds, nearly 14 seconds faster than the $\Delta = 10^{-4}$ case. This is due to the MTA-2 parallelization overhead proportional to the number of parallel phases: for $\Delta = 10^{-3}$, the number of parallel phases is 10000, and for $\Delta = 10^{-4}$ it is close to 20000. We set the Δ value to the median normalized edge weight in the experiments on the full road networks. There is no significant parallel speedup, as the average relax request size per phase is low and there is insufficient parallelism in each phase to saturate multiple processors of the MTA-2.

6. Conclusions

In this paper, we experimentally evaluate the parallel Δ-stepping NSSP algorithm for the 9th DIMACS Shortest Paths Challenge. We study the algorithm performance for core challenge graph instances on the Cray MTA-2, and observe that our implementation execution time scales impressively with number of processors for low-diameter sparse graphs. We also analyze the performance using platform-independent Δ-stepping algorithm operation counts such as the number of *phases*, and the *request set sizes*, to explain performance across graph families. For grids and road networks, we observe that the average request set size is much smaller than corresponding low-diameter graph instances of the same size. Also, the parallelization overhead is significant for these instances, as there are a higher number of parallel phases.

We also show the dependence of the bucket-width Δ on the parallel performance of the algorithm. For high diameter graphs, there is a trade-off between the number

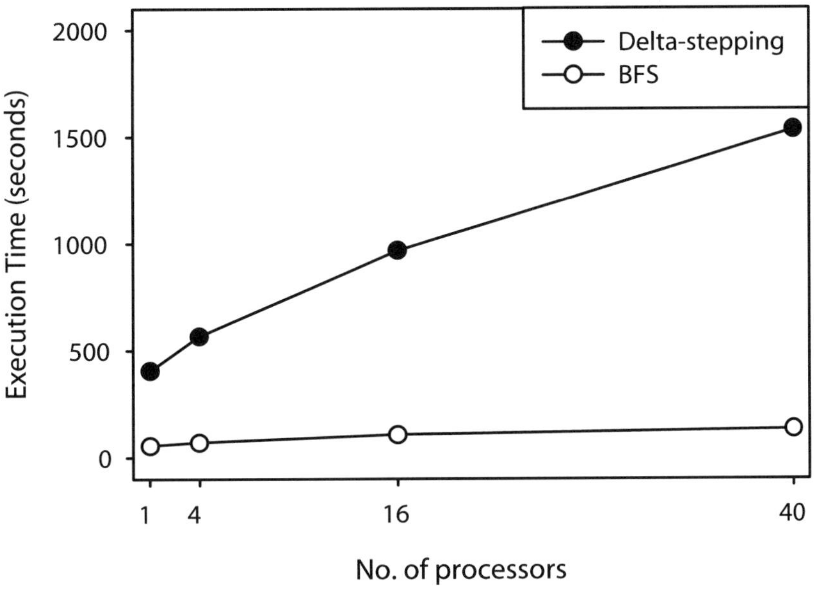

(a) Execution time for a Long-n graph instance (directed graph, n=2^{21} vertices and $m \approx 4n$ edges, random edge weights).

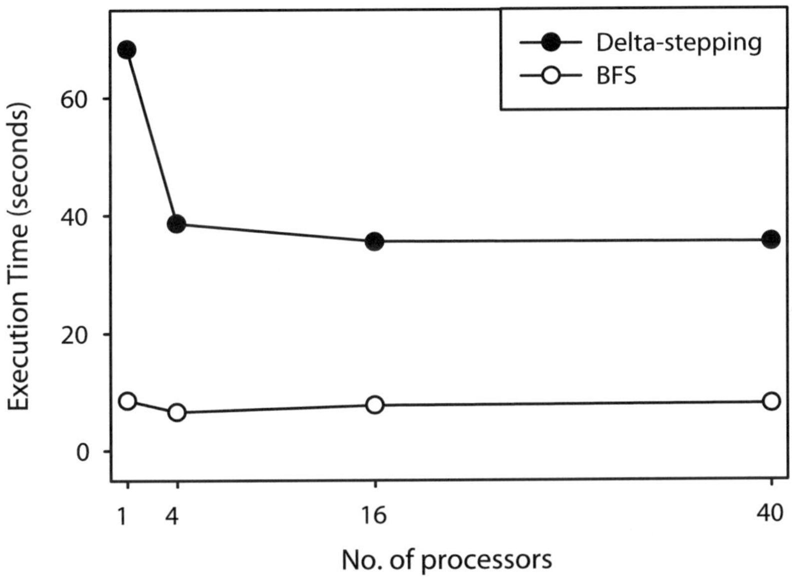

(b) Execution time for a Square-n graph instance (directed graph, n=2^{24} vertices and $m \approx 4n$ edges, random edge weights).

FIGURE 12. Δ-stepping and BFS execution times on the MTA-2 for two mesh graph instances.

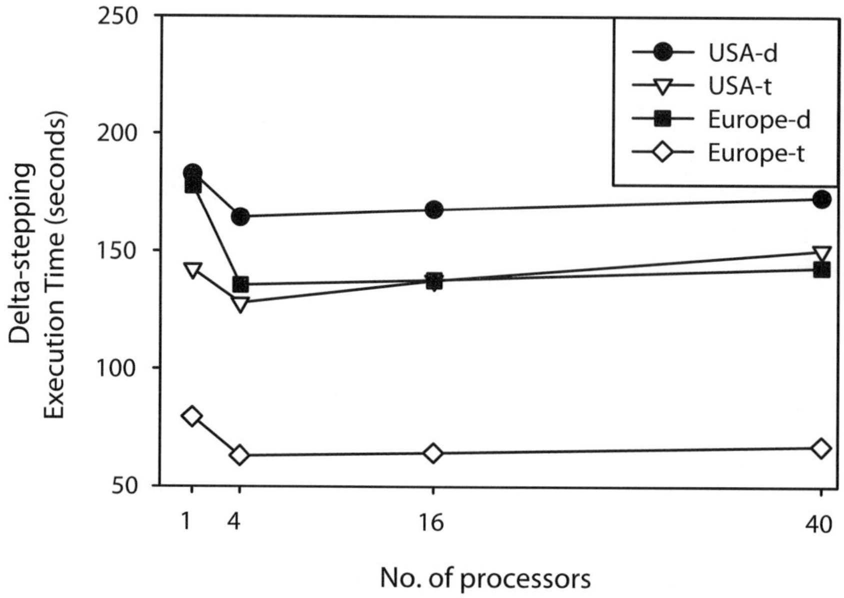

FIGURE 13. MTA-2 Parallel performance results for the Δ-stepping algorithm on the full US and Europe road networks.

of phases and the amount of work done (proportional to the number of bucket insertions). The execution time is dependent on the value of Δ as well as the number of processors. In case of road networks, where the weight distribution is not uniformly random, we have to carefully choose a value of Δ to avoid doing excessive work.

Our parallel performance studies have been restricted to the Cray MTA-2 in this paper. In future, we will extend this study to include optimized implementations of Δ-stepping on symmetric multiprocessors and multicore processors. Demonstrating scalable and efficient parallel performance for SSSP on arbitrary high-diameter graphs and road networks still remains an open challenge.

Acknowledgments

This work was supported in part by NSF Grants CAREER CCF-0611589, ACI-00-93039, NSF DBI-0420513, ITR ACI-00-81404, ITR EIA-01-21377, Biocomplexity DEB-01-20709, ITR EF/BIO 03-31654, and DARPA Contract NBCH30390004. Sandia is a multipurpose laboratory operated by Sandia Corporation, a Lockheed-Martin Company, for the United States Department of Energy under contract DE-AC04-94AL85000. We acknowledge the algorithmic inputs from Bruce Hendrickson of Sandia National Laboratories. We would also like to thank John Feo of Cray for helping us optimize the MTA-2 implementation.

Appendix A. The Cray MTA-2

This section is excerpted from [**7**].

A.1. Architecture. The Cray MTA-2 [**18**] is a novel multithreaded architecture with no data cache, and hardware support for synchronization. The computational model for the MTA-2 is *thread-centric*, not processor-centric. A thread is a logical entity comprised of a sequence of instructions that are issued in order. An MTA-2 processor consists of 128 hardware *streams* and one instruction pipeline. A stream is a physical resource (a set of 32 registers, a status word, and space in the instruction cache) that hold the state of one thread. An instruction is three-wide: a memory operation, a fused multiply-add, and a floating point add or control operation. Each stream can have up to 8 outstanding memory operations. Threads from the same or different programs are mapped to the streams by the runtime system. A processor switches among its streams every cycle, executing instructions from non-blocked streams. As long as one stream has a ready instruction, the processor remains fully utilized. No thread is bound to any particular processor. System memory size and the inherent degree of parallelism within the program are the only limits on the number of threads used by a program. The interconnection network is a partially connected 3-D torus capable of delivering one word per processor per cycle. The system has 4 GBytes of memory per processor. Logical memory addresses are hashed across physical memory to avoid stride-induced hot spots. Each memory word is 68 bits: 64 data bits and 4 tag bits. One tag bit (the full-empty bit) is used to implement synchronous load and store operations. A thread that issues a synchronous load or store remains blocked until the operation completes; but the processor that issued the operation continues to issue instructions from non-blocked streams.

The MTA-2 is closer to a theoretical PRAM machine than a shared memory symmetric multiprocessor system. Since the MTA-2 uses parallelism to tolerate latency, algorithms must often be parallelized at very fine levels to expose sufficient parallelism. However, it is not necessary that all parallelism in the program be expressed such that the system can exploit it; the goal is simply to saturate the processors. The programs that make the most effective use of the MTA-2 are those which express the parallelism of the problem in a way that allows the compiler to best exploit it.

A.2. Expressing Parallelism on the MTA-2. The MTA-2 compiler automatically parallelizes *inductive* loops of three types: parallel loops, linear recurrences and reductions. A loop is inductive if it is controlled by a variable that is incremented by a loop-invariant stride during each iteration, and the loop-exit test compares this variable with a loop-invariant expression. An inductive loop has only one exit test and can only be entered from the top. If each iteration of an inductive loop can be executed completely independently of the others, then the loop is termed parallel. To attain the best performance, we need to write code (and thus design algorithms) such that most of the loops are implicitly parallelized.

There are several compiler directives that can be used to parallelize various sections of a program. The three major types of parallelization schemes available are

- single-processor (*fray*) parallelism: The code is parallelized in such a way that just the 128 streams on the processor are utilized.
- multi-processor (*crew*) parallelism: This has higher overhead than single-processor parallelism. However, the number of streams available is much

larger, bounded by the size of the whole machine rather than the size of a single processor. Iterations can be statically or dynamically scheduled.

- future parallelism: The *future* construct (detailed below) is used in this form of parallelism. This does not require that all processor resources used during the loop be available at the beginning of the loop. The runtime growth manager increases the number of physical processors as needed. Iterations are always dynamically scheduled.

A *future* is a powerful construct to express user-specified explicit parallelism. It packages a sequence of code that can be executed by a newly created thread running concurrently with other threads in the program. Futures include efficient mechanisms for delaying the execution of code that depends on the computation within the future, until the future completes. The thread that spawns the future can pass information to the thread that executes the future via parameters. Futures are best used to implement task-level parallelism and the parallelism in recursive computations.

A.3. Synchronization support on the MTA-2. Synchronization is a major limiting factor to scalability in the case of practical shared memory implementations. The software mechanisms commonly available on conventional architectures for achieving synchronization are often inefficient. However, the MTA-2 provides hardware support for fine-grained synchronization through the full-empty bit associated with every memory word. The compiler provides a number of generic routines that operate atomically on scalar variables. We list a few useful constructs that appear in the algorithm pseudo-codes in subsequent sections.

- The `int_fetch_add` routine (`int_fetch_add`$(\&v, i)$) atomically adds integer i to the value at address v, stores the sum at v, and returns the original value at v (setting the full-empty bit to full). If v is an empty sync or future variable, the operation blocks until v becomes full.
- `readfe`$(\&v)$ returns the value of variable v when v is full and sets v empty. This allows threads waiting for v to become empty to resume execution. If v is empty, the read blocks until v becomes full.
- `writeef`$(\&v, i)$ writes the value i to v when v is empty, and sets v back to full. The thread waits until v is set empty.
- `purge`$(\&v)$ sets the state of the full-empty bit of v to empty.

Appendix B. Tables

B.1. Sequential performance of Δ-stepping implementation on the reference platform.

TABLE 1. Sequential performance (execution time in seconds, and normalized performance with reference to the baseline BFS) of our implementation for the core random graph families.

(a) Random4-n core family. Problem instance denotes the log of the number of vertices. A directed random graph of n vertices, $m = 4n$ edges, and maximum weight $C = n$.

Problem Instance	**11**	**12**	**13**	**14**	**15**	**16**
BFS	.0001	.0003	.0006	.001	.004	.02
Δ-stepping	.0007	.002	.004	.01	.03	.09
Normalized to BFS	7.00	6.67	6.67	10.00	7.50	4.50
DIMACS Reference	.0003	.0008	.002	.008	.02	.06
Normalized to BFS	3.00	2.67	3.33	8.00	5.00	3.00

Problem Instance	**17**	**18**	**19**	**20**	**21**
BFS	.05	.14	.32	.69	1.45
Δ-stepping	.23	.52	1.12	2.54	5.42
Normalized to BFS	4.60	3.71	3.50	3.68	3.74
DIMACS Reference	.13	.30	0.65	1.39	3.19
Normalized to BFS	2.60	2.14	2.03	2.01	2.20

(b) Random4-C core family. Problem instance denotes the log of the maximum edge weight. $n = 2^{20}$ vertices and $m = 4n$ edges.

Problem Instance	**0**	**1**	**2**	**3**	**4**	**5**	**6**	**7**
BFS	0.69	0.69	0.69	0.69	0.69	0.69	0.69	0.69
Δ-stepping	2.31	2.55	2.53	2.55	2.54	2.54	2.54	2.54
Normalized to BFS	3.35	3.70	3.67	3.70	3.68	3.67	3.67	3.67
DIMACS Reference	0.87	0.89	0.92	1.21	1.26	1.31	1.38	1.36
Normalized to BFS	1.26	1.29	1.33	1.75	1.83	1.90	2.00	1.97

Problem Instance	**8**	**9**	**10**	**11**	**12**	**13**	**14**	**15**
BFS	0.69	0.69	0.69	0.69	0.69	0.69	0.69	0.69
Δ-stepping	2.55	2.54	2.54	2.55	2.54	2.54	2.54	2.54
Normalized to BFS	3.70	3.68	3.68	3.70	3.68	3.68	3.68	3.68
DIMACS Reference	1.37	1.37	1.38	1.37	1.37	1.38	1.37	1.38
Normalized to BFS	1.98	1.98	2.00	1.98	1.98	2.00	1.98	2.00

TABLE 2. Sequential performance (execution time in seconds, and normalized performance with reference to the baseline BFS) of our implementation for the core long grid graph families.

(a) Long-n core family. Problem instance i denotes a grid with $x = 2^i$ and $y = 16$. $n = xy$ and $\frac{m}{n} \approx 4$.

Problem Instance	**6**	**7**	**8**	**9**	**10**	**11**
BFS	.0001	.0002	.0003	.0007	.001	.004
Δ-stepping	.0005	.001	.002	.005	.01	.03
Normalized to BFS	5.00	5.00	6.67	7.14	10.00	7.50
DIMACS Reference	.0002	.0003	.0007	.002	.006	0.01
Normalized to BFS	2.00	1.50	2.33	2.86	6.00	2.50

Problem Instance	**12**	**13**	**14**	**15**	**16**
BFS	.02	.04	.09	.19	.41
Δ-stepping	.07	.17	.35	.76	1.54
Normalized to BFS	3.50	4.25	3.89	4.00	3.76
DIMACS Reference	.03	0.06	.13	.27	.60
Normalized to BFS	1.50	1.50	1.44	1.42	1.46

(b) Long-C core family. Problem instance denotes the log of the maximum edge weight. The grid dimensions are set to $x = 2^{16}$ and $y = 16$.

Problem Instance	**0**	**1**	**2**	**3**	**4**	**5**	**6**	**7**
BFS	0.25	0.25	0.25	0.25	0.25	0.25	0.25	0.25
Δ-stepping	0.68	0.75	0.78	0.88	1.02	1.07	1.09	1.09
Normalized to BFS	2.75	3.00	3.12	3.52	4.08	4.28	4.36	4.36
DIMACS Reference	0.50	0.54	0.57	0.59	0.57	0.58	0.60	0.60
Normalized to BFS	2.00	2.16	2.28	2.36	2.28	2.32	2.40	2.40

Problem Instance	**8**	**9**	**10**	**11**	**12**	**13**	**14**	**15**
BFS	0.25	0.25	0.25	0.25	0.25	0.25	0.25	0.25
Δ-stepping	1.08	1.09	1.09	1.09	1.08	1.09	1.08	1.09
Normalized to BFS	4.32	4.36	4.36	4.36	4.32	4.36	4.32	4.36
DIMACS Reference	0.59	0.60	0.61	0.59	0.61	0.60	0.60	0.60
Normalized to BFS	2.36	2.40	2.44	2.36	2.44	2.40	2.40	2.40

TABLE 3. Sequential performance (execution time in seconds, and normalized performance with reference to the baseline BFS) of our implementation for the core square grid graph families.

(a) Square-n core family. Problem instance denotes the log of the grid x dimension. $x = y$ and $\frac{m}{n} \approx 4$.

Problem Instance	**11**	**12**	**13**	**14**	**15**	**16**
BFS	.0001	.0003	.0007	.001	.003	.01
Δ-stepping	.0008	.002	.004	.008	.03	.07
Normalized to BFS	8.00	6.67	5.71	8.00	10.00	7.00
DIMACS Reference	.0003	.0007	.002	.006	.01	.03
Normalized to BFS	3.00	2.33	2.86	6.00	3.33	3.00

Problem Instance	**17**	**18**	**19**	**20**	**21**
BFS	.04	.08	.20	.42	.93
Δ-stepping	.20	.36	.81	2.05	4.38
Normalized to BFS	5.00	4.00	4.05	4.88	4.71
DIMACS Reference	.06	0.14	.36	.84	2.01
Normalized to BFS	1.50	1.75	1.80	2.00	2.16

(b) Square-C core family. Problem instance denotes the log of the edge weight. The grid dimensions are set to $x = y = 2^{10}$, and $n = xy$.

Problem Instance	**0**	**1**	**2**	**3**	**4**	**5**	**6**	**7**
BFS	0.42	0.42	0.42	0.42	0.42	0.42	0.42	0.42
Δ-stepping	1.99	2.06	2.03	2.09	2.05	2.07	2.06	2.01
Normalized to BFS	4.74	4.90	4.83	4.89	4.88	4.93	4.90	4.79
DIMACS Reference	0.56	0.68	0.71	0.79	0.78	0.76	0.81	0.80
Normalized to BFS	1.33	1.62	1.69	1.88	1.86	1.81	1.93	1.90

Problem Instance	**8**	**9**	**10**	**11**	**12**	**13**	**14**	**15**
BFS	0.42	0.42	0.42	0.42	0.42	0.42	0.42	0.42
Δ-stepping	2.04	2.09	2.05	2.06	2.01	2.08	2.09	2.08
Normalized to BFS	4.86	4.98	4.88	4.90	4.78	4.95	4.98	4.95
DIMACS Reference	0.82	0.80	0.83	0.79	0.77	0.79	0.78	0.77
Normalized to BFS	1.95	1.90	1.98	1.88	1.83	1.88	1.86	1.83

TABLE 4. Sequential performance (execution time in seconds, and normalized performance with reference to the baseline BFS) of our implementation for the core road networks.

(a) Core graphs from the USA road network, with the transit time as the length function.

Problem Instance	**CTR**	**W**	**E**	**LKS**	**CAL**	**NE**
BFS	4.16	1.49	.65	.39	.26	.17
Δ-stepping	25.24	9.87	4.97	2.52	1.95	1.43
Normalized to BFS	6.07	6.63	7.65	6.46	7.50	8.41
DIMACS Reference	9.06	3.12	1.65	1.14	.72	.58
Normalized to BFS	2.18	2.09	2.54	2.92	2.77	3.41

Problem Instance	**NW**	**FLA**	**COL**	**BAY**	**NY**
BFS	.16	.13	.04	0.03	.02
Δ-stepping	.89	.97	.30	.21	.15
Normalized to BFS	5.56	7.46	7.5	7.00	7.50
DIMACS Reference	.45	.36	.13	.09	.07
Normalized to BFS	2.81	2.77	3.25	3.00	3.50

(b) Core graphs from the USA road network, with the distance as the length function.

Problem Instance	**CTR**	**W**	**E**	**LKS**	**CAL**	**NE**
BFS	4.32	1.89	1.05	.80	.54	.34
Δ-stepping	21.63	10.34	7.02	3.52	3.67	1.06
Normalized to BFS	5.01	5.47	6.69	4.40	6.80	3.11
DIMACS Reference	15.52	4.91	3.12	2.24	1.41	0.86
Normalized to BFS	3.59	2.60	2.97	2.80	2.61	2.53

Problem Instance	**NW**	**FLA**	**COL**	**BAY**	**NY**
BFS	.31	.28	.05	.03	.02
Δ-stepping	1.26	1.17	0.15	0.11	0.08
Normalized to BFS	4.06	4.18	3.00	3.67	4.00
DIMACS Reference	0.71	0.55	0.13	0.08	0.07
Normalized to BFS	2.29	1.96	2.60	2.67	3.50

B.2. Algorithm performance as a function of Δ.

TABLE 5. Performance of the Δ-stepping algorithm as a function of the bucket width Δ for random and scale-free networks. K denotes 10^3, M denotes 10^6 and B denotes 10^9.

(a) Random4-n graph instance ($n = C = 2^{28}$, $m = 4n$).

Δ	0.1	0.5	1	5	10
No. of phases	328	122	89	58	50
Last non-empty bucket	93	19	9	1	0
Average distance	4.94	4.94	4.94	4.94	4.94
Avg. no. of light relax requests per phase	161K	1.84M	4.43M	8.28M	20.00M
Avg. no. of heavy relax requests per bucket	3.12M	5.46M	0	0	0
Total number of relaxations	343.20M	328.70M	394.30M	480.30M	1.00B
Execution Time (40 processors MTA-2, seconds)	14.03	11.64	13.57	16.15	27.14

(b) ScaleFree4-n graph instance ($n = C = 2^{25}$, $m = 4n$).

Δ	0.1	0.5	1	5	10
No. of phases	312	117	83	51	39
Last non-empty bucket	131	26	13	2	1
Average distance	1.68	1.68	1.68	1.68	1.68
Avg. no. of light relax requests per phase	22.40K	267.00K	667.00K	2.40M	3.15M
Avg. no. of heavy relax requests per bucket	278.00K	455.80K	0	0	0
Total number of relaxations	43.78M	43.63M	55.40M	122.68M	122.76M
Execution Time (40 processors MTA-2, seconds)	4.23	2.55	2.79	5.48	6.38

(c) RandomLogUnif4-n instance ($n = C = 2^{20}$, $m = 4n$).

Δ	0.001	0.05	0.1	0.5	1
No. of phases	460	115	93	77	71
Last non-empty bucket	134	17	8	4	2
Average distance	0.04	0.04	0.04	0.04	0.04
Avg. no. of light relax requests per phase	1.50K	50.80K	84.80K	132.00K	150.01K
Avg. no. of heavy relax requests per bucket	6.50K	3.47K	3.97K	2.18K	1.42K
Total number of relaxations	1.59M	5.91M	7.92M	10.17M	10.74M
Execution Time (40 processors MTA-2, seconds)	2.15	1.18	0.96	0.80	0.75

TABLE 6. Performance of the Δ-stepping algorithm as a function of the bucket width Δ for mesh networks. K denotes 10^3, M denotes 10^6 and B denotes 10^9.

(a) Long grid instance ($n = C = 2^{20}$).

Δ	0.1	0.5	1	5	10
No. of phases	295.27K	151.43K	124.76K	97.38K	92.70K
Last non-empty bucket	216.38K	43.28K	21.64K	4.33K	2.16K
Average distance	10805	10805	10805	10805	10805
Avg. no. of light relax requests per phase	0.65	5.49	11.29	22.99	37.22
Avg. no. of heavy relax requests per bucket	5.21	9.65	0	0	0
Total number of relaxations	1.32M	1.25M	1.40M	2.23M	3.45M
Execution Time (40 processors MTA-2, seconds)	858.75	465.14	443.05	369.57	274.23

(b) Square grid instance ($n = C = 2^{20}$).

Δ	0.1	0.5	1	5	10
No. of phases	12795	5489	4188	2769	2504
Last non-empty bucket	4691	938	469	93	46
Average distance	251.86	251.86	251.86	251.86	251.86
Avg. no. of light relax requests per phase	15.51	155.11	340.50	785.18	1248.72
Avg. no. of heavy relax requests per bucket	242.22	437.44	0	0	0
Total number of relaxations	1.33M	1.26M	1.43M	2.17M	3.13M
Execution Time (40 processors MTA-2, seconds)	48.77	20.04	13.92	9.17	8.32

TABLE 7. Performance of the Δ-stepping algorithm as a function of the bucket width Δ for road networks. K denotes 10^3, M denotes 10^6 and B denotes 10^9.

(a) Central USA road instance (distance).

Δ	0.1	0.5	1	5
No. of phases	3129	2017	1669	1300
Last non-empty bucket	105	21	10	2
Average distance	3.93	3.93	3.93	3.93
Avg. no. of light relax requests per phase	6.34K	26.04K	57.89K	82.60K
Avg. no. of heavy relax requests per bucket	320.60	1.27	0	0
Total number of relaxations	19.87M	52.50M	96.60M	107.00M
Execution Time (40 processors MTA-2, seconds)	7.83	5.84	5.62	8.88

(b) NE USA road instance (transit time).

Δ	0.1	0.5	1	5
No. of phases	437	3542	3126	2220
Last non-empty bucket	315	63	31	6
Average distance	14.06	14.06	14.06	14.06
Avg. no. of light relax requests per phase	369.90	783.80	1.38K	8.66K
Avg. no. of heavy relax requests per bucket	168.40	1.85	0	0
Total number of relaxations	1.76M	2.78M	4.31M	19.21M
Execution Time (40 processors MTA-2, seconds)	12.92	9.25	8.39	6.84

B.3. Parallel Performance on the Cray MTA-2.

TABLE 8. MTA-2 performance (execution time in seconds, normalized performance with reference to the baseline BFS, relative speedup) of our implementation on Random4-n graphs. Problem instance denotes log of the number of vertices. p denotes the number of processors. $m = 4n$ edges, and maximum weight $C = n$.

p	Problem Instance	**21**	**22**	**23**	**24**	**25**	**26**	**27**	**28**
1	BFS (sec)	0.62	1.24	3.39	4.91	9.70	18.90	37.30	73.94
	Δ-stepping (sec)	3.21	6.34	12.05	23.61	46.63	93.77	187.84	371.27
	Ratio to BFS	5.18	5.11	3.55	4.81	4.81	4.96	5.04	5.02
2	BFS (sec)	0.31	0.61	1.19	2.34	4.65	9.29	18.66	37.06
	Δ-stepping (sec)	1.92	3.44	6.57	12.72	24.88	48.40	96.15	187.99
	Ratio to BFS	6.25	5.66	5.52	5.43	5.35	5.21	5.15	5.07
	Relative Speedup	1.67	1.84	1.83	1.86	1.87	1.94	1.95	1.99
4	BFS (sec)	0.16	0.31	0.61	1.19	2.37	4.71	9.38	19.59
	Δ-stepping (sec)	1.23	2.07	3.77	7.07	13.63	25.40	50.08	96.89
	Ratio to BFS	7.69	6.68	6.18	5.94	5.75	5.39	5.34	4.95
	Relative Speedup	2.61	3.06	3.20	3.34	3.42	3.69	3.75	3.83
8	BFS (sec)	0.09	0.16	0.31	0.60	1.18	2.35	4.73	9.37
	Δ-stepping (sec)	0.96	1.40	2.39	4.28	8.04	13.81	27.29	49.18
	Ratio to BFS	10.67	8.48	7.74	7.13	6.81	6.88	5.77	5.25
	Relative Speedup	3.34	4.53	5.04	5.52	5.80	6.79	6.88	7.55
16	BFS (sec)	0.06	0.10	0.17	0.32	0.62	1.20	2.39	4.73
	Δ-stepping (sec)	0.84	1.24	1.84	3.06	5.45	8.34	15.91	25.47
	Ratio to BFS	7.55	12.40	10.60	9.22	8.83	6.95	6.66	5.38
	Relative Speedup	3.82	5.11	6.55	7.71	8.55	11.24	11.81	14.58
32	BFS (sec)	0.05	0.07	0.11	0.19	0.36	0.69	1.36	2.68
	Δ-stepping (sec)	0.78	1.047	1.52	2.42	4.12	5.70	10.31	13.90
	Ratio to BFS	15.60	15.00	13.81	12.74	11.44	15.83	7.58	5.19
	Relative Speedup	4.12	6.04	7.93	9.76	11.32	16.45	18.22	26.71
40	BFS (sec)	0.04	0.06	0.10	0.17	0.32	0.61	1.20	2.37
	Δ-stepping (sec)	0.81	1.05	1.53	2.35	3.98	5.15	9.51	11.96
	Ratio to BFS	18.41	16.41	15.30	13.82	12.44	8.44	7.92	5.04
	Relative Speedup	3.96	6.04	7.88	10.05	11.72	11.11	19.75	31.04

TABLE 9. MTA-2 performance (execution time in seconds, normalized performance with reference to the baseline BFS, relative speedup) of our implementation for RandomLogUnif4-n graphs. Problem instance denotes the log of the number of vertices. p denotes the number of processors. $m = 4n$ edges and maximum weight $C = n$.

p	Problem Instance	**21**	**22**	**23**	**24**	**25**	**26**	**27**
1	BFS (sec)	0.62	1.24	3.39	4.91	9.70	18.90	37.30
	Δ-stepping (sec)	20.43	41.72	85.10	173.96	378.80	878.86	1687.59
	Ratio to BFS	32.95	33.64	25.10	35.43	39.05	46.50	45.24
4	BFS (sec)	0.16	0.31	0.61	1.19	2.37	4.71	9.38
	Δ-stepping (sec)	6.03	11.17	22.90	45.38	97.63	224.46	426.02
	Ratio to BFS	37.69	36.03	37.54	38.13	41.19	47.65	45.52
	Relative Speedup	3.38	3.73	3.72	3.83	3.88	3.91	3.96
16	BFS (sec)	0.06	0.10	0.17	0.32	0.62	1.20	2.39
	Δ-stepping (sec)	2.47	3.94	7.43	13.96	26.50	60.82	113.12
	Ratio to BFS	41.17	39.40	43.70	43.62	42.74	50.68	47.33
	Relative Speedup	8.27	10.59	11.45	12.46	14.29	14.45	14.92
40	BFS (sec)	0.04	0.06	0.10	0.17	0.32	0.61	1.20
	Δ-stepping (sec)	1.99	2.61	4.27	7.23	12.86	29.58	51.89
	Ratio to BFS	49.17	43.50	42.70	42.53	40.19	48.49	43.24
	Relative Speedup	10.27	15.98	19.93	24.06	29.46	29.71	32.52

TABLE 10. MTA-2 performance (execution time in seconds, normalized performance with reference to the baseline BFS, relative speedup) of our implementation on Random4-C graphs. Problem instance denotes the log of the maximum edge weight. p denotes the number of processors. $n = 2^{26}$ vertices, $m = 4n$ edges.

p	Problem Instance	**0**	**3**	**6**	**9**	**12**	**15**
1	BFS (sec)	19.07	19.07	19.07	19.07	19.07	19.07
	Δ-stepping (sec)	93.66	93.74	94.34	93.22	95.76	94.11
	Ratio to BFS	4.91	4.91	4.89	4.89	5.01	4.83
2	BFS (sec)	9.38	9.38	9.38	9.38	9.38	9.38
	Δ-stepping (sec)	48.24	48.15	48.78	48.5	49.25	48.63
	Ratio to BFS	5.14	5.13	5.20	5.17	5.25	5.18
	Relative Speedup	1.94	1.95	1.91	1.92	1.94	1.93
4	BFS (sec)	4.73	4.73	4.73	4.73	4.73	4.73
	Δ-stepping (sec)	25.81	25.43	25.47	25.81	25.39	25.35
	Ratio to BFS	5.46	5.38	5.38	5.46	5.37	5.36
	Relative Speedup	3.63	3.69	3.66	3.61	3.77	3.71
8	BFS (sec)	2.36	2.36	2.36	2.36	2.36	2.36
	Δ-stepping (sec)	14.06	13.67	13.86	13.85	14.07	13.85
	Ratio to BFS	5.96	5.79	5.87	5.87	5.96	5.87
	Relative Speedup	6.66	6.86	6.73	6.73	6.80	6.79
16	BFS (sec)	1.21	1.21	1.21	1.21	1.21	1.21
	Δ-stepping (sec)	8.37	8.38	8.4	8.37	8.42	8.38
	Ratio to BFS	6.92	6.92	6.94	6.92	6.96	6.92
	Relative Speedup	11.19	11.19	11.11	11.14	11.37	11.23
32	BFS (sec)	0.69	0.69	0.69	0.69	0.69	0.69
	Δ-stepping (sec)	5.66	5.65	5.66	5.68	5.66	5.67
	Ratio to BFS	8.20	8.19	8.20	8.23	8.20	8.21
	Relative Speedup	11.42	11.45	11.38	11.32	11.67	11.45
40	BFS (sec)	0.61	0.61	0.61	0.61	0.61	0.61
	Δ-stepping (sec)	5.23	5.27	5.22	5.23	5.21	5.26
	Ratio to BFS	8.52	8.58	8.50	8.52	8.48	8.57
	Relative Speedup	17.91	17.79	17.88	17.82	18.38	17.89

TABLE 11. MTA-2 performance (execution time in seconds, normalized performance with reference to the baseline BFS) of our implementation on Long-n graphs. Problem instance denotes the log of the rectangular grid x dimension. p denotes the number of processors, $y = 16$, $n = xy$, $m \approx 4n$ edges, and maximum weight $C = n$.

p	Problem Instance	**10**	**11**	**12**	**13**
1	BFS (sec)	0.54	1.22	1.54	4.19
	Δ-stepping (sec)	3.99	8.57	13.77	32.11
	Ratio to BFS	7.39	7.02	8.94	7.66
4	BFS (sec)	0.74	1.43	2.12	4.52
	Δ-stepping (sec)	5.36	11.20	17.92	42.06
	Ratio to BFS	7.24	7.83	8.45	9.30
16	BFS (sec)	1.04	1.85	3.09	6.72
	Δ-stepping (sec)	7.10	15.07	23.50	56.08
	Ratio to BFS	6.83	8.14	7.60	8.34
40	BFS (sec)	1.31	2.43	4.00	8.29
	Δ-stepping (sec)	12.53	23.64	40.02	90.59
	Ratio to BFS	9.56	9.73	10.00	10.97

p	Problem Instance	**14**	**15**	**16**	**17**
1	BFS (sec)	7.60	14.30	34.90	55.62
	Δ-stepping (sec)	57.16	123.73	243.53	404.91
	Ratio to BFS	7.52	8.65	6.97	7.28
4	BFS (sec)	9.27	19.80	39.48	71.49
	Δ-stepping (sec)	73.93	158.72	306.69	567.63
	Ratio to BFS	7.97	8.01	7.77	7.94
16	BFS (sec)	13.56	25.44	57.71	107.00
	Δ-stepping (sec)	97.99	212.51	503.33	967.70
	Ratio to BFS	7.23	8.35	8.72	9.04
40	BFS (sec)	18.14	32.33	72.99	132.36
	Δ-stepping (sec)	171.13	330.72	812.02	1534.05
	Ratio to BFS	9.43	10.23	11.12	11.59

TABLE 12. MTA-2 performance (execution time in seconds, normalized performance with reference to the baseline BFS) of our implementation on Long-C graphs. Problem instance denotes the log of the maximum edge weight. p denotes the number of processors. The grid dimensions are given by $x = 2^{14}$ and $y = 16$.

p	Problem Instance	**0**	**3**	**6**	**9**	**12**	**15**
1	BFS (sec)	7.60	7.60	7.60	7.60	7.60	7.60
	Δ-stepping (sec)	57.24	56.88	57.13	57.89	58.11	56.97
	Ratio to BFS	7.53	7.48	7.52	7.62	7.65	7.50
4	BFS (sec)	9.27	9.27	9.27	9.27	9.27	9.27
	Δ-stepping (sec)	74.02	73.88	73.92	74.68	75.17	75.49
	Ratio to BFS	7.98	7.97	7.97	8.06	8.10	8.14
16	BFS (sec)	13.56	13.56	13.56	13.56	13.56	13.56
	Δ-stepping (sec)	96.76	97.11	97.45	98.82	98.30	98.61
	Ratio to BFS	7.14	7.16	7.19	7.24	7.25	7.27
40	BFS (sec)	18.14	18.14	18.14	18.14	18.14	18.14
	Δ-stepping (sec)	172.00	171.34	173.43	172.84	172.49	173.19
	Ratio to BFS	9.48	9.44	9.56	9.53	9.51	9.55

TABLE 13. MTA-2 performance (execution time in seconds, normalized performance with reference to the baseline BFS) of our implementation on Square-n graphs. Problem instance denotes the log of the number of the grid dimension x. p denotes the number of processors. $x = y$, $n = xy$, $m \approx 4n$ edges, and maximum weight $C = n$.

p	Problem Instance	**6**	**7**	**8**	**9**	**10**	**11**	**12**
1	BFS (sec)	0.05	0.12	0.24	0.57	1.32	3.22	8.55
	Δ-stepping (sec)	0.20	0.52	1.28	2.80	7.84	20.56	68.33
	Ratio to BFS	4.00	4.33	5.33	4.91	5.94	6.38	7.99
4	BFS (sec)	0.09	0.16	0.33	0.72	1.51	3.19	6.59
	Δ-stepping (sec)	0.23	0.60	1.41	2.84	6.85	14.29	38.62
	Ratio to BFS	2.55	3.75	4.27	3.94	4.54	4.48	5.86
16	BFS (sec)	0.11	0.22	0.41	0.95	1.99	3.93	7.68
	Δ-stepping (sec)	0.28	0.73	1.64	3.3	7.83	14.93	35.51
	Ratio to BFS	2.54	3.32	4.00	3.47	3.93	3.80	4.62
40	BFS (sec)	0.12	0.23	0.44	1.00	2.05	4.01	7.90
	Δ-stepping (sec)	0.35	0.84	1.91	3.59	8.35	15.29	35.46
	Ratio to BFS	2.92	3.65	4.34	3.59	4.07	3.81	4.49

TABLE 14. MTA-2 performance (execution time in seconds) of the baseline BFS, and our shortest path implementation on the USA core road networks with distance (road-d) and transit times (road-t) as the length function.

p	Instance	**CTR**	**W**	**E**	**LKS**	**CAL**	**NE**
1	BFS time (sec)	7.69	5.19	3.95	3.38	2.39	2.01
	road-d time (sec)	49.89	32.91	23.46	15.08	13.09	14.33
	road-t time (sec)	37.06	24.01	15.12	14.51	11.32	6.66
4	BFS time (sec)	6.48	4.95	4.09	3.6	2.53	2.14
	road-d time (sec)	48.58	30.12	23.29	15.59	13.92	14.21
	road-t time (sec)	34.38	23.75	15.73	15.36	12.03	7.18
16	BFS time (sec)	7.26	5.85	4.94	4.32	3.02	2.53
	road-d time (sec)	52.83	36.74	26.91	18.94	15.96	15.03
	road-t time (sec)	39.95	27.86	18.53	18.21	14.25	8.54
40	BFS time (sec)	7.50	6.56	5.47	4.43	3.37	2.92
	road-d time (sec)	55.19	39.84	33.32	21.66	18.15	16.23
	road-t time (sec)	42.94	30.24	21.15	20.09	16.29	9.96

p	Instance	**NW**	**FLA**	**COL**	**BAY**	**NY**
1	BFS time (sec)	1.52	1.98	1.51	0.72	0.67
	road-d time (sec)	13.80	12.76	6.52	3.16	2.70
	road-t time (sec)	7.06	9.22	5.03	2.34	1.69
4	BFS time (sec)	1.57	2.15	1.76	0.83	0.76
	road-d time (sec)	13.41	12.28	7.90	3.70	3.39
	road-t time (sec)	8.07	10.45	5.95	2.73	1.91
16	BFS time (sec)	1.86	2.56	2.13	1.01	0.94
	road-d time (sec)	14.06	15.28	8.81	4.62	4.08
	road-t time (sec)	9.64	12.41	7.14	3.25	2.31
40	BFS time (sec)	2.18	2.95	2.19	1.05	0.95
	road-d time (sec)	15.11	16.44	9.93	5.31	5.06
	road-t time (sec)	12.05	14.54	9.09	3.95	2.82

TABLE 15. MTA-2 performance (execution time in seconds) of our implementation on the full USA and Europe road graphs.

p	Instance	**USA-d**	**USA-t**	**Europe-d**	**Europe-t**
1	BFS (sec)	10.04	9.93	7.04	7.05
	Δ-stepping (sec)	182.84	142.19	177.84	79.40
4	BFS (sec)	8.23	7.96	5.18	5.20
	Δ-stepping (sec)	164.29	127.81	135.49	63.04
16	BFS (sec)	10.21	10.14	5.89	6.01
	Δ-stepping (sec)	167.83	137.52	137.67	64.36
40	BFS (sec)	10.39	10.69	6.06	5.94
	Δ-stepping (sec)	173.11	150.63	143.13	67.16

References

1. P. Adamson and E. Tick, *Greedy partitioned algorithms for the shortest path problem*, Int'l Journal of Parallel Programming **20** (1991), no. 4, 271–298.
2. D. Ajwani, R. Dementiev, and U. Meyer, *A computational study of external-memory BFS algorithms*, Proc. 17th Annual ACM-SIAM Symposium on Discrete Algorithms (SODA 2006) (Miami, FL), ACM Press, January 2006, pp. 601–610.
3. D.A. Bader, G. Cong, and J. Feo, *On the architectural requirements for efficient execution of graph algorithms*, Proc. 34th Int'l Conf. on Parallel Processing (ICPP 2005) (Oslo, Norway), IEEE Computer Society, June 2005, pp. 547–556.
4. D.A. Bader, A.K. Illendula, B. M.E. Moret, and N. Weisse-Bernstein, *Using PRAM algorithms on a uniform-memory-access shared-memory architecture*, Proc. 5th Int'l Workshop on Algorithm Engineering (WAE 2001) (Århus, Denmark) (G.S. Brodal, D. Frigioni, and A. Marchetti-Spaccamela, eds.), Lecture Notes in Computer Science, vol. 2141, Springer-Verlag, 2001, pp. 129–144.
5. D.A. Bader and K. Madduri, *Designing multithreaded algorithms for breadth-first search and st-connectivity on the Cray MTA-2*, Proc. 35th Int'l Conf. on Parallel Processing (ICPP 2006) (Columbus, OH), IEEE Computer Society, August 2006, pp. 523–530.
6. ———, *Parallel algorithms for evaluating centrality indices in real-world networks*, Proc. 35th Int'l Conf. on Parallel Processing (ICPP 2006) (Columbus, OH), IEEE Computer Society, August 2006, pp. 539–550.
7. D.A. Bader, K. Madduri, G. Cong, and J. Feo, *Design of multithreaded algorithms for combinatorial problems*, Handbook of Parallel Computing: Models, Algorithms, and Applications (S. Rajasekaran and J. Reif, eds.), CRC Press, 2007.
8. A.-L. Barabási and R. Albert, *Emergence of scaling in random networks*, Science **286** (1999), no. 5439, 509–512.
9. J.W. Berry, B. Hendrickson, S. Kahan, and P. Konecny, *Software and algorithms for graph queries on multithreaded architectures*, Proc. Workshop on Multithreaded Architectures and Applications (MTAAP 2007) (Long Beach, CA), IEEE Computer Society, March 2007, pp. 1–14.
10. U. Brandes, *A faster algorithm for betweenness centrality*, Journal of Mathematical Sociology **25** (2001), no. 2, 163–177.
11. P. Briggs and L. Torczon, *An efficient representation for sparse sets*, ACM Letters on Programming Languages and Systems **2** (1993), no. 1-4, 59–69.
12. G.S. Brodal, J.L. Träff, and C.D. Zaroliagis, *A parallel priority queue with constant time operations*, Journal of Parallel and Distributed Computing **49** (1998), no. 1, 4–21.
13. A. Broder, R. Kumar, F. Maghoul, P. Raghavan, S. Rajagopalan, R. Stata, A. Tomkins, and J. Wiener, *Graph structure in the web*, Computer Networks **33** (2000), no. 1-6, 309–320.
14. D. Chakrabarti, Y. Zhan, and C. Faloutsos, *R-MAT: A recursive model for graph mining*, Proc. 4th SIAM Intl. Conf. on Data Mining (SDM 2004) (Orlando, FL), SIAM, April 2004, pp. 1–5.
15. K.M. Chandy and J. Misra, *Distributed computation on graphs: Shortest path algorithms*, Communications of the ACM **25** (1982), no. 11, 833–837.
16. B.V. Cherkassky, A.V. Goldberg, and T. Radzik, *Shortest paths algorithms: theory and experimental evaluation*, Mathematical Programming **73** (1996), 129–174.
17. E. Cohen, *Using selective path-doubling for parallel shortest-path computation*, Journal of Algorithms **22** (1997), no. 1, 30–56.
18. Cray, Inc., *The MTA-2 multithreaded architecture*, `http://www.cray.com/products/systems/mta/`, 2006.
19. J.R. Crobak, J.W. Berry, K. Madduri, and D.A. Bader, *Advanced shortest path algorithms on a massively-multithreaded architecture*, Proc. Workshop on Multithreaded Architectures and Applications (MTAAP 2007) (Long Beach, CA), IEEE Computer Society, March 2007, pp. 1–8.
20. C. Demetrescu, A. Goldberg, and D. Johnson, *9th DIMACS implementation challenge – Shortest Paths*, `http://www.dis.uniroma1.it/~challenge9/`, 2006.
21. ———, *9th DIMACS implementation challenge – Shortest Paths: Reference benchmark package*, `http://www.dis.uniroma1.it/~challenge9/download.shtml`, 2006.

22. R.B. Dial, *Algorithm 360: Shortest path forest with topological ordering*, Communications of the ACM **12** (1969), 632–633.
23. R.B. Dial, F. Glover, D. Karney, and D. Klingman, *A computational analysis of alternative algorithms and labeling techniques for finding shortest path trees*, Networks **9** (1979), 215–248.
24. E.W. Dijkstra, *A note on two problems in connexion with graphs*, Numerische Mathematik **1** (1959), 269–271.
25. J.R. Driscoll, H.N. Gabow, R. Shrairman, and R.E. Tarjan, *Relaxed heaps: An alternative to fibonacci heaps with applications to parallel computation*, Communications of the ACM **31** (1988), no. 11, 1343–1354.
26. M. Faloutsos, P. Faloutsos, and C. Faloutsos, *On power-law relationships of the Internet topology*, Proc. ACM SIGCOMM 1999 (Cambridge, MA), ACM, August 1999, pp. 251–262.
27. M.L. Fredman and R.E. Tarjan, *Fibonacci heaps and their uses in improved network optimization algorithms*, Journal of the ACM **34** (1987), 596–615.
28. M.L. Fredman and D.E. Willard, *Trans-dichotomous algorithms for minimum spanning trees and shortest paths*, Journal of Computer and System Sciences **48** (1994), 533–551.
29. L.C. Freeman, *A set of measures of centrality based on betweenness*, Sociometry **40** (1977), no. 1, 35–41.
30. A.M. Frieze and L. Rudolph, *A parallel algorithm for all-pairs shortest paths in a random graph*, Proc. 22nd Allerton Conf. on Communication, Control and Computing (Monticello, IL), 1984, pp. 663–670.
31. G. Gallo and P. Pallottino, *Shortest path algorithms*, Annals of Operations Research **13** (1988), 3–79.
32. F. Glover, R. Glover, and D. Klingman, *Computational study of an improved shortest path algorithm*, Networks **14** (1984), 23–37.
33. A.V. Goldberg, *Shortest path algorithms: Engineering aspects*, Proc. 12th Int'l Symposium on Algorithms and Computation (ISAAC 2001) (London, UK), Springer-Verlag, 2001, pp. 502–513.
34. ———, *A simple shortest path algorithm with linear average time*, 9th Annual European Symposium on Algorithms (ESA 2001) (Aachen, Germany), Lecture Notes in Computer Science, vol. 2161, Springer, 2001, pp. 230–241.
35. D. Gregor and A. Lumsdaine, *Lifting sequential graph algorithms for distributed-memory parallel computation*, Proc. 20th ACM SIGPLAN Conf. on Object Oriented Programming, Systems, Languages, and Applications (OOPSLA 2005) (New York, NY), ACM Press, 2005, pp. 423–437.
36. R. Guimerà, S. Mossa, A. Turtschi, and L.A.N. Amaral, *The worldwide air transportation network: Anomalous centrality, community structure, and cities' global roles*, Proceedings of the National Academy of Sciences USA **102** (2005), no. 22, 7794–7799.
37. T. Hagerup, *Improved shortest paths on the word RAM*, 27th Colloquium on Automata, Languages and Programming (ICALP 2000) (Geneva, Switzerland), Lecture Notes in Computer Science, vol. 1853, Springer-Verlag, 2000, pp. 61–72.
38. Y. Han, V. Pan, and J. Reif, *Efficient parallel algorithms for computing the all pair shortest paths in directed graphs*, Algorithmica **17** (1997), no. 4, 399–415.
39. M.R. Hribar and V.E. Taylor, *Performance study of parallel shortest path algorithms: Characteristics of good decomposition*, Proc. 13th Annual Conf. of Intel Supercomputers Users Group (Albuquerque, NM), 1997, pp. 1–27.
40. M.R. Hribar, V.E. Taylor, and D.E. Boyce, *Parallel shortest path algorithms: Identifying the factors that affect performance*, Report CPDC-TR-9803-015, Northwestern University, Evanston, IL, 1998.
41. ———, *Reducing the idle time of parallel shortest path algorithms*, Report CPDC-TR-9803-016, Northwestern University, Evanston, IL, 1998.
42. ———, *Termination detection for parallel shortest path algorithms*, Journal of Parallel and Distributed Computing **55** (1998), 153–165.
43. H. Jeong, S.P. Mason, A.-L. Barabási, and Z.N. Oltvai, *Lethality and centrality in protein networks*, Nature **411** (2001), 41–42.
44. P.N. Klein and S. Subramanian, *A randomized parallel algorithm for single-source shortest paths*, Journal of Algorithms **25** (1997), no. 2, 205–220.
45. F. Liljeros, C.R. Edling, L.A.N. Amaral, H.E. Stanley, and Y. Åberg, *The web of human sexual contacts*, Nature **411** (2001), 907–908.

46. K. Madduri, *9th DIMACS implementation challenge: Shortest Paths. Δ-stepping C/MTA-2 code*, `http://www.cc.gatech.edu/~kamesh/research/DIMACS-ch9`, 2006.
47. K. Madduri, D.A. Bader, J. Berry, and J.R. Crobak, *An experimental study of a parallel shortest path algorithm for solving large-scale graph instances*, Proc. The 9th Workshop on Algorithm Engineering and Experiments (ALENEX 2007) (New Orleans, LA), SIAM, January 2007, pp. 23–35.
48. U. Meyer, *Heaps are better than buckets: parallel shortest paths on unbalanced graphs*, Proc. 7th International Euro-Par Conference (Euro-Par 2001) (Manchester, United Kingdom), Springer-Verlag, 2000, pp. 343–351.
49. ———, *Buckets strike back: Improved parallel shortest-paths*, Proc. 16th Int'l Parallel and Distributed Processing Symp. (IPDPS 2002) (Fort Lauderdale, FL), IEEE Computer Society, April 2002, pp. 1–8.
50. ———, *Design and analysis of sequential and parallel single-source shortest-paths algorithms*, Ph.D. thesis, Universität Saarlandes, Saarbrücken, Germany, October 2002.
51. ———, *Average-case complexity of single-source shortest-paths algorithms: lower and upper bounds*, Journal of Algorithms **48** (2003), no. 1, 91–134.
52. U. Meyer and P. Sanders, *Parallel shortest path for arbitrary graphs*, Proc. 6th International Euro-Par Conference (Euro-Par 2000) (Munich, Germany), Lecture Notes in Computer Science, vol. 1900, Springer-Verlag, 2000, pp. 461–470.
53. ———, *Δ-stepping: a parallelizable shortest path algorithm*, J. Algs. **49** (2003), no. 1, 114–152.
54. M.E.J. Newman, *Scientific collaboration networks: II. shortest paths, weighted networks and centrality*, Phys. Rev. E **64** (2001), 016132.
55. ———, *The structure and function of complex networks*, SIAM Review **45** (2003), no. 2, 167–256.
56. M. Papaefthymiou and J. Rodrigue, *Implementing parallel shortest-paths algorithms*, DIMACS Series in Discrete Mathematics and Theoretical Computer Science **30** (1997), 59–68.
57. J. Park, M. Penner, and V.K. Prasanna, *Optimizing graph algorithms for improved cache performance*, Proc. Int'l Parallel and Distributed Processing Symposium (IPDPS 2002) (Fort Lauderdale, FL), IEEE Computer Society, April 2002.
58. PTV Europe, *European road graphs*, `http://i11www.iti.uni-karlsruhe.de/resources/roadgraphs/index.php`, 2006.
59. R. Raman, *Recent results on single-source shortest paths problem*, SIGACT News **28** (1997), 61–72.
60. H. Shi and T.H. Spencer, *Time-work tradeoffs of the single-source shortest paths problem*, Journal of Algorithms **30** (1999), no. 1, 19–32.
61. M. Thorup, *Undirected single-source shortest paths with positive integer weights in linear time*, Journal of the ACM **46** (1999), no. 3, 362–394.
62. J. L. Träff, *An experimental comparison of two distributed single-source shortest path algorithms*, Parallel Computing **21** (1995), no. 9, 1505–1532.
63. J. Ullman and M. Yannakakis, *High-probability parallel transitive closure algorithms*, Proc. 2nd Annual Symposium on Parallel Algorithms and Architectures (SPAA 1990) (Crete, Greece), ACM, July 1990, pp. 200–209.
64. F.B. Zhan and C.E. Noon, *Shortest path algorithms: an evaluation using real road networks*, Transportation Science **32** (1998), 65–73.

College of Computing, Georgia Institute of Technology
Current address: Computational Research Division, Lawrence Berkeley National Laboratory
E-mail address: `kamesh@gatech.edu`

College of Computing, Georgia Institute of Technology
E-mail address: `bader@cc.gatech.edu`

Computer Science Research Institute, Sandia National Laboratories
E-mail address: `jberry@sandia.gov`

Computer Science Department, Rutgers University
E-mail address: `crobakj@cs.rutgers.edu`

DIMACS Series in Discrete Mathematics
and Theoretical Computer Science
Volume **74**, 2009

Breadth first search on massive graphs

Deepak Ajwani, Ulrich Meyer, and Vitaly Osipov

ABSTRACT. We consider the problem of breadth first search (BFS) traversal on massive sparse undirected graphs in external memory. Engineering the algorithm of Munagala and Ranade [**28**] (MR_BFS) and the randomized and deterministic variants of the $o(n)$ I/O algorithm of Mehlhorn and Meyer [**26**] (MM_BFS) coupled with a heuristic, we discuss the effect of various implementation design choices on the actual running time of the BFS traversal. Demonstrating the viability of our BFS implementations on various synthetic and real world benchmarks, we show that BFS level decompositions for large graphs (around a billion edges) can be computed on a cheap machine in a *few hours.*

1. Introduction

Breadth first search (BFS) is a fundamental graph traversal strategy. It can also be viewed as computing single source shortest paths on unweighted graphs. It decomposes the input graph $G = (V, E)$ of n nodes and m edges into at most n levels where level i comprises all nodes that can be reached from a designated source s via a path of i edges, but cannot be reached using less than i edges.

Typical real-world applications of BFS on large graphs (and some of its generalizations like shortest paths or A^*) include crawling and analyzing the WWW [**29, 30**], route planning using small navigation devices with flash memory cards [**21**], state space exploration [**19**], etc. Since most of the large real world graphs are sparse, we mainly concentrate on the problem of computing a BFS level decomposition for massive sparse undirected graphs.

While modern processor speeds are measured in GHz, average hard disk latencies are in the range of a few milliseconds [**22**]. Hence, the cost of accessing a data element from the hard-disk (an I/O) is around a million times more than the cost of an instruction. Therefore, it comes as no surprise that the I/Os dominate the runtimes of even basic graph traversal strategies like BFS on large graphs, making their standard implementations non-viable. Since the traditional RAM model,

---

This work was partially supported by the DFG grants SA 933/1-3, ME 2088/1-3, ME 3250/1-1 and by MADALGO - Center for Massive Data Algorithmics, a Center of the Danish National Research Foundation.

which assumes an unbounded amount of memory with unit cost access to any location, does not capture the I/Os into the performance metric, we need to look at alternative models of computation.

1.1. Computation models. We consider the commonly accepted external memory model of Aggarwal and Vitter [**2**] and the cache-oblivious model [**20**]. They both assume a two level memory hierarchy with faster internal memory having a capacity to store M vertices/edges. In an I/O operation, one block of data, which can have B vertices/edges is transferred between disk and internal memory. The measure of performance of an algorithm is the number of I/Os it performs. The number of I/Os needed to read N contiguous items from disk is $\mathrm{scan}(N) = \Theta(N/B)$. The number of I/Os required to sort N items is $\mathrm{sort}(N) = \Theta((N/B)\log_{M/B}(N/B))$. For all realistic values of N, B, and M, $\mathrm{scan}(N) < \mathrm{sort}(N) \ll N$. The difference between the two models is that the values of B and M are not known to the algorithm in the cache-oblivious model, allowing the algorithms designed under this model to be simultaneously efficient on all levels of the memory hierarchy.

1.2. Algorithms. BFS is well-understood in the RAM model. There exists a simple linear time algorithm [**15**] (hereafter refered as IM_BFS) for the BFS traversal in a graph. IM_BFS keeps a set of appropriate candidate nodes for the next vertex to be visited in a FIFO queue Q. Furthermore, in order to find out the unvisited neighbors of a node from its adjacency list, it marks the nodes as either visited or unvisited.

Unfortunately, as the storage requirements of the graph starts approaching the size of the internal memory, the running time of this algorithm deviates significantly from the predicted $O(n+m)$ asymptotic performance of the RAM model. Figure 1 displays the results of our experiments with the commonly used BFS routine of the LEDA [**27**] graph package on random graph $G(n,m)$ with $m = 4n$. These experiments were done on a machine with Intel Xeon 2.0 GHz processor, 1 GB RAM and 2 GB swap space on a Seagate Baracuda hard-disk [**32**]. On random graphs with 3.6 million nodes (and 14.4 million edges), it takes around 10 *hours* as compared to just 10 *seconds* for graphs with 1.8 million nodes (and 7.2 million edges). On massive graphs (with a billion or more edges), IM_BFS is simply non-viable as it requires many *months* for the requisite graph traversal. As discussed before, the main cause for such a poor performance of this algorithm on massive graphs is the number of I/Os it incurs. Remembering visited nodes needs $\Theta(m)$ I/Os in the worst case and the unstructured indexed access to adjacency lists may result in $\Theta(n)$ I/Os.

The algorithm by Munagala and Ranade [**28**] (refered as MR_BFS) ignores the second problem but addresses the first by exploiting the fact that the neighbors of a node in BFS level $t-1$ are all in BFS levels $t-2$, $t-1$ or t. Let $L(t)$ denote the set of nodes in BFS level t, and let $A(t)$ be the multi-set of neighbors of nodes in $L(t-1)$. Given $L(t-1)$ and $L(t-2)$, MR_BFS builds $L(t)$ (depicted in Figure 2) as follows: Firstly, $A(t)$ is created by $|L(t-1)|$ random accesses to get the adjacency lists of all nodes in $L(t-1)$. Thereafter, duplicates are removed from $A(t)$ to get a sorted set $A'(t)$. This is done by sorting $A(t)$ according to node indices, followed by a scan and compaction phase. The set $L(t) := A'(t)/\{L(t-1) \cup L(t-2)\}$ is computed by scanning "in parallel" the sorted sets of $A'(t)$, $L(t-1)$, and $L(t-2)$ to

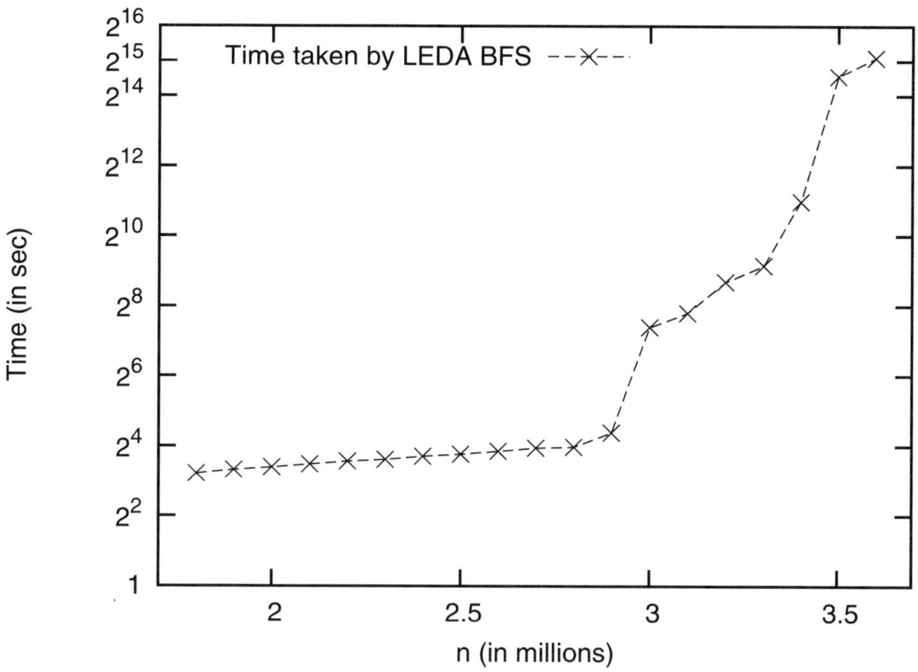

FIGURE 1. Time (in seconds) required by BFS from the LEDA graph package on random graphs with $m = 4n$ edges

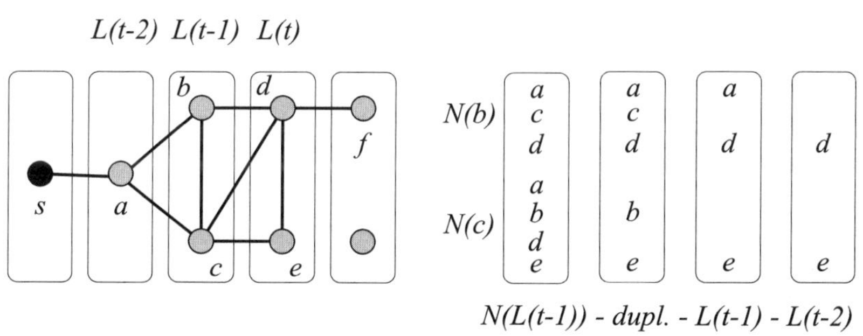

FIGURE 2. A phase in the BFS algorithm of Munagala and Ranade

filter out the nodes already present in $L(t-1)$ or $L(t-2)$. The resulting worst-case I/O-bound is $O\left(\sum_t L(t) + \sum_t \operatorname{sort}(A(t))\right) = O\left(n + \operatorname{sort}(n+m)\right)$.

Mehlhorn and Meyer suggested another approach [**26**] (MM_BFS) which involves a preprocessing phase to restructure the adjacency lists of the graph representation. It groups the vertices of the input graph into disjoint clusters of small diameter and stores the adjacency lists of the nodes in a cluster contiguously on the disk. Thereafter, an appropriately modified version of MR_BFS is run. MM_BFS exploits the fact that whenever the first node of a cluster is visited then the remaining nodes of this cluster will be reached soon after. By spending only one random access (and possibly, some sequential accesses depending on cluster size) in order to load the whole cluster and then keeping the cluster data in some efficiently

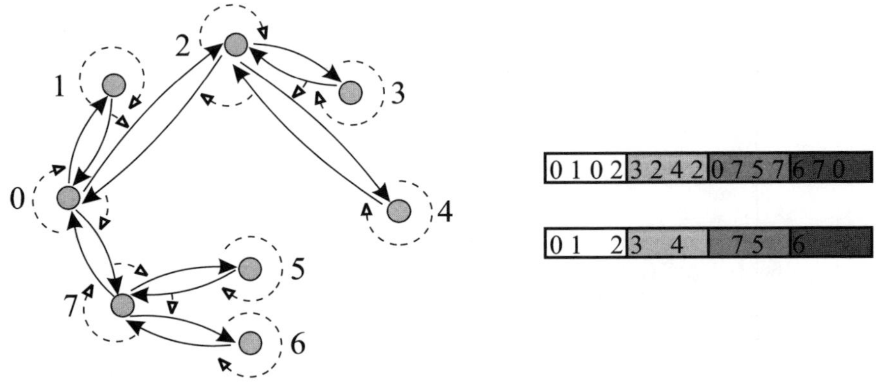

FIGURE 3. The bi-directed tree (shaded circles and solid lines) and the closed linked list of its edges (dashed lines) on the left. The order of the vertices and their partitioning before and after the duplicates removal on the right

accessible data structure (pool) until it is all used up, on sparse graphs the total amount of I/O can be reduced by a factor of up to $\sqrt{B}$. The neighboring nodes of a BFS level can be computed simply by scanning the pool and not the whole graph. Though some edges may be scanned more often in the pool, unstructured I/O in order to fetch adjacency lists is considerably reduced, thereby saving the total number of I/Os. The preprocessing of MM_BFS comes in two variants: randomized and deterministic (refered as MM_BFS_R and MM_BFS_D, respectively).

In the randomized variant, the input graph is partitioned by choosing master nodes independently and uniformly at random with a probability μ. If we have no apriori information about the structure of the graph, $\mu = \min\{1, \sqrt{\frac{(n+m)\cdot\log n}{n\cdot B}}\}$ minimizes the total expected number of I/Os. The partitioning is generated "in parallel": in each round, each master node tries to capture all unvisited neighbors of its current sub-graph into its partition, with ties being resolved arbitrarily. At the beginning of a round, the nodes lying on the boundaries of the current partitions carry the label of its corresponding master node. A scan through the set of adjacency lists collects the neighbors of all the boundary nodes together with the corresponding labels. These are then sorted, duplicates target nodes removed (with labels being resolved arbitrarily) and added to the labelled partition. Note that the diameter of such a cluster is bounded by $\max\{1, \sqrt{\frac{n\cdot B\cdot\log n}{n+m}}\}$ w.h.p and thus we need $O(\sqrt{\frac{n\cdot B\cdot\log n}{n+m}})$ rounds w.h.p. to partition the whole graph.

The deterministic variant first builds a spanning tree T_s for the connected component of G that contains the source node. Each undirected edge is then replaced by two oppositely directed edges. Note that a bi-directed tree always has at least one Euler tour. In order to construct the Euler tour around this bi-directed tree, each node chooses a cyclic order [**16**] of its neighbors. As depicted in Figure 3, the successor of an incoming edge is defined to be the outgoing edge to the next node in the cyclic order. The tour is then broken at the source node and the elements of the resulting list are then stored in consecutive order using an external memory list-ranking algorithm (e.g. [**13, 31**]). Thereafter, we chop the Euler tour into chunks of $\max\{1, \sqrt{\frac{n \cdot B}{n+m}}\}$ nodes and remove the duplicates with a couple of sorting steps. The adjacency lists are then re-ordered based on the position of their corresponding nodes in the chopped duplicate-free Euler tour. Note that the diameter of the clusters obtained by the deterministic preprocessing can not exceed $\max\{1, \sqrt{\frac{n \cdot B}{n+m}}\}$.

The randomized variant incurs an expected number of $O(\sqrt{n \cdot (n+m) \cdot \log(n)/B} + \text{sort}(n+m))$ I/Os, while the deterministic variant incurs $O(\sqrt{n \cdot (n+m)/B} + \text{sort}(n+m) + ST(n,m))$ I/Os, where $ST(n,m)$ is the number of I/Os required for computing a spanning tree of a graph with n nodes and m edges. Arge et al. [**4**] show an upper bound of $O((1+\log\log(B \cdot n/m)) \cdot \text{sort}(n+m))$ I/Os for computing such a spanning tree.

Brodal et al. [**9**] gave a cache-oblivious algorithm for BFS achieving the same worst case I/O bounds as MM_BFS_D. Their preprocessing is similar to that in MM_BFS_D, except that it produces a hierarchical clustering using the cache-oblivious algorithms for sorting, spanning tree, Euler tour and list ranking. The BFS phase uses a data-structure that maintains a hierarchy of pools and provides the set of neighbors of the nodes in a BFS level efficiently.

The other known external-memory algorithms for BFS are restricted to special graphs classes like trees [**12**], grid graphs [**5**], planar graphs [**25**], outer-planar graphs [**23**], and graphs of bounded tree width [**24**].

1.3. Related Work. Ajwani et al. [**3**] showed that the usage of the two external memory algorithms MR_BFS and MM_BFS_R along with disk parallelism and pipelining can alleviate the I/O bottleneck of BFS on many large sparse graph classes, thereby making the BFS viable for these graphs. Even with just a single disk, they computed a BFS level decomposition of small diameter large graphs (around 256 million nodes and a billion edges) in a few *hours* and moderate and large diameter graphs in a few *days*, which otherwise would have taken a few *months* with IM_BFS. As for their relative comparison, MR_BFS performs better than MM_BFS_R on small-diameter random graphs saving a few *hours*. However, the better asymptotic worst-case I/O complexity of MM_BFS helps it to outperform MR_BFS for large diameter sparse graphs (computing in a few *days* versus a few *months*), where MR_BFS incurs close to its worst case of $\Omega(n)$ I/Os.

Independently, Christiani [**14**] gave a prototypical implementation of MR_BFS, MM_BFS_R as well as MM_BFS_D and reached similar conclusions regarding the comparative performance between MR_BFS and MM_BFS_R. Though their implementation of MR_BFS and MM_BFS_R is competetive and on some graph classes even better than [**3**], their experiments were mainly carried out on smaller graphs (up to 50 million nodes). Since their main goal was to design cache-oblivious BFS, they used cache-oblivious algorithms for sorting, minimum spanning tree and list ranking even for the cache-aware algorithm MM_BFS_D. As we discuss later, these algorithms slow down the deterministic preprocessing in practice, even though they have the same asymptotic I/O complexity as their external memory counterparts.

1.4. Our Contribution. Our contributions in this paper are the following:

- We improve upon the MR_BFS and MM_BFS_R implementation described in [**3**] by reducing the computational overhead associated with each BFS level, thereby improving the results for large diameter graphs.
- We discuss the various choices made for a fast MM_BFS_D implementation. This involved experimenting with various available external memory connected component and minimum spanning tree algorithms. Our partial re-implementation of the list ranking algorithm of [**31**] adapting it to the STXXL framework outperforms the other list ranking algorithms for the sizes of our interest. As for the Euler tour in the deterministic preprocessing, we compute the cyclic order of edges around the nodes using the STXXL sorting.
- We conduct a comparative study of MM_BFS_D with other external memory BFS algorithms and show that for most graph classes, MM_BFS_D outperforms MM_BFS_R. Also, we compare our BFS implementations with Christiani's implementations [**14**], which have some cache-oblivious subroutines. This gives us some idea of the loss factor that we will have to face for the performance of cache-oblivious BFS.
- We propose a heuristic for maintaining the pool in the BFS phase of MM_BFS. This heuristic improves the runtime of MM_BFS in practice, while preserving the worst case I/O bounds of MM_BFS.
- Putting everything together, we show that the BFS traversal can also be done on moderate and large diameter graphs in a few *hours*, which would have taken the implementations of [**3**] and [**14**] several *days* and IM_BFS several *months*. Also, on low diameter graphs, the time taken by our improved MR_BFS is around one-third of that in [**3**]. Towards the end, we summarize our results (Table 13) by giving the state of the art implementations of external memory BFS on different graph classes.

2. Improvements over the previous implementations of MR_BFS and MM_BFS_R

The computation of each level of MR_BFS involves sorting and scanning of neighbors of the nodes in the previous level. Even if there are very few elements to be sorted, there is a certain overhead associated with initializing the external sorters. In particular, while the STXXL stream sorter (with the flag STXXL_SMALL_INPUT_PSORT_OPT) does not incur an I/O for sorting less than B elements, it still requires to allocate some memory and does some computation for initialization. This overhead accumulates over all levels and for large diameter graphs, it dominates the running time. This problem is also inherited by the BFS phase of MM_BFS. Since in the pipelined implementation of [**3**], we do not know in advance the exact number of elements to be sorted, we can't switch between the external and the internal sorter so easily. In order to get around this problem, we first buffer the first B elements and initialize the external sorter only when the buffer is full. Otherwise, we sort it internally.

In addition to this, we make the graph representation for MR_BFS more compact. Except the source and the destination node pair, no other information is stored with the edges.

TABLE 1. Timing in minutes for sorting n elements using either CO_SORT or STXXL_SORT

n	CO_SORT	STXXL_SORT
256×10^6	21	8
512×10^6	46	13
1024×10^6	96	25

3. Deterministic preprocessing

As discussed in Section 1.2, the key components of the deterministic preprocessing of MM_BFS include sorting, minimum spanning tree and list ranking the Euler tour. In this section, we discuss the various design choices for each of these components.

3.1. External memory sorting. Implementations of many different algorithms for sorting large data sets in memory hierarchies are available, e.g. the cache-oblivious sorting algorithm (CO_SORT) given in [**10**] and its external memory counterpart provided by different external memory libraries like STXXL (STXXL_SORT) [**18**], and TPIE [**6**]. While CO_SORT provides tight asymptotic guarantees on all levels of memory hierarchy, it is a factor three to four slower than STXXL_SORT in practice for data-sizes that do not fit in the main memory. Our results shown in Table 1 are in conformity with that of Brodal et el. [**10**], where it is shown that the external memory sorting algorithm in the library TPIE [**6**] is better than their carefully implemented cache-oblivious sorting algorithm, when run on disk. We therefore use STXXL_SORT for our implementations.

3.2. External memory spanning forest. The problem of computing a minimum spanning tree (MST) of an undirected graph with nonnegative edge weights can be solved in $O(\text{sort}(m))$ expected I/O steps [**1**]. We use the implementation by Dementiev et al. (EM_MST) [**17**], which is based on contracting spanning forest edges. In each iteration it chooses a random node u and an edge $e = (u, v)$, where v has the smallest index among the edges incident to u. The algorithm replaces all edges (v, w) by (u, w), outputs e as one of the spanning forest edges and contracts e by "merging" u and v. When the number of nodes is reduced to $O(M)$ the spanning forest is computed by the semi-external adaptation of Kruskal's algorithm. Though the expected $O(\text{sort}(m)\lceil\log(n/M)\rceil)$ I/O complexity of this algorithm is inferior to the algorithm by Abello et al. [**1**], it uses at least a factor of four less I/Os [**17**] on most inputs on a "well-behaved" machine.

The deterministic preprocessing of Christiani [**14**] uses the cache-oblivious MST (CO_MST) algorithm [**1**]. Table 2 shows the total time required by Christiani's deterministic preprocessing [**14**] using CO_MST and the one in which CO_MST is replaced by EM_MST.

3.3. External memory list ranking and Euler tour. List ranking in external memory can be solved in $O(\text{sort}(n))$ I/Os [**13**]. We found the algorithm by Sibeyn [**31**] promising as it has low constant factors in its I/O complexity. The algorithm splits the input list into sublists of size $O(M)$ and goes through the data in a wavelike manner. For all elements of the current sublist, it follows the links running through the elements of the same sublist and updates the information on

TABLE 2. Timing in hours required by deterministic preprocessing of Christiani's implementation using either CO_MST or EM_MST.

Graph class	CO_MST	EM_MST
Random graph; $n = 2^{28}$, $m = 2^{30}$	107	35
Line graph with contiguous disk layout (Simple Line); $n = 2^{28}$	38	16
Line graph with random disk layout (Random Line); $n = 2^{28}$	47	22

their final element and the number of links to it. For all elements with links running outside the current sublist, the required information is requested from the sublists containing the elements to which they are linked. *Bucketing* and *lazy processing* of the requests and the answers to the sublists, i.e., storing them in one common stack and processing them only when the wave through the data hits the corresponding sublist, make the implementation [**31**] superior to algorithms based on independent set removal by a factor of about four. Unfortunately, Sibeyn's implementation relies on the operating system for I/Os and does not guarantee that the top blocks of all the stacks remain in the internal memory, which is a necessary assumption for the asymptotic analysis of the algorithm. Besides, its reliance on internal arrays and swap space puts a restriction on the size of the lists it can rank. The deeper integration of the algorithm in the STXXL framework, using the STXXL stacks and vectors in particular, makes it possible to obtain a scalable solution, which could handle graph instances of the size we require while keeping the theoretical worst case bounds.

The cache-oblivious implementation [**14**] uses the algorithm based on independent set removal [**13**] for list ranking. While it takes around 14.3 *hours* for ranking 2^{29} element random list using 3 GB RAM, our adaptation of Sibeyn's algorithm takes less than 40 *minutes* in the same setting.

Recall that in order to construct the Euler tour around the bi-directional minimum spanning tree, each node chooses a cyclic order of its neighbors. For every edge (u, v), its successor is defined to be the edge (v, w) (u may be the same as w) such that in the cyclic order of neighors of v, u is followed by w. In one scan of the edges of the bi-directional tree, each edge is linked to its successor. The linear ordering induced by the successor function constitutes the Euler tour. The position of an edge in this tour is computed using our adaptation of Sibeyn's list ranking algorithm in the STXXL framework. The Euler tour is then subdivided into chunks of size $\max\{1, \sqrt{\frac{n \cdot B}{n+m}}\}$, duplicates eliminated using STXXL_SORT and then used for partitioning the graph.

3.4. Remark on the shape of the spanning tree. The shape of the computed spanning tree can have a significant impact on the clustering and the disk layout of the adjacency list after the deterministic preprocessing, and consequently on the BFS phase. For instance, in the case of the square grid graphs, a spanning tree containing a list with elements in a snake-like row major order produces long and narrow clusters, while a "random" spanning tree is likely to result in clusters with low diameters. Such a "random" spanning tree can be attained by assigning

TABLE 3. Time taken (in hours) by the BFS phase of MM_BFS_D with long and random clustering

Graph class	n	m	Long clusters	Random clusters
Grid($2^{14} \times 2^{14}$)	2^{28}	2^{29}	51	28

random weights to the edges of the graph and then computing a minimum spanning tree or by randomly permuting the indices of the nodes. The nodes in the long and narrow clusters tend to stay longer in the pool and therefore, their adjacency lists are scanned more often. This causes the pool to grow external and results in larger I/O volume. On the other hand, low diameter clusters are evicted from the pool sooner and are scanned less often reducing the I/O volume of the BFS phase. Consequently as Table 3 shows, the BFS phase of MM_BFS_D takes only 28 hours with clusters produced by "random" spanning tree, while it takes 51 hours with long and narrow clusters.

4. A heuristic for maintaining the pool

As noted in Section 1.2, the asymptotic improvement and the performance gain in MM_BFS over MR_BFS is obtained by decomposing the graph into low diameter clusters and maintaining an efficiently accessible pool of adjacency lists that will be required in the next few levels. Whenever the first node of a cluster is visited during the BFS, the remaining nodes of this cluster will be reached soon after and hence, this cluster is loaded into the pool. For computing the neighbors of the nodes in the current level, we just need to scan the pool and not the entire graph. Efficient management of this pool is thus, crucial for the performance of MM_BFS. In this section, we propose a heuristic for efficient management of the pool, while keeping the worst case I/O bounds of MM_BFS.

For many large diameter graphs, the pool fits into the internal memory most of the time. However, even if the number of edges in the pool is not so large, scanning all the edges in the pool for each level can be computationally quite expensive. Hence, we keep a portion of the pool that fits in the internal memory as a multimap hash table. Given a node as a key, it returns all the nodes adjacent to the current node. Thus, to get the neighbors of a set of nodes we just query the hash function for those nodes and delete them from the hash table. For loading the cluster, we just insert all the adjacency lists of the cluster in the hash table, unless the hash table has already $O(M)$ elements.

Recall that after the deterministic preprocessing, the elements are stored on the disk in the order in which they appear on the Euler tour around a spanning tree of the input graph. The Euler tour is then chopped into clusters with $\max\{1, \sqrt{\frac{n \cdot B}{n+m}}\}$ elements (before the duplicate removal) ensuring that the maximum distance between any two nodes in the cluster is at most $\max\{1, \sqrt{\frac{n \cdot B}{n+m}}\} - 1$. However, the fact that the contiguous elements on the disk are also closer in terms of BFS levels is not restricted to intra-cluster adjacency lists. The adjacency lists that come alongside the requisite cluster will also be required soon and by caching these other adjacency lists, we can save some I/Os in the future. This caching is particularly beneficial when the pool fits in the internal memory. Note that we still load the

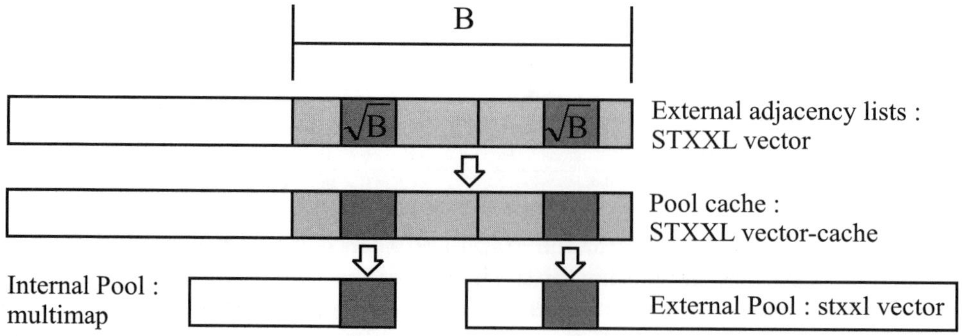

FIGURE 4. Schema depicting the implementation of our heuristic

$\max\{1, \sqrt{\frac{n \cdot B}{n+m}}\}$ node clusters in the pool, but keep the remaining elements of the block in the pool-cache. For line graphs, this means that we load $O(\sqrt{B})$ nodes in the internal pool, while keeping the remaining $O(B)$ adjacency lists which we get in the same block, in the pool-cache, thereby reducing the I/O complexity for the BFS traversal on line graphs to the computation of a spanning tree.

We represent the adjacency lists of nodes in the graph as a STXXL vector. STXXL already provides a fully associative vector-cache with every vector. Before doing an I/O for loading a block of elements from the vector, it first checks if the block is already there in the vector-cache. If so, it avoids the I/O loading the elements from the cache instead. Increasing the vector-cache size of the adjacency list vector with a layout computed by the deterministic preprocessing and choosing the replacement policy to be LRU provides us with an implementation of the pool-cache. Figure 4 depicts the implementation of our heuristic.

5. Experiments

Configuration. We have implemented the algorithms in C++ using the g++ 4.02 compiler (optimization level -O3) on the *GNU/Linux* distribution with a 2.6 *kernel* and the external memory library STXXL version 0.77. Our experimental platform has two 2.0 GHz Opteron processors, 3 GB of RAM, 1 MB cache and 250 GB Seagate Baracuda hard-disks [**32**]. These hard-disks have 8 MB buffer cache. The average seek time for read and write is 8.0 and 9.0 msec, respectively, while the sustained data transfer rate for outer zone (maximum) is 65 MByte/s. This means that for graphs with 2^{28} nodes, n random read and write I/Os will take around 600 and 675 hours, respectively. In order to compare better with the results of [**3**], we restrict the available memory to 1 GB for our experiments and use only one processor and one disk.

First, we show the comparison between improved MM_BFS_R and MR_BFS with the corresponding implementations in [**3**]. Then we compare our implementation of MM_BFS_D (without our heuristic) with Christiani's implementation based on cache-oblivious routines. Finally, we look at the relative performance of improved versions of MR_BFS, MM_BFS_R and MM_BFS_D. We summarize this section by highlighting the best algorithms for each graph class and its run-time. Note that some of the results shown in this section have been interpolated using the symmetry in the graph structure.

TABLE 4. Timing in hours taken for BFS by the two MM_BFS_R implementations

Graph class	n	m	MM_BFS_R of [3]		Improved	
			Phase 1	Phase 2	Phase 1	Phase 2
Random	2^{28}	2^{30}	5.1	4.5	5.2	3.8
MM_worst	$\sim 4.3 \cdot 10^7$	$\sim 4.3 \cdot 10^7$	6.7	26	5.2	18
MR_worst	2^{28}	2^{30}	5.1	45	4.3	40
Grid ($2^{14} \times 2^{14}$)	2^{28}	2^{29}	7.3	47	4.4	26
Simple Line	2^{28}	$2^{28}-1$	85	191	55	2.9
Random Line	2^{28}	$2^{28}-1$	81	203	64	25
Webgraph	$\sim 1.4 \cdot 10^8$	$\sim 1.2 \cdot 10^9$	6.2	3.2	5.8	2.8

Graph classes. We consider the same graph classes as in [**3**] - Random, Grid, MR_worst graph, MM_worst graph, line graphs with different layouts and the webgraph. They cover a broad spectrum of different performances of external memory BFS algorithms.

Random graph: On a n node graph, we randomly select m edges with replacement (i.e., m times selecting a source and target node such that source $\neq$ target) and remove duplicate edges to obtain random graphs.

MR_worst graph: This graph consists of B levels, each having $\frac{n}{B}$ nodes, except the level 0 which contains only the source node. The edges are randomly distributed between consecutive levels, such that these B levels approximate the BFS levels. The initial layout of the nodes on the disk is random. This graph causes MR_BFS to incur its worst case of $\Omega(n)$ I/Os.

Grid graph ($x \times y$): It consists of a $x \times y$ grid, with edges joining the neighbouring nodes in the grid.

MM_BFS worst graph: This graph causes MM_BFS_R to incur its worst case of $\Theta(n \cdot \sqrt{\frac{\log n}{B}} + \text{sort}(n))$ I/Os ($m = O(n)$ for this graph).

Line graphs: A line graph consists of n nodes and $n-1$ edges such that there exist two nodes u and v, with the path from u to v consisting of all the $n-1$ edges. We took two different initial layouts:

- Simple, in which all blocks consists of B consecutively lined nodes
- Random, in which the arrangement of nodes on disk is given by a random permutation.

Web graph: As an instance of a real world graph, we consider an actual crawl of a part of the world wide web in 2001 [**33**], where an edge represents a hyperlink between two sites. This graph has around 130 million nodes and 1.4 billion edges. It has a core which consists of most of its nodes and behaves like random graph.

Comparing MM_BFS_R. Table 4 shows the improvement that we achieved in MM_BFS_R. As Table 5 shows, these improvements are achieved by reducing the computation time per level in the BFS phase. On I/O bound random graphs, the improvement is just around 15%, while on computation bound line graphs with random disk layout, we improve the running time of the BFS phase from around 200 hours to 25 hours. Our implementation of the randomized preprocessing in the case of simple line graphs additionally benefits from the way clusters are laid out on the disk as this layout reflects the order in which the nodes are visited by the BFS. This reduces the total running time for the BFS phase of MM_BFS_R on

TABLE 5. I/O wait time and the total time in hours for the BFS phase of the two MM_BFS_R implementations on moderate to large diameter graphs

Graph class	n	m	MM_BFS_R of [3]		Improved	
			I/O wait	Total	I/O wait	Total
MM_worst	$\sim 4.3 \cdot 10^7$	$\sim 4.3 \cdot 10^7$	13	26	16	18
Grid ($2^{14} \times 2^{14}$)	2^{28}	2^{29}	46	47	24	26
Simple Line	2^{28}	$2^{28} - 1$	0.5	191	0.05	2.9
Random Line	2^{28}	$2^{28} - 1$	21	203	21	25

TABLE 6. Timing in hours taken for BFS by the two MR_BFS implementations

Graph class	n	m	MR_BFS of [3]		Improved	
			I/O wait	Total	I/O wait	Total
Random	2^{28}	2^{30}	2.4	3.4	1.2	1.4
Webgraph	$\sim 135 \times 10^6$	$\sim 1.18 \times 10^9$	3.7	4.0	2.5	2.6
MM_worst	$\sim 42.6 \times 10^6$	$\sim 42.6 \times 10^6$	25	25	13	13
Simple line	2^{28}	$2^{28} - 1$	0.6	10.2	0.06	0.4

TABLE 7. Timing in hours for computing the deterministic preprocessing of MM_BFS by the two implementations of MM_BFS_D

Graph class	n	m	Christiani's implementation	Our implementation
Random graph	2^{28}	2^{30}	107	5.2
Random Line	2^{28}	$2^{28} - 1$	47	3.2

TABLE 8. Timing in hours for the BFS phase of MM_BFS by the two implementations of MM_BFS_D (without heuristic)

Graph class	n	m	Christiani's implementation	Our implementation
Random graph	2^{28}	2^{30}	16	3.4
Random Line	2^{28}	$2^{28} - 1$	0.5	2.8

simple line graphs from around 190 hours to 2.9 hours. The effects of caching are also seen in the I/O bound BFS phase on the grid ($2^{14} \times 2^{14}$) graphs, where the I/O wait time decreases from 46 hours to 24 hours.

Comparing MR_BFS. Improvements in MR_BFS are shown in the Table 6. On random graphs where MR_BFS performs better than the other algorithms, we improve the runtime from 3.4 hours to 1.4 hours. Similarly for the web-crawl based graph, the running time reduces from 4.0 hours to 2.6 hours. The other graph class where MR_BFS outperforms MM_BFS_R is the MM_worst graph and here again, we improve the performance from around 25 hours to 13 hours.

TABLE 9. Timing in hours taken by our implementations of different external memory BFS algorithms.

Graph class	MR_BFS	MM_BFS_R	MM_BFS_D
Random graph	1.4	8.9	8.7
Random Line	4756	89	3.6

TABLE 10. I/O volume (in GB) required in the preprocessing phase by the two variants of MM_BFS

Graph class	n	m	Randomized	Deterministic
Random graph	2^{28}	2^{30}	500	630
Random Line	2^{28}	$2^{28}-1$	10500	480

TABLE 11. Preprocessing time (in hours) required by the two variants of MM_BFS, with the heuristic

Graph class	n	m	Randomized	Deterministic
Random graph	2^{28}	2^{30}	5.2	5.2
Random Line	2^{28}	$2^{28}-1$	64	3.2

Penalty for cache-obliviousness. We compared the performance of our implementation of MM_BFS_D (without the heuristic) with Christiani's implementation [**14**] based on cache-oblivious subroutines. Table 7 and 8 show the results of the comparison on the two extreme graph classes - random graphs and line graphs with random layout on disk - for the preprocessing and the BFS phase respectively. We observed that on both graph classes, the preprocessing time required by our implementation is significantly less than the one by Christiani. While pipelining helps the BFS phase of our implementation on random graphs, it becomes a liability on line graphs as it brings extra computation cost per level.

We suspect that these performance losses are inherent in cache-oblivious algorithms to a certain extent and will be carried over to the cache-oblivious BFS implementation.

Comparing MM_BFS_D with other external memory BFS algorithm implementations. Table 9 shows the performance of our implementations of different external memory BFS algorithms with the heuristic for maintaining the pool. While MR_BFS performs better than the other two on random graphs saving a few *hours*, our implementation of MM_BFS_D with the heuristic outperforms MR_BFS and MM_BFS_R on line graphs with random layout on disk saving a few *months* and a few *days*, respectively. Random line graphs are an example of a tough input for external memory BFS as they not only have a large number of BFS levels, but also their layout on the disk makes the random accesses to adjacency lists very costly. Also, on moderate diameter grid graphs, MM_BFS_D, which takes 21 hours, outperforms MM_BFS_R and MR_BFS. It is interesting to note that Christiani [**14**] reached a different conclusion regarding the relative performance of MM_BFS_D and MM_BFS_R. As noted before, this is because of the cache-oblivious routines used in his implementation.

TABLE 12. Time taken (in hours) by the two phases of MM_BFS_D with our heuristic

Graph class	n	m	MM_BFS_D	
			Phase1	Phase2
Random	2^{28}	2^{30}	5.2	3.4
Webgraph	$\sim 1.4 \cdot 10^8$	$\sim 1.2 \cdot 10^9$	3.3	2.4
Grid ($2^{21} \times 2^7$)	2^{28}	$\sim 2^{29}$	3.6	0.4
Grid ($2^{27} \times 2$)	2^{28}	$\sim 2^{28} + 2^{27}$	3.2	0.6
Simple Line	2^{28}	$2^{28} - 1$	2.6	0.4
Random Line	2^{28}	$2^{28} - 1$	3.2	0.5

On large diameter sparse graphs such as line graphs, the randomized preprocessing scans the graph $\Omega(\sqrt{B})$ times, and thus incurring an expected number of $O(\sqrt{n \cdot (n+m) \cdot \log(n)/B})$ I/Os. On the other hand, the I/O complexity of the deterministic preprocessing is $O((1 + \log\log(B \cdot n/m)) \cdot \text{sort}(n+m))$, dominated by the spanning tree computation. Note that the Euler tour computation followed by list ranking only requires $O(\text{sort}(m))$ I/Os. This asymptotic difference shows in the I/O volume of the two preprocessing variants (Table 10), thereby explaining the better performance of the deterministic preprocessing over the randomized one (Table 11). On low diameter random graphs, the diameter of the clusters is small and consequently, the randomized variant scans the graph fewer times leading to less I/O volume. As compared to MM_BFS_R, MM_BFS_D provides dual advantages: First, the preprocessing itself is faster and second, for most graph classes, the partitioning is also more robust, thus leading to better worst-case runtimes in the BFS phase. The later is because the clusters generated by the deterministic preprocessing are of diameter at most $\max\{1, \sqrt{\frac{n \cdot B}{n+m}}\}$, while the ones by randomized preprocessing can have a larger diameter $(O(\sqrt{\frac{n \cdot B \cdot \log n}{n+m}}))$ causing adjacency lists to be scanned more often. Also, MM_BFS_D benefits much more from our caching heuristic than MM_BFS_R as the deterministic preprocessing gathers neighboring clusters of the graph on contigous locations in the disk.

Results with heuristic. Table 12 shows the results of MM_BFS_D with our heuristic on different graph classes. On moderate diameter grid graphs as well as large diameter random line graphs, MM_BFS_D with our heuristic provides the fastest implementation of BFS in the external memory.

Summary. Table 13 gives the current state of the art implementations of external memory BFS on different graph classes.

Our improved MR_BFS implementation outperforms the other external memory BFS implementations on low diameter graphs or when the nodes of a graph are arranged on the disk in the order required for BFS traversal. For random graphs with 256 million nodes and a billion edges, our improved MR_BFS performs BFS in just 1.4 hours. Similarly, improved MR_BFS takes only 2.6 hours on webgraphs (whose runtime is dominated by the short diameter core) and 0.4 hours on line graph with contigous layout on disk. On moderate diameter square grid graphs, the total time for BFS is brought down from 54.3 hours for MM_BFS_R implementation in [**3**] to 21 hours for our implementation of MM_BFS_D with heuristics, an

TABLE 13. The best total running time (in hours) for BFS traversal on different graphs with the best external memory BFS implementations; Entries like $> 25x$ denote that this algorithm takes more than 25 times the time taken by the best algorithm for this input instance

Graph class	n	m	MR_BFS	MM_BFS_R	MM_BFS_D
Random	2^{28}	2^{30}	**1.4**	7×	6×
Webgraph	$\sim 1.4 \cdot 10^8$	$\sim 1.2 \cdot 10^9$	**2.6**	3.5×	2×
Grid ($2^{14} \times 2^{14}$)	2^{28}	2^{29}	2.5×	1.25×	**21**
Grid ($2^{21} \times 2^7$)	2^{28}	$\sim 2^{29}$	>100×	>10×	**4.0**
Grid ($2^{27} \times 2$)	2^{28}	$\sim 2^{28} + 2^{27}$	>500×	>25×	**3.8**
Simple Line	2^{28}	$2^{28} - 1$	**0.4**	7×	7×
Random Line	2^{28}	$2^{28} - 1$	>1300×	>75×	**3.6**
Max			~ **1/2 year**	~ **1 week**	1 **day**

improvement of more than 60%. For large diameter graphs like random line graphs, MM_BFS_D along with our heuristic computes the BFS in just about 3.6 *hours*, which would have taken the MM_BFS_R implementation in [**3**] around 12 *days* and MR_BFS and IM_BFS a few *months*, an improvement by a factor of more than 75 and 1300, respectively.

6. Discussion

We implemented the deterministic variant of MM_BFS and showed its comparative analysis with other external memory BFS algorithms. Together with the improved implementations of MR_BFS and MM_BFS_R and our heuristic for maintaining the pool, it provides viable BFS traversal on different classes of massive sparse graphs. In particular, we obtain an improvement factor between 75 and 1300 for line graphs with random disk layout over the previous external memory implementations of BFS.

Acknowledgements

We are grateful to Rolf Fagerberg and Frederik Juul Christiani for providing us their code. Also thanks are due to Dominik Schultes and Roman Dementiev for their help in using the external MST implementation and STXXL, respectively. The authors also acknowledge the usage of the computing resources of the University of Karlsruhe.

References

[1] J. Abello, A. Buchsbaum, and J. Westbrook. A functional approach to external graph algorithms. *Algorithmica 32(3)*, pages 437–458, 2002.

[2] A. Aggarwal and J. S. Vitter. The input/output complexity of sorting and related problems. *Communications of the ACM, 31(9)*, pages 1116–1127, 1988.

[3] D. Ajwani, R. Dementiev, and U. Meyer. A computational study of external-memory BFS algorithms. *SODA*, pages 601–610, 2006.

[4] L. Arge, G. Brodal, and L. Toma. On external-memory MST, SSSP and multi-way planar graph separation. *SWAT*, volume 1851 of *LNCS*, pages 433–447. Springer, 2000.

[5] L. Arge, L. Toma, and J. S. Vitter. I/O-efficient algorithms for problems on grid-based terrains. *ALENEX*, 2000.

[6] L. Arge et.al. http://www.cs.duke.edu/TPIE/.

[7] D. K. Blandford, G. E. Blelloch, and I. A. Kash. Compact representations of separable graphs. *SODA*, pages 679–688, 2003.

[8] D. K. Blandford, G. E. Blelloch, and I. A. Kash. An experimental analysis of a compact graph representation. *ALENEX*, 2004.

[9] G. Brodal, R. Fagerberg, U. Meyer, and N. Zeh. Cache-oblivious data structures and algorithms for undirected breadth-first search and shortest paths. *SWAT*, volume 3111 of *LNCS*, pages 480–492. Springer, 2004.

[10] G. Brodal, R. Fagerberg, and K. Vinther. Engineering a cache-oblivious sorting algorithm. *ALENEX*, pages 4–17. SIAM, 2004.

[11] G. S. Brodal and R. Fagerberg. Cache oblivious distribution sweeping. *ICALP*, pages 426–438, 2002.

[12] A. Buchsbaum, M. Goldwasser, S. Venkatasubramanian, and J. Westbrook. On external memory graph traversal. *SODA*, pages 859–860. ACM-SIAM, 2000.

[13] Y. J. Chiang, M. T. Goodrich, E. F. Grove, R. Tamasia, D. E. Vengroff, and J. S. Vitter. External memory graph algorithms. *SODA*, pages 139–149. ACM-SIAM, 1995.

[14] Frederik Juul Christiani. Cache-oblivious graph algorithms, 2005. Master's thesis, Department of Mathematics and Computer Science, University of Southern Denmark.

[15] T. H. Cormen, C.E. Leiserson, and R.L. Rivest. *Introduction to Algorithms*. McGraw-Hill, 1990.

[16] S. Baase. Introduction to parallel connectivity, list ranking, and euler tour techniques. In J. H. Reif, editor, Synthesis of Parallel Algorithms, chapter 2, pages 61-114. Morgan Kaufmann, 1993.

[17] R. Dementiev, P. Sanders, D. Schultes, and J. Sibeyn. Engineering an external memory minimum spanning tree algorithm. *TCS*, pages 195–208. Kluwer, 2004.

[18] R. Dementiev, L. Kettner, P. Sanders. STXXL: Standard Template Library for XXL Data Sets. *ESA*, volume 3669 of *LNCS*, pages 640–651. Springer, 2005.

[19] S. Edelkamp, S.. Jabbar, and S. Schrödl. External A*. *KI*, volume 3238 of *LNAI*, pages 226–240. Springer, 2004.

[20] M. Frigo, C. E. Leiserson, H. Prokop, and S. Ramachandran. Cache-oblivious algorithms. *FOCS*, pages 285–297. IEEE Computer Society Press, 1999.

[21] A. Goldberg and R. Werneck. Computing point-to-point shortest paths from external memory. *ALENEX*. SIAM, 2005.

[22] P.C. Guide. Disk Latency. http://www.pcguide.com/ref/hdd/perf/perf/spec/posLatency-c.html.

[23] A. Maheshwari and N. Zeh. External memory algorithms for outerplanar graphs. *ISAAC*, volume 1741 of *LNCS*, pages 307–316. Springer, 1999.

[24] A. Maheshwari and N. Zeh. I/O-efficient algorithms for graphs of bounded treewidth. *SODA*, pages 89–90. ACM-SIAM, 2001.

[25] A. Maheshwari and N. Zeh. I/O-optimal algorithms for planar graphs using separators. *SODA*, pages 372–381. ACM-SIAM, 2002.

[26] K. Mehlhorn and U. Meyer. External-memory breadth-first search with sublinear I/O. *ESA*, volume 2461 of *LNCS*, pages 723–735. Springer, 2002.

[27] K. Mehlhorn and S. Naher. The LEDA Platform of Combinatorial and Geometric Computing. Cambridge University Press, 1999.

[28] K. Munagala and A. Ranade. I/O-complexity of graph algorithms. *SODA*, pages 687–694. ACM-SIAM, 1999.

[29] M. Najork and J. Wiener. Breadth-first search crawling yields high-quality pages. *WWW*, pages 114–118, 2001.

[30] V. Shkapenyuk and T. Suel. Design and implementation of a high-performance distributed web crawler. *ICDE*. IEEE, 2002.

[31] J. F. Sibeyn. From parallel to external list ranking, 1997. Technical report, Max Planck Institut für Informatik, Saarbrücken, Germany.

[32] Seagate Technology. http://www.seagate.com/cda/products/discsales/marketing/detail/0,1081,628,00.html.

[33] The stanford webbase project. http://www-diglib.stanford.edu/~testbed/doc2/WebBase/.

Deepak Ajwani, Max-Planck-Institut für Informatik, Stuhlsatzenhausweg 85, 66123 Saarbrücken, Germany
E-mail address: ajwani@mpi-inf.mpg.de

Ulrich Meyer, Johann Wolfgang Goethe-Universität Frankfurt, 304, Robert-Mayer-Str. 11-15, 60325 Frankfurt/Main, Germany
E-mail address: umeyer@ae.cs.uni-frankfurt.de

Vitaly Osipov, Universität Karlsruhe (TH), Am Fasanengarten 5, 76131 Karlsruhe, Germany
E-mail address: osipov@ira.uka.de

Engineering Label-Constrained Shortest-Path Algorithms

Chris Barrett, Keith Bisset, Martin Holzer,
Goran Konjevod, Madhav Marathe, and Dorothea Wagner

Dept. of Computer Science and Virginia Bioinformatics Institute, Virginia Tech
{cbarrett, kbisset, mmarathe}@vbi.vt.edu
Fakultät für Informatik, Universität Karlsruhe (TH)
{mholzer, wagner}@ira.uka.de
Dept. of Computer Science and Engineering, Arizona State University
goran@asu.edu

Abstract. We consider a generalization of the shortest-path problem: given an alphabet Σ, a graph G whose edges are weighted and Σ-labeled, and a regular language $L \subseteq \Sigma^*$, the *L-constrained shortest-path problem* consists of finding a shortest path p in G such that the concatenated labels along p form a word of L. This definition allows to model, e. g., many traffic-planning problems. We present extensions of well-known speed-up techniques for the standard shortest-path problem, and conduct an extensive experimental study of their performance with various networks and language constraints. Our results show that depending on the network type, both goal-directed and bidirectional search speed up the search considerably, while combinations of these do not.

1 Introduction

Consider a *multimodal* road network, with roads differentiated by categories (highways, local streets etc.) and find in that network a shortest path from a given start to a destination point that uses highway at most once (thus avoiding on- and off-ramps). Another example is the *k-similar-path* problem, where we want to compute two shortest paths between the same pair of vertices such that the second path reuses at most k edges of the first (this can be useful to avoid traffic jams in vehicle routing).

To formalize such problems, we augment the network edges with appropriate labels and model the given restriction as a formal language. The labels of the edges on a shortest path concatenated must then form an element of the language. A detailed theoretical study of this *(formal-) language-constrained shortest-path problem* (LCSP) was undertaken in [4], where also a generalization of Dijkstra's algorithm to solve this problem is given, and in [3] an implementation of this algorithm was tested for the special case of linear regular languages (LINLCSP).

Building on this earlier work, we now consider the LCSP with *arbitrary* regular expressions (REGLCSP): we propose a concrete implementation of an algorithm solving this extended problem, and present adaptations of speed-up

R. Fleischer and J. Xu (Eds.): AAIM 2008, LNCS 5034, pp. 27–37, 2008.

techniques designed for the standard, or *unimodal*, shortest-path problem to our setting. In a systematic experimental study on realistic transportation networks we investigate the applicability of *goal-directed search*, *Sedgewick-Vitter heuristic*, and *bidirectional search* as well as combinations of these. We also explore the scalability of our implementation by applying them to instances of increasing size as well as involving languages of varying complexity (both linear and general regular expressions).

Our experiments show that goal-directed search and the Sedgewick-Vitter heuristic yield substantial speed-up for RegLCSP on all European and some US road networks, while bidirectional search performs well especially on railway networks. Unlike in the unimodal case, combinations of bidirectional search with one of the other techniques do not perform any better than each variant applied separately. Experiments with k-similar paths confirm growing speed-up factors with increasing NFA sizes.

1.1 Related Work

Research on generalizations of the standard shortest-path problem has traditionally focused on the extension of Dijkstra's algorithm [5] to time-dependent cost functions (cf. [8]), while comparatively little work has been done on constraints restricting the set of feasible paths. There are reports on studies of *multimodal*, or *intermodal*, shortest paths in transportation science literature, however, generally of limited applicability. Regular languages as a model for constrained-shortest-path problems were first suggested in [9], and applications to database queries described in [13,7]. Our initial motivation for studying LCSP problems comes from the TRANSIMS project [1,2]. A theoretical study on algorithmic and complexity-related issues can be found in [4].

In [3], an algorithm for the RegLCSP problem with time-dependent edge weights obeying the FIFO property is described, where an *implicit* representation (cf. Section 3.1) is used; experimental results are presented only for LinLCSP. Also for time-dependent RegLCSP, [11] gives an implicit algorithm, running in linear time for FIFO weight functions. This algorithm is extended in [12] to allow for turn penalties; some experimental results are reported in both these papers. The present work deals with RegLCSP, where the focus is on an extensive experimental evaluation of speed-up techniques with diverse classes of larger networks.

2 Foundation

In this section we formally define the regular-language-constrained shortest-path problem, and describe two out of many algorithmic problems, *multimodal plans* and *k-similar paths*, that can be tackled using our problem formulation.

2.1 Problem Statement

The *regular-language-constrained shortest-path problem* (RegLCSP) is defined as follows. Given a finite alphabet Σ, a graph—also referred to by network—$G = (V, E, c, \ell)$ with cost function $c : E \to \mathbb{R}_+$ and label function $\ell : E \to \Sigma$, and a regular language $L \subseteq \Sigma^*$. For a *query* $(s, d) \in V \times V$ find a shortest s-d-path $p = (e_1, e_2, \dots, e_k)$ in G such that $\ell(p) \in L$, where $\ell(p)$ is the concatenation $\ell(e_1) \cdot \ell(e_2) \cdots \ell(e_k)$ of the labels of p's edges. The cost—or length—of p is the sum of costs of p's edges. By Kleene's theorem, a regular language can be represented through a nondeterministic finite automaton (NFA), allowing for a concise description.

2.2 Applications

We describe two applications of the RegLCSP problem, which are also respected in our experiments (other examples include turn complexity, counting constraints, and trip chaining; cf. [3] for an overview).

Multimodal Plans. Consider a traveler who wants to take a bus from a start s to a destination point d and suppose transfers are undesirable, while walks from s to a bus stop and from a bus stop to d be allowed. To solve such a task, add to the given road network a vertex for every bus stop and an edge between each consecutive pair of stops. Label the edges according to the modes of travel allowed (e. g., c for car travel; w for walking on sidewalks and pedestrian bridges; b for bus transit). Now the traveler's restriction can be modeled as $w^*b^*w^*$, if we make sure that the network contains a zero-length w-edge for each change of bus.

k-Similar Paths. We want to consecutively route two (or more) vehicles from s to d such that the second uses at most k of the edges passed by the first one. This can be useful, e. g., to plan for a travel group different transfers between two fixed points. Note that the second path thus found may, depending on the network and the choice of k, be of greater length. To do this, find a shortest s-d-path p in the given network, label p's edges by t (for *taken*), the remaining ones by f (for *free*), and solve the s-d-query again for the expression $f^*(t \cup f^*)^k f^*$.

3 Algorithms

We now show how RegLCSP can be solved through a product network constructed from given network and NFA, and present some adaptations of unimodal speed-up techniques to our multimodal algorithm.

3.1 Product Network

Consider the direct product of a weighted, Σ-labeled network $G = (V, E, c, \ell_G)$ and an NFA $A = (Q, \Sigma, \delta, q_0, F)$ with set Q of vertices/states, alphabet Σ, transition function δ, start vertex q_0, and set F of final vertices; let T be the set of state transitions $t = (q_1, q_2)$ with $\delta(q_1) = q_2$ and labels $\ell_A(t) \in \Sigma$. The product network $P = G \times A$ is defined to have vertex set $\{(v, q) \mid v \in V,\ q \in Q\}$

and edge set $\{(e,t) \mid e \in E,\, t \in T,\, \ell_G(e) = \ell_A(t)\}$. The cost of an edge $(e,t) \in P$ corresponds to $c(e)$.

Theorem 1. *Finding an L-constrained shortest path for some $L \subseteq \Sigma^*$ and $(s,d) \in V \times V$ is equivalent to finding a shortest path in the product network $P = G \times A$ from vertex (s, q_0) to (d, f) for some $f \in F$.*

For the proof, one need only observe that there is a one-to-one correspondence between paths in P starting at (s, q_0) and ending at some vertex (d, f) and s-d-paths in G whose labeling belongs to L (for details please refer to [4]). Theorem 1 immediately yields the **RegLCSP Algorithm**, which performs an s-d-shortest-path search in P.

Implementation. Obviously, a direct implementation of this algorithm would require $\Theta(|G| \cdot |A|)$ space. We therefore propose a more practical way without having to compute an explicit representation of P: the algorithm considers pairs $(v,q) \in P$ but to iterate over (v,q)'s outgoing edges, simultaneously accesses the adjacency lists of v and q. To do this efficiently, we store the outgoing edges of both G and A bundled by their labels and keep pointers to the first edge of each bundle. Now we need only iterate over all labels $l \in \Sigma$ and consider each combination of vertices v' reachable from v via an edge labeled l and q' reachable from q via an edge labeled l. This *implicit representation* reduces storage space to $\Theta(|G| + |A|)$, while time complexity does not increase by more than a constant factor.

3.2 Speed-Up Techniques

In order to improve the above-described RegLCSP Algorithm, we adopt several approaches designed to speed up unimodal Dijkstra's algorithm: the key idea is to apply the genuine technique to the product network.

Goal-Directed Search (go). For given source and destination vertices s and d, goal-directed search, or A^* *search*, modifies the edge costs so that during the search, edges pointing roughly towards d are preferred to those pointing away from it. The effect is that potentially fewer vertices (and edges) have to be *touched* before d is found. With networks featuring *distance metric*, this modification, $\bar{c}$, can typically be achieved through Euclidean distances $\underline{\text{dist}}(v,w)$ between vertices v and w:

$$\bar{c}(v,w) = c(v,w) - \underline{\text{dist}}(v,d) + \underline{\text{dist}}(w,d)$$

(in this case, c accounts for curves, bridges, etc., so that $\underline{\text{dist}}$ provides a lower bound). When using *travel*, or *time*, *metric*, letting $\underline{\text{dist}}'(\cdot,d) = \underline{\text{dist}}(\cdot,d)/v_{\max}$ with $v_{\max} = \max_{\{v,w\}\in E} \underline{\text{dist}}(v,w)/\, c(v,w)$ yields a feasible lower bound.

Sedgewick-Vitter Heuristic (sv). If we do not insist on exact shortest paths, a canonical extension of go is to bias the search to d even further: the Sedgewick-Vitter heuristic [10] uses as modified cost function $\bar{c}(v,w) = c(v,w) - \alpha \cdot \underline{\text{dist}}(v,d) + \alpha \cdot \underline{\text{dist}}(w,d)$ for some $\alpha \geq 1$, influencing the trade-off between gain in running time and path length increase. For previous work exploring this technique, cf. [6].

Bidirectional Search (`bi`). To reduce the search space, we run two simultaneous searches, *forward* and *backward*, starting from s and d, respectively; the expected improvement is a halving of the number of touched vertices[1]. A shortest s-d-path has been found when a vertex u is about to be scanned that has already been settled (i. e., its distance from the search's origin has become permanent) by the search in the other direction (note that the shortest path such found need not pass by u).

Combinations (`bi+`). We provide two variants for the combination of `bi` with either `go` or `sv`: *(1)* the cost function used for the backward search corresponds to that for the forward search; *or (2)* the cost function for both searches is the average of modified cost functions with respect to s and d, respectively:

$$\bar{c}(v,w) = c(v,w) + 1/2 \cdot (-\underline{\text{dist}}(v,d) + \underline{\text{dist}}(w,d) - \underline{\text{dist}}(w,s) + \underline{\text{dist}}(v,s)) .$$

4 Experimental Study

The empiric part of this work systematically investigates our implementation of the RegLCSP Algorithm and the speed-up techniques described in the previous section. It can be seen as both a continuation and extension of [3] in that the restriction to linear expressions is waived and besides goal-directed search and the Sedgewick-Vitter heuristic, bidirectional search and combinations of techniques are employed. Our main focus is on the suitability of each technique for several networks and NFAs and the speed-ups attainable. Moreover, two different problems that can be tackled by RegLCSP (cf. Section 2.2) are considered.

4.1 Setup

Our experiments are conducted using realistic networks, representing various US and European road as well as European railway networks[2] (cf. Table 1). The road networks are weighted with actual distances (not necessarily Euclidean lengths) and labeled with values reflecting road category (from 1 to 4 for US and from 1 to 15 for EU networks, ranking from fast highways to local/rural streets). The railway network represents trains and other means of public transportation, where vertices mark railway stations/bus stops and edges denote non-stop connections between two embarking points, weighted with average travel times and labeled from 0 to 9 for rapid Intercity Express trains to slower local buses. One important difference between the US and the European road data collections is that the former come undirected, while the latter are directed.

We apply several specific language constraints of varying complexity, listed in Table 2: for the US networks, we distinguish between enforced use of highway

[1] The forward search will explore roughly r^2 vertices to find an r-edge shortest path, while the searches are likely to meet when each has explored roughly $(r/2)^2$ vertices.

[2] The US networks are taken from the TIGER/LINE collection, available at `http://www.dis.uniroma1.it/~challenge9/data/tiger/`. The European road and railway data were provided courtesy of PTV AG, Karlsruhe, and HaCon, Hannover.

Table 1. Network sizes. Left: US road networks; right: European road and railway networks. For each network, a short key, its type (road/rail), and the numbers n and m of vertices and edges, respectively, are indicated.

key	network	type	n	m
AZ	Arizona	road	545111	665827
DC	Distr/Col	road	9559	14909
GA	Georgia	road	738879	869890

key	network	type	n	m
LUX	Luxembourg	road	30087	70240
CHE	Switzerland	road	586025	1344496
DEU	Germany	rail	6900	24223

Table 2. Left: language constraints used with different networks (from top to bottom: US road, European road and railway). For each NFA, a short key, regular expression recognized by the NFA, informal description, and the numbers n and m of vertices and edges, respectively, are indicated. Right: complex NFA representing for a decomposition of the alphabet $\Sigma = \Sigma_1 \cup \Sigma_2$ the *unrestricted* expression, Σ^*.

	key	expression	description	n	m
US rd.	$\mathcal{H}$	$(3 \cup 4)^*(1 \cup 2)^+ \cdot (3 \cup 4)^*$	highway usage	3	10
	$\mathcal{R}$	$(2 \cup 3 \cup 4)^*$	regional transfer	1	3
	$\mathcal{L}$	4^*	local streets	1	1
EU rd.	$\mathcal{S}$	$(1 \cup \ldots \cup 15)^*$	unrestr. (simple)	1	15
	$\mathcal{C}$	$(1 \cup \ldots \cup 15)^*$	unrestr. (complex)	5	72
	$\mathcal{L}$	$(10 \cup 11 \cup 12)^*$	local streets	1	3
EU r/w	$\mathcal{S}$	$(0 \cup \ldots \cup 9)^*$	unrestr. (simple)	1	10
	$\mathcal{C}$	$(0 \cup \ldots \cup 9)^*$	unrestr. (complex)	5	50
	$\mathcal{N}$	$(1 \cup \ldots \cup 9)^*$	normal-speed trains	1	9

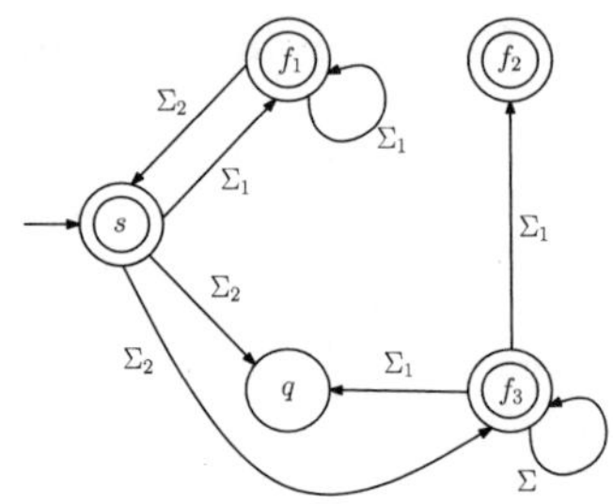

(interstate or national), regional transfer (all categories but interstates), and use of local/rural streets only; for the European networks, we employ two different NFAs imposing no restriction at all (besides the 'canonical', $\mathcal{S}$, also an 'artificially made-complex' one, $\mathcal{C}$) as well as such restricting to local streets and avoiding high-speed trains (the latter usually being a little more expensive), respectively.

To measure the performance of each speed-up technique T, we compute the ratio $\mathrm{tv}_{\mathrm{pl}}/\mathrm{tv}_T$ of *touched vertices* (product vertices added to the priority queue), where $\mathrm{tv}_{\mathrm{pl}}$ and tv_T stand for the number of touched vertices with *plain Dijkstra* (i. e., pure REGLCSP Algorithm) and with T, respectively. This definition of *speed-up* both is machine independent and proved to reflect actual running times quite precisely. Our code was compiled with the GCC (version 3.4) and executed on several 2- or 4-core AMD Opteron machines with between 8 and 32 GB of main memory. Unless otherwise noted, each series consists of 1000 queries.

4.2 Multimodal Routing

The term *multimodal* here is used in an extended sense since depending on the network type, it may refer either to multiple road categories or train classes. For comparability reasons, we explore the exact algorithms and the Sedgewick-Vitter heuristic separately.

Exact Algorithms. Assessment of our results is done in two steps, where we first provide a synopsis of the overall outcome and then detail on a few networks under the aspect of path lengths.

Synopsis. Figure 1 shows for each combination of network, NFA, and algorithm the average speed-up in terms of touched vertices in the product graph; the algorithms are distinguished along the x-axis (using the abbreviations introduced in Section 3.2, where `pl` stands for plain Dijkstra) and the NFAs are marked by their short keys (cf. Table 2). As a general result, it can be stated that both variants of the bidirectional/goal-directed combination (lumped together under `bi+`) always seem to be dominated by `bi`: there are just tiny differences in the number of vertices touched by `bi` and either one of the `bigo` variants (or even `bisv`). This is astounding insofar as in the unimodal case such combinations usually outperform both `go` and `bi`. Moreover, NFA size (mostly the number of vertices) has a direct impact on the number of touched vertices: the NFAs $\mathcal{H}$ and $\mathcal{C}$ incur considerably higher numbers of touched vertices than the others do.

One striking difference between the various US networks is that for the AZ and GA graphs, `go` search does not yield any improvement over `pl` at all, however, a speed-up factor of up to 2 (i. e., a reduction of 50 % of touched vertices) can be achieved for DC. Similar improvement of a factor of 2 is reached with European road networks, while `go` accelerates the DEU railway graph only marginally. On the other hand, `bi` gives good speed-up of around 2 for the railway network; little improvement in general for the US networks; and no speed-up, or even a slow-down (especially with NFA $\mathcal{L}$), for the European road networks.

Overall, the performance of each algorithm is strongly dependent on the network properties, such as density or the metric used. It is also noteworthy that some NFAs are so much restrictive that a larger number of queries cannot be answered: e. g., with $\mathcal{L}$, no feasible path is found for 34 and 53 % of the queries in the LUX and CHE networks, respectively.

Dijkstra Rank. To get a finer picture, we now consider, exemplarily for the LUX network, the speed-up values categorized by the lengths of the belonging shortest paths found, also called *Dijkstra rank.* Figure 2 shows in the form of standard box plots the average speed-up with the algorithms `go` and `bi` and NFA $\mathcal{C}$. The best factors are obtained when the Dijkstra rank lies somewhere in the middle of the complete range: a certain minimal distance between start and destination seems to be required for the speed-up technique to kick in; with higher ranks (both vertices are located near opposite borders of the network), however, the `pl` search is naturally bounded already, so that the speed-up factors decrease again.

Sedgewick-Vitter. Performance of the `sv` heuristic can be measured in terms of both reduction in the number of touched vertices and path length increase: the bigger the choice of α, i. e., the greater the distortion towards the target, the smaller gets the search space; however, with increasing α, accuracy of the found paths drops. For the LUX network, we observed that an α of 1.2 reduces the number of touched vertices by well over 20 % on average while the path

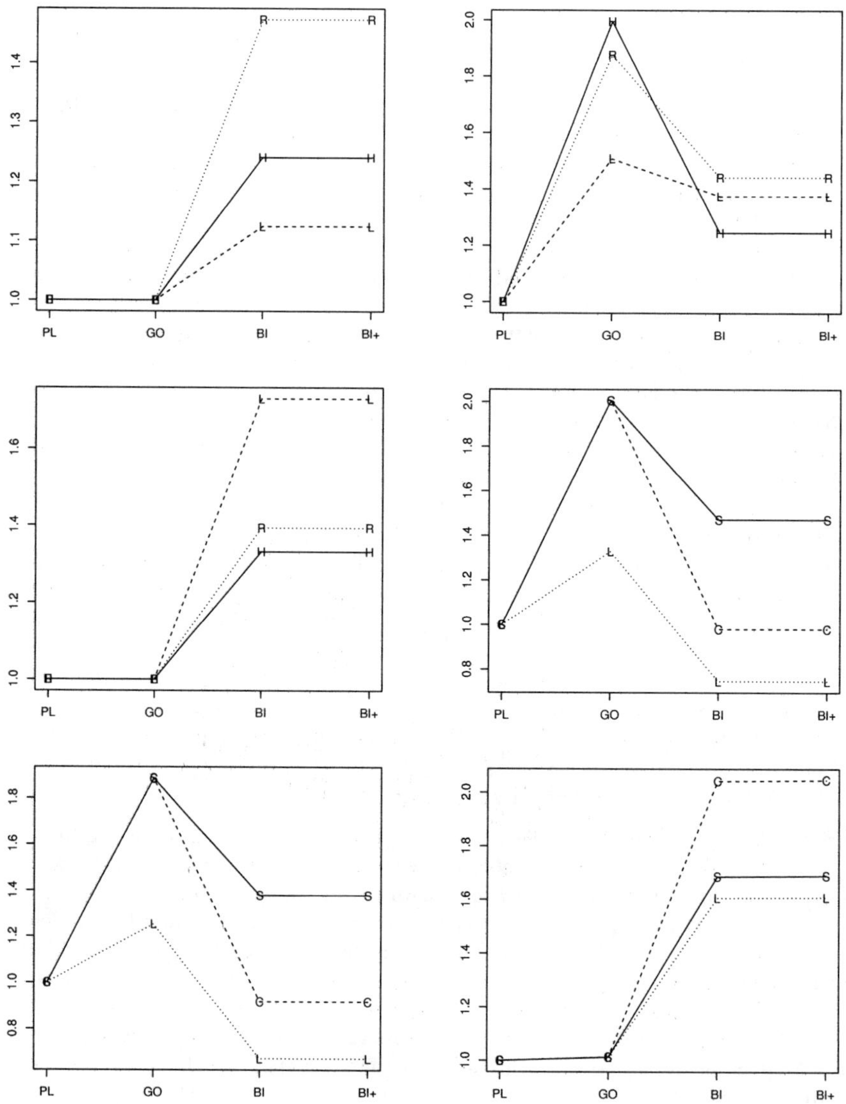

Fig. 1. Average speed-up in terms of touched vertices with each of the algorithms plain (`pl`), goal-directed (`go`), bidirectional (`bi`), and bidirectional/goal-directed combinations (`bi+`), applied to different networks (from top to bottom and left to right: AZ, DC; GA, LUX; CHE, DEU); the NFAs used are indicated by the characters on the lines (cf. Table 2).

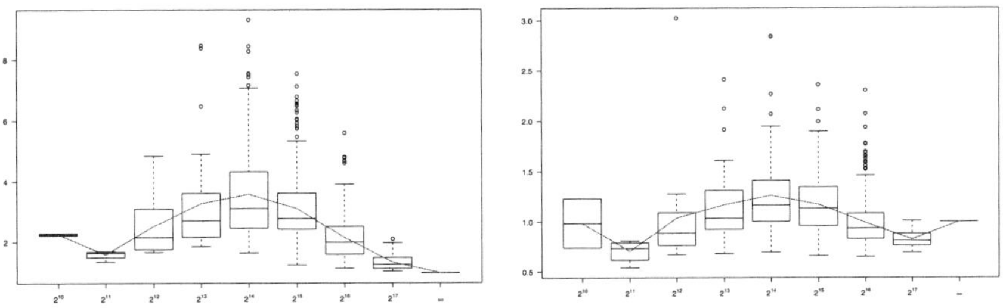

Fig. 2. Speed-up with go (left) and bi (right) applied to the LUX network and $\mathcal{C}$ NFA, categorized by Dijkstra rank. The x-axis denotes the (approximate) length of a path (∞ comprises infeasible queries); in each plot, the curve joins the mean values.

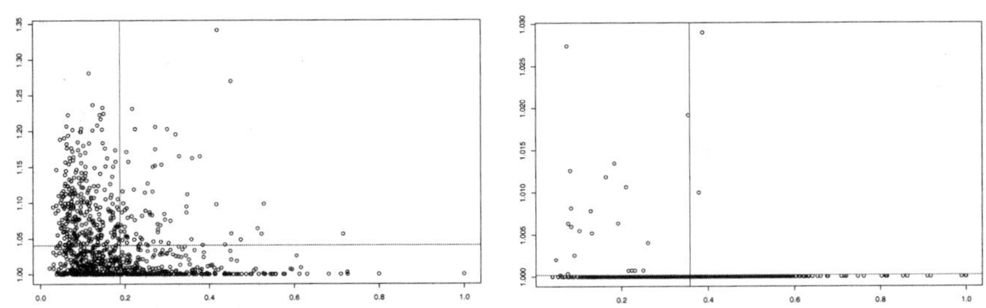

Fig. 3. Reduction in the number of touched vertices and path length increase with sv, applied to the LUX (left) and DEU (right) networks and the NFA $\mathcal{S}$ with α-parameters of 2 and 50, respectively. The x-axis denotes the share of touched vertices with sv in the number of vertices touched by go, while the y-axis denotes the path length increase. The horizontal and vertical lines mark the respective mean values.

lengths remain exact for all but a few queries. When raising α to 2, we save just over 80 % of touched vertices on average, but path lengths increase by around 4 % (cf. Figure 3). The picture for the DEU network looks similar, although much higher factors of α are needed to cause some effect: a reasonable choice seems to be 50 (the number of touched vertices diminishes to roughly a third with path quality almost unaffected).

4.3 k-Similar Paths

Besides exploring yet another practical application, the k-similar-path problem, as defined in Section 2.2, allows to construct NFAs of virtually arbitrary sizes in a natural way: the NFA restricting the second path found to $k \geq 0$ edges shared with the first one consists of $k+1$ vertices and $2k+1$ edges. Figure 4 shows for the DC network and increasing values of k both the number of vertices touched and speed-ups achieved with each algorithm. As can be predicted from theory, the curves joining the numbers of touched vertices exhibit to be almost

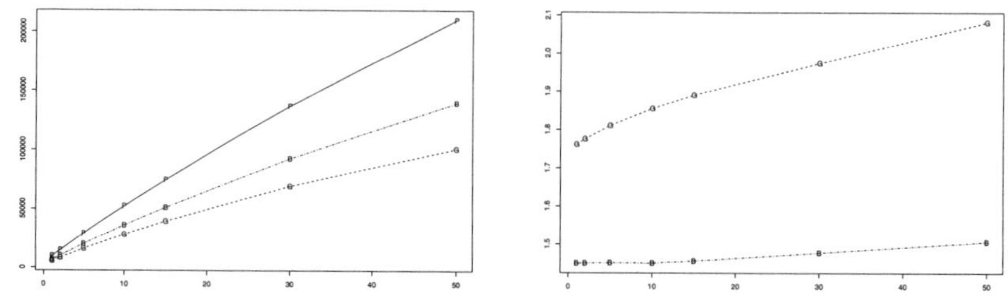

Fig. 4. k-similar-path computation: number of touched vertices (left) and speed-up (right) with the DC network and the algorithms `pl`, `go`, and `bi` for different choices of k (denoted along the x-axis)

linear (in fact, they appear slightly sublinear). With increasing NFA sizes (and hence bigger product networks), speed-ups also rise: with `bi` search, only a small growth is noticeable while with the `go` variant, the increase ranges between 1.75 (with $k = 1$) and 2 (with $k = 50$).

5 Conclusion

We have shown how to solve the regular-language-constrained shortest-path problem using product networks composed of a graph and an NFA, and proposed techniques to speed up point-to-point queries. In a practical implementation, all variants were empirically tested with a variety of real-world networks and constraints. Goal-directed and bidirectional search are found to give good speed-ups also for the multimodal setting. However, their performance greatly depends on the network properties. The Sedgewick-Vitter heuristic can, especially for railway graphs, take high α-parameters, while the paths found remain near-optimal. Surprisingly, combinations with bidirectional search do not perform better than the individual techniques. Investigating the application of k-shortest-path problems, it could be shown that the speed-up attainable increases with the size of the NFA.

In our opinion, the most interesting questions for future research include: adaptation and implementation of further speed-up techniques and heuristics; comparison between implicit and explicit product network representations; and integration of time-dependent cost functions.

Acknowledgments

The author Martin Holzer wants to give special thanks to Valentin Mihaylov for his assistance with parts of the implementation and execution of the experiments.

References

1. Barrett, C., Birkbigler, K., Smith, L., Loose, V., Beckman, R., Davis, J., Roberts, D., Williams, M.: An operational description of TRANSIMS. Technical report, Los Alamos National Laboratory (1995)
2. Barrett, C.L., Bisset, K., Holzer, M., Konjevod, G., Marathe, M.V., Wagner, D.: Engineering the label-constrained shortest-path algorithm. Technical report, NDSSL, Virginia Tech. (2007)
3. Barrett, C.L., Bisset, K., Jacob, R., Konjevod, G., Marathe, M.V.: Classical and contemporary shortest path problems in road networks: Implementation and experimental analysis of the TRANSIMS router. In: Möhring, R.H., Raman, R. (eds.) ESA 2002. LNCS, vol. 2461, pp. 126–138. Springer, Heidelberg (2002)
4. Barrett, C.L., Jacob, R., Marathe, M.V.: Formal-language-constrained path problems. SIAM J. Comput. 30(3), 809–837 (2000)
5. Dijkstra, E.W.: A note on two problems in connexion with graphs. Numerische Mathematik 1, 269–271 (1959)
6. Jacob, R., Marathe, M.V., Nagel, K.: A computational study of routing algorithms for realistic transportation networks. ACM Journal of Experimental Algorithms 4(6) (1999)
7. Mendelzon, A.O., Wood, P.T.: Finding regular simple paths in graph databases. SIAM J. Comput. 24(6), 1235–1258 (1995)
8. Orda, A., Rom, R.: Shortest-path and minimum-delay algorithms in networks with time-dependent edge-length. J. ACM 37(3), 607–625 (1990)
9. Romeuf, J.-F.: Shortest path under rational constraint. Information Processing Letters 28, 245–248 (1988)
10. Sedgewick, R., Vitter, J.S.: Shortest paths in euclidean graphs. Algorithmica 1(1), 31–48 (1986)
11. Sherali, H.D., Jeenanunta, C., Hobeika, A.G.: Time-dependent, label-constrained shortest path problems with applications. Transportation Science 37(3), 278–293 (2003)
12. Sherali, H.D., Jeenanunta, C., Hobeika, A.G.: The approach-dependent, time-dependent, label-constrained shortest path problems. Networks 48(2), 57–67 (2006)
13. Yannakakis, M.: Graph-theoretic methods in database theory. In: PODS, pp. 230–242 (1990)

Titles in This Series

74 **Camil Demetrescu, Andrew V. Goldberg, and David S. Johnson, Editors,** The Shortest Path Problem: Ninth DIMACS Implementation Challenge

73 **Paul H. Siegel, Emina Soljanin, Adriaan J. van Wijngaarden, and Bane Vasić, Editors,** Advances in Information Recording

72 **Regina Y. Liu, Robert Serfling, and Diane Souvaine, Editors,** Data Depth: Robust Multivariate Analysis, Computational Geometry and Applications

71 **Zhilan Feng, Ulf Dieckmann, and Simon Levin, Editors,** Disease Evolution: Models, Concepts, and Data Analyses

70 **James Abello and Graham Cormode, Editors,** Discrete Methods in Epidemiology

69 **Siemion Fajtlowicz, Patrick W. Fowler, Pierre Hansen, Melvin F. Janowitz, and Fred S. Roberts, Editors,** Graphs and Discovery

68 **A. Ashikhmin and A. Barg, Editors,** Algebraic Coding Theory and Information Theory

67 **Ravi Janardan, Michiel Smid, and Debasish Dutta, Editors,** Geometric and Algorithmic Aspects of Computer-Aided Design and Manufacturing

66 **Piyush Gupta, Gerhard Kramer, and Adriaan J. van Wijngaarden, Editors,** Advances in Network Information Theory

65 **Santosh S. Vempala,** The Random Projection Method

64 **Melvyn B. Nathanson, Editor,** Unusual Applications of Number Theory

63 **J. Nešetřil and P. Winkler, Editors,** Graphs, Morphisms and Statistical Physics

62 **Gerard J. Foschini and Sergio Verdú, Editors,** Multiantenna Channels: Capacity, Coding and Signal Processing

61 **M. F. Janowitz, F.-J. Lapointe, F. R. McMorris, B. Mirkin, and F. S. Roberts, Editors,** Bioconsensus

60 **Saugata Basu and Laureano Gonzalez-Vega, Editors,** Algorithmic and Quantitative Real Algebraic Geometry

59 **Michael H. Goldwasser, David S. Johnson, and Catherine C. McGeoch, Editors,** Data Structures, Near Neighbor Searches, and Methodology: Fifth and Sixth DIMACS Implementation Challenges

58 **Simon Thomas, Editor,** Set Theory: The Hajnal Conference

57 **Eugene C. Freuder and Richard J. Wallace, Editors,** Constraint Programming and Large Scale Discrete Optimization

56 **Alexander Barg and Simon Litsyn, Editors,** Codes and Association Schemes

55 **Ding-Zhu Du, Panos M. Pardalos, and Jie Wang, Editors,** Discrete Mathematical Problems with Medical Applications

54 **Erik Winfree and David K. Gifford, Editors,** DNA Based Computers V

53 **Nathaniel Dean, D. Frank Hsu, and R. Ravi, Editors,** Robust Communication Networks: Interconnection and Survivability

52 **Sanguthevar Rajasekaran, Panos Pardalos, and D. Frank Hsu, Editors,** Mobile Networks and Computing

51 **Pierre Hansen, Patrick Fowler, and Maolin Zheng, Editors,** Discrete Mathematical Chemistry

50 **James M. Abello and Jeffrey Scott Vitter, Editors,** External Memory Algorithms

49 **Ronald L. Graham, Jan Kratochvíl, Jaroslav Nešetřil, and Fred S. Roberts, Editors,** Contemporary Trends in Discrete Mathematics

48 **Harvey Rubin and David Harlan Wood, Editors,** DNA Based Computers III

47 **Martin Farach-Colton, Fred S. Roberts, Martin Vingron, and Michael Waterman, Editors,** Mathematical Support for Molecular Biology

46 **Peng-Jun Wan, Ding-Zhu Du, and Panos M. Pardalos, Editors,** Multichannel Optical Networks: Theory and Practice

45 **Marios Mavronicolas, Michael Merritt, and Nir Shavit, Editors,** Networks in Distributed Computing

44 **Laura F. Landweber and Eric B. Baum, Editors,** DNA Based Computers II

43 **Panos Pardalos, Sanguthevar Rajasekaran, and José Rolim, Editors,** Randomization Methods in Algorithm Design

42 **Ding-Zhu Du and Frank K. Hwang, Editors,** Advances in Switching Networks

41 **David Aldous and James Propp, Editors,** Microsurveys in Discrete Probability

40 **Panos M. Pardalos and Dingzhu Du, Editors,** Network Design: Connectivity and Facilities Location

39 **Paul W. Beame and Samuel R Buss, Editors,** Proof Complexity and Feasible Arithmetics

38 **Rebecca N. Wright and Peter G. Neumann, Editors,** Network Threats

37 **Boris Mirkin, F. R. McMorris, Fred S. Roberts, and Andrey Rzhetsky, Editors,** Mathematical Hierarchies and Biology

36 **Joseph G. Rosenstein, Deborah S. Franzblau, and Fred S. Roberts, Editors,** Discrete Mathematics in the Schools

35 **Dingzhu Du, Jun Gu, and Panos M. Pardalos, Editors,** Satisfiability Problem: Theory and Applications

34 **Nathaniel Dean, Editor,** African Americans in Mathematics

33 **Ravi B. Boppana and James F. Lynch, Editors,** Logic and Random Structures

32 **Jean-Charles Grégoire, Gerard J. Holzmann, and Doron A. Peled, Editors,** The SPIN Verification System

31 **Neil Immerman and Phokion G. Kolaitis, Editors,** Descriptive Complexity and Finite Models

30 **Sandeep N. Bhatt, Editor,** Parallel Algorithms: Third DIMACS Implementation Challenge

29 **Doron A. Peled, Vaughan R. Pratt, and Gerard J. Holzmann, Editors,** Partial Order Methods in Verification

28 **Larry Finkelstein and William M. Kantor, Editors,** Groups and Computation II

27 **Richard J. Lipton and Eric B. Baum, Editors,** DNA Based Computers

26 **David S. Johnson and Michael A. Trick, Editors,** Cliques, Coloring, and Satisfiability: Second DIMACS Implementation Challenge

25 **Gilbert Baumslag, David Epstein, Robert Gilman, Hamish Short, and Charles Sims, Editors,** Geometric and Computational Perspectives on Infinite Groups

24 **Louis J. Billera, Curtis Greene, Rodica Simion, and Richard P. Stanley, Editors,** Formal Power Series and Algebraic Combinatorics/Séries Formelles et Combinatoire Algébrique, 1994

23 **Panos M. Pardalos, David I. Shalloway, and Guoliang Xue, Editors,** Global Minimization of Nonconvex Energy Functions: Molecular Conformation and Protein Folding

22 **Panos M. Pardalos, Mauricio G. C. Resende, and K. G. Ramakrishnan, Editors,** Parallel Processing of Discrete Optimization Problems

21 **D. Frank Hsu, Arnold L. Rosenberg, and Dominique Sotteau, Editors,** Interconnection Networks and Mapping and Scheduling Parallel Computations

20 **William Cook, László Lovász, and Paul Seymour, Editors,** Combinatorial Optimization

19 **Ingemar J. Cox, Pierre Hansen, and Bela Julesz, Editors,** Partitioning Data Sets